Herbicide Resistance and World Grains

Herbicide Resistance and World Grains

Edited by
Stephen B. Powles
and **Dale L. Shaner**

CRC Press
Taylor & Francis Group
Boca Raton London New York

CRC Press is an imprint of the
Taylor & Francis Group, an **informa** business

Published 2001 by CRC Press
Taylor & Francis Group
6000 Broken Sound Parkway NW, Suite 300
Boca Raton, FL 33487-2742

First issued in paperback 2019

No claim to original U.S. Government works

ISBN-13: 978-0-367-45532-3 (pbk)
ISBN-13: 978-0-8493-2219-8 (hbk)

**Visit the Taylor & Francis Web site at
http://www.taylorandfrancis.com**

**and the CRC Press Web site at
http://www.crcpress.com**

Library of Congress Cataloging-in-Publication Data

Herbicide resistance and world grains / edited by Stephen B. Powles and Dale L. Shaner.
    p. cm.
Includes bibliographical references.
ISBN 0-8493-2219-7 (alk. paper)
1. Herbicide resistance. 2. Herbicide-resistant crops. I. Powles, Stephen B. II. Shaner,
Dale L. (Dale Lester)

SB951.4 .H437 2001
632'.954—dc21
                                                00-140113
                                                CIP

Library of Congress Card Number 00-140113

Preface

Weeds have been the bane of farmers since humanity changed from hunting and gathering to settled agriculture. Weeds compete directly with crops for space, nutrients, and light, and they are among the major limitations on crop yield. The discovery and use of selective herbicides that began with the auxinic herbicides in the late 1940s was one of the primary reasons for the spectacular increases in crop productivity over the past 50 years. Farmers throughout the major crop production areas of the world now depend on herbicides to manage weeds. However, evolution continues, and the selective pressure that herbicides exert on weed populations selects for rare resistant biotypes that can quickly proliferate and dominate the population. If a herbicide is used persistently the selection of resistance is inevitable. Australia is a model of what can happen with unbridled use of herbicides and the selection of herbicide resistant populations.

Management of herbicide-resistant weed populations has received increasing attention following the initial selection and identification of triazine-resistant weeds 30 years ago in the United States. In the 1990s industry and academia made major efforts to develop practical, cost-effective guidelines that farmers could use to achieve integrated weed management with herbicides as a major, but not exclusive, tool.

Although there are several excellent books and numerous papers on herbicide resistance, there is no compilation of the latest information on resistance and its management in the major world grain crops. Both editors of this book have been intimately involved in practical herbicide-resistance management from academic and industrial perspectives. As a result, we feel that the agricultural community needs access to the information on herbicide-resistance management from experts who have been dealing with this problem.

Farmers will adopt herbicide-resistance management techniques and tools if they accept the economic sense of these practices. Farming must be profitable in order for farmers to remain in business and resistance management has both short- and long-term costs. It is important to understand these costs and how they impact the implementation of resistance management. If resistance management is not economically acceptable, the recommendations for resistance management will not be adopted. Hence, we include chapters on resistance management in major world grain crops, the economic impact of resistance management, and the latest information on new models for comparing different resistance-management strategies. We also include regulatory affairs because major changes in the regulation of herbicides now include resistance management as a key component in the registration of new herbicidal products.

The chapter authors summarize the current state of herbicide resistance in major grain crops and present the latest information on how to manage resistant weed populations. They also indicate unanswered questions and future research needs. This information can be the springboard to new research and new solutions.

An overriding theme of this book is the need for integrated weed management. As one of the editors has frequently stated, herbicide resistance has been a primary driver for the implementation of integrated weed management in Australia. This is the future of weed management. Herbicides have played a vital role in increasing crop production throughout the world, but complete dependence on herbicides for weed management is nonsustainable. Herbicides must be part of an integrated system. Herbicide-resistance problems demonstrate the weakness of over-reliance on a single weed management tool. The editors hope that this book will help lead the way to true integrated weed management, in which herbicides play an integral, but not exclusive, role.

Acknowledgments

No book is the product of a single person. We gratefully acknowledge the good work of the chapter authors who willingly gave their time and effort to produce an excellent distillation of current information on herbicide resistance and management. We thank our co-workers and colleagues who reviewed chapters.

We also thank all of the authors for their generous offers to dedicate all royalties for this book to purchasing copies of the book to be distributed to libraries in developing nations to help them manage herbicide-resistant weed populations. We are also grateful to CRC Press for agreeing to publish this book and to the invaluable aid of its staff.

Editors

Stephen B. Powles, Ph.D., received his B.Sc. in crop science from the University of Western Sydney, his M.Sc. from Michigan State University, and his Ph.D. in plant physiology from the Australian National University. Postdoctoral research followed at Stanford University (Carnegie Plant Biology) and at C.N.R.S., Gif-sur-Yvette, France.

Dr. Powles returned to Australia in 1983 and built a large research program on fundamental and applied aspects of herbicide resistance at the University of Adelaide. From 1995 to 1997 he was director of an Australia-wide research program on weed management (CRC). In 1998 he was appointed professor at the University of Western Australia and director of a major research initiative focused on herbicide resistance and management.

Dr. Powles is the author of more than 120 research papers and reviews on herbicide resistance. He has given numerous invited lectures at international meetings. Many graduate students and postdoctoral fellows who have graduated from his herbicide-resistance program now work worldwide in industry and academia. In 1997 he was elected an International Honorary Fellow of the Weed Science Society of America and in 1999 he was elected to the Australian Academy of Technological Sciences and Engineering. Aside from family and his professional interests in the plant sciences and agriculture, he remains active in field hockey.

Dale L. Shaner, Ph.D., is a senior research fellow at BASF Corporation, Princeton, NJ. He is a member of the Agricultural Products Research Department at the Agricultural Research Center in Princeton, where he conducts research and provides technical support on imidazolinone-resistant crops, herbicide-resistance management, and herbicide mode of action. Prior to his present position, Dr. Shaner was the director of the Ag Biotech Department of American Cyanamid.

Dr. Shaner received his B.A. in botany in 1970 from DePauw University, his M.A. in plant ecology in 1972 from the University of Colorado, and his Ph.D. in plant physiology in 1976 from the University of Illinois. After obtaining his Ph.D., Dr. Shaner served as assistant professor of weed science/plant physiology at the University of California, Riverside until 1979. He then joined the Herbicide Discovery Group at American Cyanamid, where he helped conduct the early research on the greenhouse and field activity of the imidazolinones. He moved into the laboratory and was instrumental in discovering the mode of action of the imidazolinone herbicides.

In 1984 Dr. Shaner received Cyanamid's Scientific Achievement Award for his work on imazamethabenz-methyl. He has won several awards for scientific publication from Cyanamid. He was the recipient of the Weed Science Society of America's (WSSA) Industry Award in 1993. He was elected a fellow of the WSSA in

2000. Dr. Shaner has also been very active in industry's efforts to manage herbicide resistance. He is the past chairman of the Herbicide Resistance Action Committee and has given numerous presentations on industry's position in herbicide-resistance management. He helped establish herbicide-resistance groups in North and South America.

Dr. Shaner is a member of the WSSA and the American Chemical Society. He is an associate editor for *Weed Science Journal*.

Dr. Shaner is the author of more than 50 papers. He has given more than 40 invited lectures at symposia and seminars. His research interests include all aspects of the mode of action of herbicides, the effect of environmental factors on herbicidal activity, and herbicide resistance in crops and weeds.

Contributors

Art J. Diggle, Ph.D.
Agriculture Western Australia
Locked Bag 4
Bentley Delivery Centre, WA 6983
Australia

Ian Heap, Ph.D.
Weedsmart
P.O. Box 1365
Corvallis, OR 97339
U.S.A.

Kazuyuki Itoh, Ph.D.
National Institute of Agro-
  Environmental Sciences
Tsukuba Science City,
  Ibaraki 305-8604
Japan

Homer LeBaron, Ph.D.
1230 N. Cottonwood Circle
Heber, UT 84032
U.S.A.

Deirdre Lemerle, Ph.D.
NSW Agriculture and CRC
  for Weed Management Systems
Agricultural Institute
Wagga Wagga, NSW 2650
Australia

Paul Leonard, Ph.D.
BASF Corporation
Chaussee de Tirlemont
B-5030 Gembloux
Belgium

Carol A. Mallory-Smith, Ph.D.
Department of Crop and Soil Science
Oregon State University
107 Crop Science Building
Corvallis, OR 97331-3002
U.S.A.

Paul Neve, Ph.D.
Western Australian Herbicide
  Resistance Initiative
University of Western Australia
Nedlands, WA 6907
Australia

Micheal D. K. Owen, Ph.D.
Department of Agronomy
Iowa State University
Ames, IA 50011
U.S.A.

David J. Pannell, Ph.D.
Agricultural and Resource
  Economics
University of Western Australia
Nedlands, WA 6907
Australia

Stephen B. Powles, Ph.D.
Western Australian Herbicide
  Resistance Initiative
University of Western Australia
Nedlands, WA 6907
Australia

Christopher Preston, Ph.D.
University of Adelaide and CRC
  for Weed Management Systems
PMB 1
Glen Osmond, SA 5064
Australia

Dale L. Shaner, Ph.D.
BASF Corporation
Princeton, NJ 08543-0400
U.S.A.

Donald C. Thill, Ph.D.
Department of Plant, Soil,
  and Entomological Science
University of Idaho
P.O. Box 442339
Moscow, ID 83844-2339
U.S.A.

Bernal E. Valverde, Ph.D.
Department of Weed Science
The Royal Veterinary
  and Agricultural University
Agrovej 10
DK-2630 Taastrup
Denmark

David Zilberman, Ph.D.
Department of Agricultural
  and Resource Economics
University of California
207 Giannini Hall
Berkeley, CA 94720
U.S.A.

Meeting of Chapter Authors at the Third International Weed Science Congress, Iguazu, Brazil, June 2000. Left to right: Art J. Diggle, Paul Neve, Micheal D. K. Owen, Ian Heap, Bernal E. Valverde, Stephen B. Powles, Dale L. Shaner, Carol A. Mallory-Smith, Christopher Preston, and David J. Pannell.

Contents

1 Introduction and Overview of Resistance

Ian Heap and Homer LeBaron

CONTENTS

1.1 INTRODUCTION

Humans have struggled against the negative impact of weeds since the cultivation of crops commenced around 10,000 B.C.[1] Weed control technologies have evolved from hand-weeding to include primitive hoes (6000 B.C.), animal-powered implements (1000 B.C.), mechanically powered implements (1920 A.D.), biological control (1930 A.D.), and chemical (herbicide) control (1947 A.D.).[1] Since the introduction of the first selective herbicides, 2,4-D and MCPA, in 1947, herbicides have had a major positive impact on world agricultural production, initially in developed nations and more recently in developing nations.[2] Herbicides are often the most reliable and least expensive method of weed control available, and the success of herbicides is largely responsible for the abundant and sustained food production necessary to

0-8493-2219-7/01/$0.00+$.50
© 2001 by CRC Press LLC

support an increasing world population.[3] The availability of herbicides has allowed plant breeders to move from taller, more competitive crop varieties to shorter, higher-yielding crop varieties.[4] The efficacy and cost-effectiveness of herbicides has led to heavy reliance on them in the developed world. Nevertheless, there are some real and perceived problems with herbicides. In recent years there has been increased concern about residues and associated food safety issues, their adverse impact on the environment,[5] and the widespread occurrence of herbicide-resistant weeds.[6] The focus of this book is on the last of these concerns.

1.2   RESISTANCE

Pesticide resistance evolved in insects, fungi, and bacteria long before it was observed in weeds.[7] Resistance evolves following persistent selection for mutant genotypes that may be pre-existing or arise *de novo* in weed populations.[7] Herbicide-resistant weeds were predicted shortly after the introduction of herbicides.[8,9] Given the examples of pesticide resistance in insects and fungi it seemed inevitable that the continuous or frequent use of the same herbicide against the same populations of weeds would eventually result in resistant weeds. The first herbicides to be used persistently over large areas were 2,4-D and MCPA. Fortunately, these auxinic herbicides are not prone to rapid selection for resistance and, with the exception of 2,4-D-resistant *Daucus carota*,[10,11] resistance was not reported until the appearance of triazine herbicide-resistant weeds (Section 1.2.3.1) in the early 1970s.[12,13]

1.2.1   RESISTANCE DEFINED

In the context of this book, *resistance*, unless otherwise stated, denotes the evolved capacity of a previously herbicide-susceptible weed population to withstand a herbicide and complete its life cycle when the herbicide is used at its normal rate in an agricultural situation.

Target-site resistance is the result of a modification of the herbicide-binding site (usually an enzyme), which precludes a herbicide from effectively binding. This is the most common resistance mechanism.

Cross-resistance occurs where a single resistance mechanism confers resistance to several herbicides. *Target-site cross-resistance* can occur to herbicides binding to the same target site (enzyme). Good examples are the two classes of herbicide chemistry (aryloxyphenoxypropionates and cyclohexanediones). While chemically dissimilar, both inhibit the enzyme acetyl coenzyme A carboxylase and resistant biotypes frequently exhibit varying levels of target-site cross-resistance in both. Some target sites may have more than one domain, e.g., the targeted protein in photosystem II has separate domains that bind triazine-type herbicides and phenolic-type herbicides.[14]

Nontarget-site resistance is resistance due to a mechanism(s) other than a target-site modification. Nontarget-site resistance can be endowed by mechanisms such as enhanced metabolism, reduced rates of herbicide translocation, sequestration, etc. Such mechanisms reduce the amount of herbicide reaching the target site.

Nontarget-site cross-resistance occurs when a single mechanism endows resistance across herbicides with different modes of action. Such mechanisms are usually unrelated to the herbicide target site. Examples are cytochrome P450–based nontarget-site cross-resistance,[15–17] and glutathione transferase–based resistances,[18–20] which degrade a spectrum of herbicides that have different sites of action.

Multiple-resistance occurs when two or more resistance mechanisms are present within individual plants or a population. Depending on the number and type of mechanisms, a population and/or individual plants within a population may simultaneously exhibit multiple-resistance to many different herbicides.

While there has not been international standardization on terminology, this book uses the above-defined terms of resistance, target-site and nontarget-site resistance, and multiple-resistance throughout.[21]

1.2.2 LEARNING FROM THE HISTORY OF RESISTANCE

At this moment in time, and in this introductory chapter on herbicide resistance, it is informative to look back over the past 30 years. Prior to 1970, the few reports or observations of weeds exhibiting reduced levels of control with 2,4-D or other early herbicides received little notice or concern among farmers or scientists. Minor interspecific and intraspecific differences or shifts in selectivity among weed populations were very common following the use of herbicides. This was to change dramatically with the first clear evidence of evolved resistance to the triazine herbicides. The triazine herbicides provided excellent weed control; however, even with the remarkably effective and consistent triazine herbicides, missed or escaped weeds were often observed under certain soil and climatic conditions.

The first report of triazine resistance in the previously very susceptible *Senecio vulgaris* populations in a Washington State nursery (U.S.A.) in the late 1960s[12] rightfully received much attention. The weed had evolved an extremely high level of resistance that was genetically transferred to its progeny.[22,23] Many realized that this was likely a harbinger of things to come. Holm et al.,[24] in their major and classic work on world weeds, stated, "This discovery [i.e., resistance to triazines] has proven to be one of the most important events since the inception of weed science. If triazine-resistant weeds are not controlled by some other means, the resistant biotype rapidly predominates, reproduces and becomes a solid infestation that can no longer be controlled by the herbicide. If only triazine herbicides are available to control weeds in that crop, the farmer may no longer be able to produce the crop of choice."

Not long after this first report of triazine-resistant *S. vulgaris* another equally important discovery was made. In 1975, S. Radosevich, a graduate student with Dr. F. M. Ashton at the University of California, Davis, wrote a letter to Dr. Homer M. LeBaron of Ciba-Geigy in which he stated that isolated chloroplasts from triazine-resistant *S. vulgaris* were insensitive to simazine. Radosevich and Appleby[25] had earlier confirmed that there were no differences between the susceptible and resistant biotypes of *S. vulgaris* in herbicide uptake, distribution, or metabolism, whereas it was well known that maize and other tolerant crops avoided injury because they are able to metabolize atrazine.[26] In 1976, Radosevich and DeVilliers published the first report confirming that the triazine resistance in this weed was due to alteration of the target site. This evidence was the first documentation of what would later become many cases of target-site-based resistance to triazine and, ultimately, many other herbicides. They further reported that these triazine herbicide–insensitive chloroplasts were capable of continuing photosynthesis in the presence of simazine or atrazine.

Earlier research had demonstrated that chloroplasts isolated from crop species resistant to triazine herbicides were susceptible to triazines. Moreland[27] had reported that photosynthesis in isolated chloroplasts was equally inhibited by simazine, whether they came from resistant maize or susceptible spinach. Radosevich (1977) soon documented that triazine-resistant common *Chenopodium album* and *Amaranthus retroflexus*, which had recently been found in Washington State maize fields and elsewhere, also had chloroplasts that were insensitive to atrazine. These discoveries on the mechanism of triazine herbicide resistance were more unexpected, and had greater impact on weed science and management, than the original evolution of triazine-resistant weeds.

During the 1970s, many additional important weed species were reported to be resistant to triazine herbicides and several other herbicides. These were widely scattered independent outbreaks in the United States, Canada, and Europe. Some species (e.g., *Amaranthus* spp., *Chenopodium* spp., *Conyza canadensis, Kochia scoparia, Solanum nigrum, Echinochloa crus-galli, Senecio vulgaris,* and *Poa annua*) have evolved resistance to the triazine herbicides frequently and in many locations (Table 1.3). The majority of triazine-resistance cases have occurred from separate evolutionary events and have not been caused by the spread of resistant seed or propagules.[28,29] However, there has been some evolution of triazine resistance along roadsides and rights-of-way, where vehicles have dispersed seed. Triazine resistance has also appeared in orchards and roadsides throughout the world from the persistent use of simazine. In North America, *K. scoparia, Bromus tectorum,* and *S. vulgaris* have mostly evolved resistance to triazine herbicides in roadsides, railways, nurseries, or perennial crops. By 1980, 32 weed species (26 broadleafs and 6 grasses) had evolved resistance to triazine herbicides.

A significant scientific benefit flowing from the work on triazine-resistant weeds has been the increased knowledge and understanding of herbicide-binding sites and modes of action. As further discussed in Chapter 3, the availability of two biotypes identical except for herbicide resistance, due to a small change at the herbicide-binding domain on a protein in the thylakoid membrane, provided a powerful tool to study the mechanisms of photosynthesis, herbicide mechanisms of action, and other physiological and molecular genetic processes.[30–32] Indeed, the target (binding site) has been isolated and crystallized from resistant and susceptible photosynthetic bacteria,[33] leading to a Nobel Prize in medicine, not because the research dealt with herbicide resistance, but because of its universal implications for drug binding, design, and resistance.

1.2.3 WHEN AND WHERE RESISTANCE EVOLVED

Although the above-mentioned first reports of triazine-resistant weeds were of concern to weed scientists they were not initially of practical significance to most growers, as they were very limited and localized in area, and were usually well controlled with other herbicides. Even among scientists, herbicide-resistant weeds were mostly of academic interest and few were researching how they should be managed. However, it was becoming obvious that the problem deserved greater attention. During 1977, LeBaron organized an informal but enthusiastic meeting at the Weed Science Society of America (WSSA) convention held in Dallas, Texas on February 10, 1978. From this first discussion, research efforts were coordinated and LeBaron organized a formal 1-day symposium, held the following year at the WSSA meeting in San Francisco on February 8, 1979. Of the 14 papers, 8 were presented by scientists from the United States, 4 by scientists from Canada, and 1 each by scientists from Germany and Israel. Dr. J. Gressel agreed to work with LeBaron to edit a book from these papers. The landmark book, *Herbicide Resistance in Plants*, was published in 1982. Chapters in this first book on herbicide resistance included the remarkable progress on photosynthetic mechanisms and herbicide-binding sites,[30] the potential of new herbicide-resistant crops,[22,34] models for predictions of future herbicide-resistant weeds,[35] and physiological responses and fitness of susceptible and resistant weed biotypes.[36] This book was well received and brought to the attention of weed scientists, plant physiologists, and many others the practical and scientific importance of herbicide-resistant weeds.

From these beginnings catalyzed by the first cases of triazine resistance, resistance has evolved in the field to almost all herbicide chemistries and modes of action (Table 1.1) and in almost all cropping systems (Tables 1.2 and 1.3). In many cases researchers have simply stated that resistant weeds occur in "cropland" without specifying a crop. Thus, the fact that a resistant weed biotype is not listed under a particular crop does not necessarily mean that it has not been found in that crop (Table 1.2). The evolution of resistant populations has been slow to appear with some herbicides (Figure 1.1), but resistance continues to evolve to almost all herbicide modes of action being used (Figure 1.2).

In August 2000 the International Survey of Herbicide-Resistant Weeds recorded 235 herbicide-resistant weed biotypes in 47 countries.[37] A new resistant biotype refers to the first instance of a weed species evolving resistance to one or more herbicides in a herbicide group. Up-to-date information on the International Survey of Herbicide-Resistant Weeds can be found on the Web (http://www.weedscience.com).[37] There has been a relatively steady increase in the number of new cases of resistance (approximately nine new cases per year) since 1980 (Figure 1.1). In the 5-year period from 1978 to 1983, scientists around the world documented 33 new cases of triazine-resistant weeds (Figure 1.2). More recently, ALS- and ACCase-herbicide-resistant weeds have accounted for a large portion of the resistant species (Figure 1.2). The first cases of glyphosate-resistant weeds have appeared recently in Australia[38,39] and Malaysia.[40] Importantly, the Malaysian *Eleusine indica* resistant to glyphosate[41] has been reported to have target-site resistance due to a mutation in the EPSP-synthase gene.[40] Considering the recent rapid adoption of glyphosate-resistant crops, more cases of glyphosate resistance are likely; however, it is expected that resistance to glyphosate will appear less frequently than for most herbicide modes of action, following a pattern similar to that observed for phenoxy herbicides (Figure 1.2).

TABLE 1.1
The Occurrence of Herbicide-Resistant Weed Biotypes to Different Herbicide Groups

Herbicide Group	WSSA[a] Code	HRAC[b] Code	Australian Code	Example	Resistant Weed Biotypes Dicots	Resistant Weed Biotypes Monocots	Resistant Weed Biotypes Total
ALS inhibitors	2	B	B	Chlorsulfuron	44	20	64
Triazines	5	C1	C	Atrazine	42	19	61
Bipyridiliums	22	D	L	Paraquat	18	7	25
ACCase inhibitors	1	A	A	Diclofop-methyl	0	21	21
Synthetic auxins	4	O	I	2,4-D	15	5	20
Ureas/amides	7	C2	C	Chlorotoluron	6	11	17
Dinitroanilines	3	K1	D	Trifluralin	2	7	9
Triazoles	11	F3	C	Amitrole	1	3	4
Chloroacetamides	15	K3	E	Butalochlor	0	3	3
Thiocarbamates	8	N	E	Triallate	0	3	3
Glycines	9	G	M	Glyphosate	0	2	2
Benzoflurans	16	N	K	Ethofumesate	0	1	1
Chloro-carbonic-acids	26	N	J	Dalapon	0	1	1
Nitriles	6	C3	C	Bromoxynil	1	0	1
Organoarsenicals	17	Z		MSMA	1	0	1
Pyrazoliums	8	Z		Difenzoquat	0	1	1
Unknown	25	Z		Flamprop-methyl	0	1	1
				Totals	130	105	235

[a] Retzinger and Mallory-Smith[48]
[b] Schmidt [Scm]

Source: Compiled from data in Reference 37.

TABLE 1.2
The Occurrence of Herbicide-Resistant Weeds in Various Cropping Situations

Category	Crop	No. of Resistant Biotypes
Field Crops	Wheat/barley	57
	Corn	50
	Rice	24
	Soybean	22
	Canola	11
	Cotton	5
	Sugarbeet	4
	Unspecified cropland[a]	62
Vegetables	Vegetables (carrot, lettuce, potato, etc.)	16
Perennial Crops	Orchard (apple, pear, peach, ... including vineyard)	37
	Pasture (clover, alfalfa, pasture seed, etc.)	23
	Forestry	8
	Other perennial (tea, coffee, rubber, mint, etc.)	8
Noncrop	Noncrop (roadside, railway, industrial site)	35

[a] Respondents of the survey only indicated that the resistant biotype was found on "cropland" in their region in general and did not specify all the crops.

Source: Compiled from data in Reference 37.

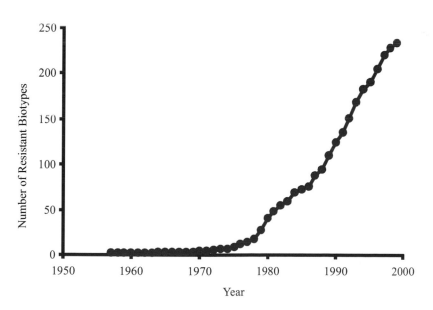

FIGURE 1.1   The chronological increase in the number of herbicide-resistant weeds worldwide.

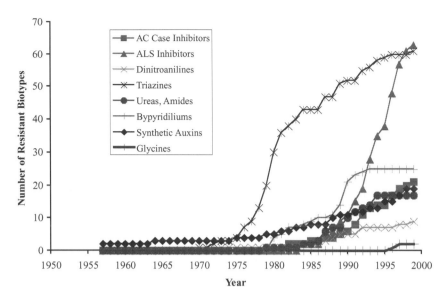

FIGURE 1.2 The chronological increase in the number of herbicide-resistant weeds for several herbicide classes.

1.2.3.1 World Grain Crops

Detailed information on the extent and spread of resistant weed biotypes in the major crops will be provided in subsequent chapters. Here we summarize briefly the resistant weed populations in some of the major crops.

1.2.3.1.1 Wheat

Fifty-seven herbicide-resistant weed biotypes have been identified in wheat and barley (Table 1.2). Grass weeds with resistance to ACCase-inhibiting herbicides and dicot weeds with resistance to ALS-inhibiting herbicides account for the majority of the wheat area infested with herbicide-resistant weeds. Target-site cross-resistance and multiple-resistance in the grass weeds *Lolium rigidum* from Australia and *Alopecurus myosuroides* from Europe are two of the most intractable problems. In some instances there are few remaining herbicides for control of these multiple-resistant biotypes. ACCase-herbicide resistance in weedy *Avena* spp. is widespread in all major wheat-producing regions of the world and multiple-resistant *Avena* spp. are beginning to appear in North America. *Avena* spp. are already among the most serious weeds of cereals, and now they are poised to become the worst herbicide-resistant weed of wheat.

    Twenty-four of the herbicide-resistant biotypes occurring in wheat and barley are resistant to ALS-inhibiting herbicides. ALS-herbicide-resistant biotypes of *Kochia scoparia* and *Salsola iberica* now infest more than 60% of wheat fields in the northern United States. In general there have been sufficient alternatives for control of ALS-herbicide-resistant weeds in wheat and barley and the present threat to production is less than that of ACCase-herbicide-resistant grasses (see Chapter 5).

While seven dicot weed species have evolved resistance to synthetic auxin-type herbicides in wheat, none has become a widespread problem and all are easily controlled with alternative modes of action.[37]

1.2.3.1.2 Rice

The *Echinochloa* spp. are major weeds wherever rice is grown and they present growers with the most serious resistance problems in rice (see Chapter 6). Propanil resistance in *E. crus-galli* and *E. colona* has evolved in South America and the United States as a result of elevated levels of the same acylamidase enzyme system that rice uses (nontarget-site-based resistance) to detoxify propanil.[42,43] Target-site cross-resistance to butachlor and thiobencarb has evolved in populations of *E. crus-galli* in China in over 2 million ha.[44] In California (U.S.A.), Fischer et al.[45] identified populations of *E. oryzoides* and *E. phyllopogon* with multiple-resistance (4- to 20-fold) to thiobencarb, molinate, fenoxaprop, and propanil. This has left growers with few alternatives to control *Echinochloa* spp. in some California rice fields (see Chapter 6).

In addition to the above-mentioned grass weed problems of rice there has been a steady increase in the number of sites and the area infested with ALS-herbicide-resistant weeds. Over half of the 24 herbicide-resistant biotypes in rice are resistant to ALS-inhibiting herbicides. With the exception of *Echinochloa* spp., they represent all of the major weed threats to the rice industry. Of particular concern are *Alisma plantago-aquatica* in Italy and Portugal, *Cyperus difformis* and *Sagittaria montevidensis* in Australia and the United States, *Scirpus mucronatus* in Italy and the United States, and *Lindernia* sp. in Asia. The lack of good alternatives in rice for control of some of these species heightens the concern of growers (see Chapter 6).

1.2.3.1.3 Maize

At least 50 herbicide-resistant weed biotypes have evolved resistance in maize production systems worldwide and all but 11 of them are resistant to triazine herbicides (see Chapter 4). The widespread adoption of atrazine for weed control in maize fields resulted in triazine-resistant weeds becoming the first widespread herbicide-resistant weed problem. Triazine-resistant *Chenopodium album* and *Amaranthus* spp. have achieved particular notoriety because of the large areas infested with these biotypes and the fact that they have been identified in 18 countries. *Amaranthus* spp. are of greater concern to growers because they have shown the potential to sequentially evolve multiple-resistance to triazine and ALS herbicides. Two dicot species (*Amaranthus rudis* and *A. palmeri*) and four grasses (*Setaria* spp. and *Sorghum bicolor*) have evolved resistance to ALS herbicides in maize, all in the United States. Even fewer ALS-herbicide-resistant species have evolved in European maize production due to far less usage of ALS herbicides in Europe. Fortunately, the majority of herbicide-resistant weeds in corn can be easily controlled by alternative herbicides (see Chapter 4).

1.2.3.1.4 Soybean

Of the 22 herbicide-resistant biotypes found in soybean, 16 are resistant to ALS herbicides and 3 are resistant to ACCase herbicides. In soybean, five ALS-herbicide-resistant

TABLE 1.3
The Occurrence of Triazine-Resistant Weeds Worldwide

	Species	Common Name	Family	Weed Group	Country and Year
1.	Abutilon theophrasti	Velvetleaf	Malvaceae	Dicot	U.S.A. (1984)
2.	Amaranthus albus	Tumble pigweed	Amaranthaceae	Dicot	Spain (1987)
3.	A. blitoides	Prostrate pigweed	Amaranthaceae	Dicot	Israel (1983), Spain (1986)
4.	A. cruentus	Smooth pigweed	Amaranthaceae	Dicot	Spain (1989)
5.	A. hybridus	Smooth pigweed	Amaranthaceae	Dicot	U.S.A. (1972), Italy (1980), France (1980), Switzerland (1982), Spain (1985), Israel (1986), South Africa (1993)
6.	A. lividus	Livid amaranth	Amaranthaceae	Dicot	Switzerland (1978), France (1981)
7.	A. palmeri	Palmer amaranth	Amaranthaceae	Dicot	U.S.A. (1993)
8.	A. powellii	Powell amaranth	Amaranthaceae	Dicot	Canada (1977), France (1982), Switzerland (1986), Czech Republic (1989), U.S.A. (1992)
9.	A. retroflexus	Redroot pigweed	Amaranthaceae	Dicot	Canada (1980), U.S.A. (1980), Germany (1980), France (1980), Switzerland (1982), Bulgaria (1984), Czech Republic (1985), Spain (1986), Poland (1991), Chile (1995)
10.	A. rudis	Common waterhemp	Amaranthaceae	Dicot	U.S.A. (1994)
11.	Ambrosia artemisiifolia	Common ragweed	Asteraceae	Dicot	Canada (1976), U.S.A. (1993)
12.	Arenaria serpyllifolia	Thymeleaf sandwort	Caryophyllaceae	Dicot	France (1980)
13.	Atriplex patula	Spreading orach	Chenopodiaceae	Dicot	Germany (1980)
14.	Bidens tripartita	Bur beggarticks	Asteraceae	Dicot	Austria (1979)
15.	Brachypodium distachyon	False brome	Poaceae	Monocot	Israel (1975)
16.	Brassica campestris	Birdsrape mustard	Brassicaceae	Dicot	Canada (1977)
17.	Bromus tectorum	Downy brome	Poaceae	Monocot	France (1981), Spain (1990)
18.	Capsella bursa-pastoris	Shepherd's-purse	Brassicaceae	Dicot	Poland (1984)
19.	Chamomilla suaveolens	Disc mayweed	Asteraceae	Dicot	United Kingdom (1989)

20.	Chenopodium album	Lambsquarters	Chenopodiaceae	Dicot	Canada (1973), Switzerland (1977), U.S.A. (1977), France (1978), New Zealand (1979), Belgium (1980), The Netherlands (1980), Germany (1980), Italy (1982), Czech Republic (1986), Spain (1987), Bulgaria (1989), United Kingdom (1989), Poland (1991), Norway (1994), Chile (1995), Slovenia (1996)
21.	C. ficifolium	Figleaved goosefoot	Chenopodiaceae	Dicot	Germany (1980), Switzerland (1986)
22.	C. polyspermum	Manyseeded goosefoot	Chenopodiaceae	Dicot	France (1980), Switzerland (1982), Germany (1988)
23.	C. strictum var. glaucophyllum	Late flowering goosefoot	Chenopodiaceae	Dicot	Canada (1976), Czech Republic (1989)
24.	Chloris inflata	Swollen fingergrass	Poaceae	Monocot	U.S.A. (1987)
25.	Conyza bonariensis	Hairy fleabane	Asteraceae	Dicot	Spain (1987), Israel (1993)
26.	C. canadensis	Horseweed	Asteraceae	Dicot	France (1981), Switzerland (1982), United Kingdom (1982), Poland (1983), Czech Republic (1987), Spain (1987), Belgium (1989), Israel (1993)
27.	Crypsis schoenoides	Swamp pricklegrass	Poaceae	Monocot	Israel (1995)
28.	Datura stramonium	Jimsonweed	Solanaceae	Dicot	U.S.A. (1992), Chile (1995)
29.	Digitaria sanguinalis	Large crabgrass	Poaceae	Monocot	France (1983), Poland (1995)
30.	Echinochloa crus-galli	Barnyardgrass	Poaceae	Monocot	U.S.A. (1978), Canada (1981), France (1982), Spain (1992), Poland (1995)
31.	Epilobium adenocaulon	American willowherb	Onagraceae	Dicot	Belgium (1980), United Kingdom (1981), Poland (1995)
32.	E. tetragonum	Square-stalked willowherb	Onagraceae	Dicot	France (1981), Germany (1988)
33.	Fallopia convolvulus	Climbing buckwheat	Polygonaceae	Dicot	Austria (1980), Germany (1988)
34.	Galinsoga ciliata	Hairy galinsoga	Asteraceae	Dicot	Germany (1980), Switzerland (1991)
35.	Kochia scoparia	Kochia	Chenopodiaceae	D.cot	U.S.A. (1976)
36.	Lolium rigidum	Rigid ryegrass	Poaceae	Monocot	Israel (1979), Australia (1988), Spain (1992)
37.	Lophochloa smyrnacea	Catstail	Poaceae	Monocct	Israel (1979)
38.	Matricaria matricarioides	Pineapple-weed	Asteraceae	Dicot	United Kingdom (1989)
39.	Panicum capillare	Witchgrass	Poaceae	Monocct	Canada (1981)

TABLE 1.3 (CONTINUED)
The Occurrence of Triazine-Resistant Weeds Worldwide

	Species	Common Name	Family	Weed Group	Country and Year
40.	*P. dichotomiflorum*	Fall panicum	Poaceae	Monocot	Spain (1981)
41.	*Phalaris paradoxa*	Hood canarygrass	Poaceae	Monocot	Israel (1979)
42.	*Plantago lagopus*	Plantain	Plantaginaceae	Dicot	Israel (1992)
43.	*Poa annua*	Annual bluegrass	Poaceae	Monocot	France (1978), Germany (1980), Belgium (1981), The Netherlands (1981), United Kingdom (1981), Japan (1982), Czech Republic (1988), U.S.A. (1994), Norway (1996)
44.	*Polygonum aviculare*	Prostrate knotweed	Polygonaceae	Dicot	The Netherlands (1987)
45.	*P. hydropiper*	Marshpepper smartweed	Polygonaceae	Dicot	France (1989)
46.	*P. lapathifolium*	Pale smartweed	Polygonaceae	Dicot	France (1979), Czech Republic (1982), Germany (1988), Spain (1991)
47.	*P. pennsylvanicum*	Pennsylvania smartweed	Polygonaceae	Dicot	U.S.A. (1990)
48.	*P. persicaria*	Ladysthumb	Polygonaceae	Dicot	New Zealand (1980), France (1980), Czech Republic (1989)
49.	*P. monspeliensis*	Rabbitfoot polypogon	Poaceae	Monocot	Israel (1979)
50.	*Raphanus raphanistrum*	Wild radish	Brassicaceae	Dicot	Australia (1999)
51.	*Senecio vulgaris*	Common groundsel	Asteraceae	Dicot	U.S.A. (1970), United Kingdom (1977), Germany (1980), Belgium (1982), Switzerland (1982), The Netherlands (1982), France (1982), Czech Republic (1988), Chile (1995), Norway (1996)
52.	*Setaria faberi*	Giant foxtail	Poaceae	Monocot	U.S.A. (1984), Spain (1987)

53.	*S. glauca*	Yellow foxtail	Poaceae	Monocot	Canada (1981), France (1981), U.S.A. (1984), Spain (1987)
54.	*S. verticillata*	Bristly foxtail	Poaceae	Monocot	Spain (1992)
55.	*S. viridis*	Green foxtail	Poaceae	Monocot	France (1982), Spain (1987)
56.	*S. viridis var. major*	Giant green foxtail	Poaceae	Monocot	France (1982)
57.	*Sinapis arvensis*	Wild mustard	Brassicaceae	Dicot	Canada (1994)
58.	*Solanum nigrum*	Black nightshade	Solanaceae	Dicot	France (1979), Italy (1980), Germany (1980), Belgium (1981), The Netherlands (1981), Switzerland (1983), United Kingdom (1983), Spain (1987), Poland (1995)
59.	*Sonchus asper*	Spiny sowthistle	Asteraceae	Dicot	France (1980)
60.	*Stellaria media*	Common chickweed	Caryophyllaceae	Dicot	Germany (1978)
61.	*Urochloa panicoides*	Liverseedgrass	Poaceae	Monocot	Australia (1996)

Source: Compiled from data in Reference 37.

Amaranthus spp. and four ALS-herbicide-resistant *Setaria* spp. have evolved in the midwestern United States. The rapid adoption of glyphosate-resistant soybean in the United States has curbed the rate of spread of many biotypes that evolved resistance to other herbicides in soybean (see Chapter 4). South American soybean growers face increasing problems of ACCase-herbicide-resistant populations of *Brachiaria plantaginea*, *Eriochloa punctata*, and *Sorghum sudanese.*

1.2.3.1.5   Canola

Eleven cases of herbicide-resistant weed biotypes have been reported in canola. The most serious of these are ACCase-herbicide-resistant *Avena fatua* and *Setaria viridis* and the ALS-herbicide (ethametsulfuron-methyl)-resistant *Sinapis arvensis*. While the ALS-herbicide-resistant *S. arvensis* is not currently widespread (approximately 200 ha) it is particularly troublesome, as ethametsulfuron-methyl is the only solution for removal of *Sinapis* from conventional canola crops. Fortunately, glyphosate- and glufosinate-resistant canola varieties have alleviated this problem for some growers.

1.2.3.2   Resistance in Other Ecosystems

1.2.3.2.1   Cotton

Given the intensive use of herbicides in cotton production it is surprising that only five weeds have been recorded as having evolved resistance to herbicides in cotton (Table 1.2). This low number, despite the propensity of monoculture cotton, is probably due to the multiplicity of different herbicides typically used throughout the cotton-growing season and the continued use of mechanical weed control, often including hand-removal of survivors. In the cotton-growing regions of the southern United States, *Sorghum halepense* has evolved ACCase-herbicide resistance; *Amaranthus palmeri*, *Eleusine indica*, and *S. halepense* have evolved dinitroaniline resistance; and *Xanthium strumarium* has resistance to MSMA/DSMA.

1.2.3.2.2   Perennial crops

Thirty-seven herbicide-resistant weed biotypes have evolved in orchards and vineyards worldwide (Table 1.2). The majority are triazine-resistant weeds of orchards (19 species) and 11 species with resistance to bipyridilium herbicides. On the whole, these have been easily controlled with alternative herbicides or cultural controls, but typically at a substantially increased cost of weed control to the grower. The economic impact of herbicide-resistant weeds in these crops has been far less than in annual cropping systems because of the alternatives available and the lower impact of weeds in general on perennial crops. Two recent and important cases of resistance in orchards are those of glyphosate-resistant *Lolium rigidum* in an apple orchard in Australia[39] and in an almond orchard in the United States and glyphosate-resistant *Eleusine indica* in a fruit orchard in Malaysia.[41] These cases are of interest not because of their economic importance in orchards but because they show that glyphosate-resistant weeds will evolve given sufficient selection pressure, a point that was debated in the early and mid-1990s by industry and academia.

1.2.3.2.3   Noncrop

There have been 35 herbicide-resistant weed biotypes recorded for roadside, railway, and industrial sites. Nineteen of these species have evolved resistance to triazine herbicides and eight to ALS herbicides. Considering that only about 10% of the total herbicide usage is for these purposes, this is well above average. A major reason is that, traditionally, only the least expensive herbicides are used, leading to repeated use of the same herbicide at high rates without other herbicides, or other control measures such as tillage or competition from crops. Resistance evolved to the most persistent herbicides that exert the greatest selection pressure. Persistence of the herbicide can derive from the inherent slow degradation of a herbicide used often at high rates or from the propensity to make multiple applications of a herbicide throughout the year.

It is notable that many of the same resistant biotypes occur on railway lines/roadsides and in adjacent fields, which may indicate that the spread of resistant weeds in either direction has been a factor in exacerbating resistance management strategies. This is certainly the case for ALS-herbicide-resistant *Kochia scoparia* where resistant populations selected through usage of ALS herbicides along railway lines and roadsides have spread and have subsequently been detected in potato fields of Idaho where no ALS herbicides had been used.

There are many alternatives for control of weeds in noncrop situations and undoubtedly many resistant weeds of noncrop situations have gone undetected due to the ease of changing herbicide mixtures when weed populations survive treatment.

1.2.4   Weeds That Are Resistance Generalists

Some weed species have a far greater propensity to evolve resistance than other species. Of the 152 weed species that have evolved resistance to one or more herbicide modes of action (MOA), 106 have evolved resistance to only 1 MOA, 28 species to 2 MOA, 10 species to 3 MOA, 2 species to 4 MOA, 2 species to 5 MOA, 3 species to 6 MOA, and 1 species (*Lolium rigidum*) has evolved resistance to 8 MOA. Grass weed species account for all the cases of weed species that have evolved multiple-resistance to five or more herbicide modes of action. These grasses are *L. rigidum* (8 MOA and 16 countries), *Echinochloa crus-galli* (6 MOA and 15 countries), *Avena fatua* (6 MOA and 16 countries), *Poa annua* (6 MOA and 15 countries), *Alopecurus myusoroides* (5 MOA and 8 countries), and *Eleusine indica* (5 MOA and 5 countries). These grasses have evolved resistance to many of the herbicides used against them. The first three account for hundreds of thousands of hectares of herbicide-resistant weed infestations throughout the world, representing the most intractable cases of herbicide resistance worldwide.

Few dicot weed species have evolved resistance to a wide range of herbicide modes of action. The most widespread herbicide-resistant dicot biotypes are *Chenopodium album* (18 countries), *Amaranthus retroflexus* (15 countries), *Conyza canadensis* (14 countries), *Senecio vulgaris* (12 countries), and *Solanum nigrum* (10 countries). *Amaranthus* spp. in general appear to be the most resistance prone of the dicot weeds. *A. retroflexus* has evolved resistance to 3 MOA; *A. hybridus*,

A. lividus, and *A. palmeri* to 3 MOA; *A. powellii, A. blitoides,* and *A. rudis* to 2 MOA; and *A. albus* and *A. cruentus* to 1 MOA. The *Amaranthus* spp. have shown the capacity to evolve resistance to ALS, triazine, urea, bypyridilium, and dinitro-aniline herbicides. It is also feared that *A. rudis* may have evolved resistance to glyphosate in the midwestern United States.[46]

Initially, triazine-resistant dicot weeds accounted for the greatest area infested by herbicide-resistant weeds. This has changed over the last 10 years and now ALS-herbicide-resistant weeds account for the greatest number of resistant species and probably the largest area affected by resistance. ALS-herbicide-resistant populations of *Kochia scoparia* and *Salsola iberica* are now so widespread throughout the northern United States that few growers consider using an ALS herbicide for control of these species.

The area infested by a given species provides the most agronomically significant description of the extent of resistance. Unfortunately, there are few estimates of areas, and most are only based on educated guesses. The 2 million ha of butachlor/thiobencarb cross-resistant *Echinochloa crus-galli* is a local scientist's estimation.[44] The millions of hectares of resistant *L. rigidum* in Australia are based on limited data. It may be more appropriate to see which weed species have evolved in the greatest numbers of locations, with the greatest numbers of reports. Using these criteria, the 25 most prevalent resistant species are summarized in Table 1.4. By analyzing the data amassed in the resistance Web site it is evident that some weed species are on the list due to biological factors, as they have the ability to evolve resistance to many herbicide groupings or in diverse agroecosystems. Some weed species appear on the list because they exist in many cropping systems, and a single herbicide group is used in all those systems. Many weed species could be there because, based on educated guesswork, they are in the top 76[47] or the top 180[24] listing of the world's worst weeds, and their distribution is widespread. Still, some of the top 76 weeds have not evolved herbicide resistance, including some of the top 10.

1.3   RESISTANCE IN THE DEVELOPED VS. DEVELOPING WORLD

Based on the survey by Heap,[37] the following countries are reporting the greatest number of herbicide-resistant biotypes: United States, 80; Australia, 32; Canada, 32; France, 30; Spain, 24; U.K., 19; Israel, 18; Germany, 15; Belgium, 15; Switzerland, 14; and Japan, 12. These and many other developed countries already have serious infestations of herbicide-resistant weeds, especially in major crops, and in the most productive and fertile areas where herbicides are most essential. The comparison of countries based on their present economies and high technology agriculture, how-ever, presents some important ironies and paradoxical challenges. The developing countries have not depended as much on herbicides due to economic limitations and the availability of cheap labor. They therefore have fewer problems with herbicide-resistant weeds, yet these developing countries suffer the greatest losses from weeds. Where food production is least efficient and subsistence farming is commonly practiced, the needs are acute for better weed control tools, including herbicides. As economies and farming efficiency improve, the farm workers presently used for hand-hoeing find more satisfying and profitable occupations elsewhere. Conversely, the more extensive use of herbicides will bring with it greater problems from resistant weeds and require better management of weeds and herbicides. Over the next decade this will be most evident for rice-growing regions in developing nations (see Chapter 6).

Resistance is widespread throughout the developed world. The control of weeds that have a propensity to evolve resistance (Table 1.4) is most likely to result in the rapid evolution of resistance. While resistance is inevitable wherever herbicides are persistently used, the preemptive practices outlined throughout this book can slow the evolutionary processes leading to resistance. Herbicide use is sporadic in the developing world, and thus resistance is found only in pockets. Where there is resistance (mainly in high-value plantation crops and in dwarf or semidwarf wheat and rice), the areas affected are quite extensive. As developing countries industrialize and herbicide use replaces farm labor, more cases can be expected of the same resistances as have occurred in the grass weeds of wheat and rice in the developed countries. It is typical of developing countries to develop and produce a single herbicide for a single weed problem, and farmers have fewer alternatives and are less likely to rotate herbicide chemistries than in the developed world. This happened in Hungary, where almost all the corn land was infested with triazine-resistant weeds because of monoculture and the local production of cheap atrazine. For reasons such as these we anticipate a rapid increase in the number of herbicide-resistance problems appearing in the developing nations.

1.4 SUMMARY

As will be elaborated in the following chapters, there have been considerable advances in our understanding of the causes, nature, genetics, mechanisms, and solutions for herbicide-resistant weeds since the first triazine-resistant *Senecio vulgaris* was documented 30 years ago. Understanding these factors is a necessary step in devising effective herbicide-resistance management strategies. However, implementing these resistance management strategies has proven to be the most difficult step. Cooperation among academia, industry, and growers is necessary in devising management strategies that are attractive for growers to adopt (see Chapter 7). Most growers still consider herbicide-resistance avoidance a low priority and do not change their weed control programs to avoid resistance because of financial or logistic constraints (see Chapter 7). New biotechnology-derived technologies, such as herbicide-resistant crops, will provide us with opportunities for management of existing herbicide-resistance problems, but in the long run may themselves cause future resistance problems. As outlined throughout this text the solutions to achieve sustainable weed management practices will differ between regions and agroecosystems but will, inevitably, involve more diversity in weed control technologies than is currently evident in many developed nations. It is hoped that this chapter and book can help catalyze more diverse weed control systems in world agriculture.

TABLE 1.4
The Top 25 Worst Herbicide-Resistant Weeds Weighted by Propensities in Countries, MOA, Sites, Hectares, and Cropping Systems

	Common Name	Species	No. Countries	No. MOA	No. Sites	No. Hectares	No. Cropping Regimes
1.	Rigid ryegrass	*Lolium rigidum*	16	8	7,000	836,400	6
2.	Wild oat	*Avena fatua*	16	6	22,100	2,941,200	4
3.	Redroot pigweed	*Amaranthus retroflexus*	15	3	11,500	31,900	10
4.	Lambsquarters	*Chenopodium album*	18	2	19,700	463,600	5
5.	Green foxtail	*Setaria viridis*	6	4	3,800	1,220,900	5
6.	Barnyardgrass	*Echinochloa crus-galli*	15	6	1,200	817,600	4
7.	Goosegrass	*Eleusine indica*	5	5	6,300	20,100	6
8.	Kochia	*Kochia scoparia*	4	3	50,400	189,200	4
9.	Horseweed	*Conyza canadensis*	14	4	1,400	7,300	7
10.	Palmer amaranth	*Amaranthus palmeri*	3	3	12,000	356,200	5
11.	Common groundsel	*Senecio vulgaris*	12	3	1,900	6,800	6
12.	Smooth pigweed	*Amaranthus hybridus*	8	2	10,200	32,900	4
13.	Annual bluegrass	*Poa annua*	15	6	1,100	5,200	4

14.	Blackgrass	*Alopecurus myosuroides*	13	4	1,900	9,300	3
15.	Black nightshade	*Solanum nigrum*	10	2	1,600	4,500	6
16.	Italian ryegrass	*Lolium multiflorum*	7	2	4,200	26,200	3
17.	Common waterhemp	*Amaranthus rudis*	2	2	8,300	25,800	5
18.	Common ragweed	*Ambrosia artemisiifolia*	3	2	2,200	15,900	3
19.	Prostrate pigweed	*Amaranthus blitoides*	3	2	800	2,500	4
20.	Powell amaranth	*Amaranthus powellii*	7	2	100	700	5
21.	Little seed canary grass	*Phalaris minor*	3	2	55,800	609,300	1
22.	Sumatran fleabane	*Conyza sumatrensis*	4	1	900	2,200	5
23.	Wild poinsettia	*Euphorbia heterophylla*	3	2	800	24,600	2
24.	American willowherb	*Epilobium adenocaulon*	5	2	900	300	2
25.	Hood canary grass	*Phalaris paradoxa*	3	2	800	2,500	2

Note: These 25 weeds were chosen by cycling through the whole database[37] five times summing the ranks for each of the 150 weed species. The weeds were then sorted and ranked separately by the number of countries, MOA, etc. for each of the categories. The cumulative rank for each species for each of the five categories was determined and the 25 with the highest ranks are shown. The rest may be seen on the Web site.[37] Despite being performed by numeric criteria, there is an arbitrariness of having large and small countries equalized. There is good genetic reason to consider the *Amaranthus* spp. as a single complex, which would enhance its position.

Source: Compiled from data in Reference 37.

REFERENCES

1. Hay, J. R., Gains to the grower from weed science, *Weed Sci.*, 22, 439, 1974.
2. Freed, V. H., Weed science: the emergence of a vital technology, *Weed Sci.*, 28, 621, 1980.
3. Avery, D.T., Saving the Planet with Pesticides and Plastic, Hudson Institute, Indianapolis, IN, 1995.
4. Herms, D. A. and Mattson, W. J., The dilemma of plants, to grow or to defend, *Q. Rev. Biol.*, 67, 283, 1992.
5. Pimentel, D. and Greiner, A., Environmental and socio-economic costs of pesticide use, in *Techniques for Reducing Pesticide Use: Economic and Environmental Benefits*, Pimentel, D., Ed., John Wiley & Sons, New York, 1997, chap. 4.
6. Heap, I. M., International survey of herbicide-resistant weeds: lessons and limitations, *Proc. Brit. Weed Cont. Conf.*, 3, 769, 1999.
7. Roush, R. T. and McKenzie, J. A., Ecological genetics of insecticide and acaricide resistance, *Annu. Rev. Entomol.*, 32, 361, 1987.
8. Harper, J. L., The evolution of weeds in relation to resistance to herbicides, *Proc. Brit. Weed Cont. Conf.*, 3, 179, 1956.
9. Abel, A. L., The rotation of weed killers, *Proc. Brit. Weed Cont. Conf.*, 1, 249, 1954.
10. Switzer, C. M., The existence of 2,4-D resistant strains of wild carrot, *Proc. North Eastern Weed Cont. Conf.*, 11, 315, 1957.
11. Whitehead, C. W. and Switzer, C. M., The differential response of strains of wild carrot to 2,4-D and related herbicides, *Can. J. Plant Sci.*, 43, 255, 1963.
12. Ryan, G. F., Resistance of common groundsel to simazine and atrazine, *Weed Sci.*, 18, 614, 1970.
13. LeBaron, H. and Gressel, J., Eds., *Herbicide Resistance in Plants*, Wiley, New York, 1982.
14. Gressel, J. and Evron, Y., Pyridate is not a two-site inhibitor, and may be more prone to evolution of resistance than other phenolic herbicides, *Pestic. Biochem. Physiol.*, 44, 140, 1992.
15. Kemp, M. S., Moss, S. R., and Thomas, T. H., Herbicide resistance in *Alopecurus myosuroides*, in *Managing Resistance to Agrichemicals: From Fundamental Research to Practical Strategies*, Green, B. M., LeBaron, H. M., and Moberg, W. K., Eds., ACS Symp. Ser. 421, American Chemical Society, Washington, D.C., 1990, 376.
16. Hall, L. M., Moss, S. R., and Powles, S. B., Mechanisms of resistance to aryloxyphenoxypropionate herbicides in two resistant biotypes of *Alopecurus myosuroides* (black-grass): herbicide metabolism as a cross-resistance mechanism, *Pestic. Biochem. Physiol.*, 57, 87, 1997.
17. Hyde, R. J., Hallahan, D. L., and Bowyer, J. R., Chlorotoluron metabolism in leaves of resistant and susceptible biotypes of the grass weed *Alopecurus myosuroides*, *Pestic. Sci.*, 47, 185, 1996.
18. Cummins, I., Moss, S., Cole, D. J., and Edwards, R., Glutathione transferases in herbicide-resistant and herbicide-susceptible black-grass *(Alopecurus myosuroides)*, *Pestic. Sci.*, 51, 244, 1997.
19. Cummins, I., Cole, D. J., and Edwards, R., Purification of multiple glutathione transferases involved in herbicide detoxification from wheat (*Triticum aestivum* L.) treated with the safener fenchlorazole-ethyl, *Pestic. Biochem. Physiol.*, 59, 35, 1997.
20. Cummins, I., Cole, D. J., and Edwards, R., A role for glutathione transferases functioning as glutathione peroxidases in resistance to multiple herbicides in black-grass, *Plant J.*, 18, 285, 1999.

21. Hall, L. M., Holtum, J. A. M., and Powles, S. B., Mechanisms responsible for cross resistance and multiple resistance, in *Herbicide Resistance in Plants, Biology and Biochemistry*, Powles, S. B. and Holtum, J. A. M., Eds., Lewis Publishers, Boca Raton, FL, 1994, 243.

22. Souza-Machado, V., Inheritance and breeding potential of triazine tolerance and resistance in plants, in *Herbicide Resistance in Plants*, LeBaron, H. M. and Gressel, J., Eds., John Wiley & Sons, New York, 1982, 257.

23. Scott, K. R. and Putwain, P. D., Maternal inheritance of simazine resistance in a population of *Senecio vulgaris, Weed Res.*, 21, 137, 1981.

24. Holm, L., Doll, J., Holm, J., Pancho, E., and Herberger, J., *World's Weeds: Natural Histories and Distribution*, John Wiley & Sons, New York, 1997, 1129.

25. Radosevich, S. R. and Appleby, A. P., Studies on the mechanism of resistance to simazine in common groundsel, *Weed Sci.*, 21, 497, 1973.

26. Shimabukuro, R. H., Detoxication of herbicides, in *Weed Physiology, Vol. II: Herbicide Physiology*, Duke, S. O., Ed., CRC Press, Boca Raton, FL, 1985, 215.

27. Moreland, D. E., Inhibitors of chloroplast electron transport: structure activity relations, in *Progress in Photosynthesis Research*, Vol. 3, Metzner, H., Ed., Lichtenstern, Tübingen, Germany, 1969, 1693.

28. Gasquez, J. and Darmency, H., Variation for chloroplast properties between two triazine resistant biotypes of *Poa annua* L., *Plant Sci. Lett.*, 30, 99, 1983.

29. Gasquez, J., Mouemar, A., and Darmency, H., Triazine resistance in *Chenopodium album* L. occurrence and characteristics of an intermediate biotype, *Pestic. Sci.*, 16, 390, 1985.

30. Arntzen, C. J., Pfister, K., and Steinback, K. E., The mechanism of chloroplast triazine resistance: alterations in the site of herbicide action, in *Herbicide Resistance in Plants*, LeBaron, H. M. and Gressel, J., Eds., John Wiley & Sons, New York, 1982, 185.

31. Arntzen, C. J. and Duesing, J. H., Chloroplast-encoded herbicide resistance, in *Advances in Gene Technology: Molecular Genetics of Plants and Animals*, Downey, K., Voellmy, R. W., Ahmad, F., and Schultz, J., Eds., Academic Press, New York, 1983, 273.

32. Hirschberg, J. and McIntosh, L., Molecular basis of herbicide resistance in *Amaranthus hybridis* L., *Science*, 222, 1346, 1983.

33. Michel, H. and Deisenhofer, J., Relevance of the photosynthetic reaction center from purple bacteria to the structure of photosystem II, *Biochemistry*, 27, 1, 1988.

34. Faulkner, J. S. D., Breeding herbicide-tolerant crop cultivars by conventional methods, in *Herbicide Resistance in Plants*, LeBaron, H. M. and Gressel, J., Eds., John Wiley & Sons, New York, 1982, 235.

35. Gressel, J. and Segel, L. A., Interrelating factors controlling the rate of appearance of resistance: the outlook for the future, in *Herbicide Resistance in Plants*, LeBaron, H. M. and Gressel, J., Eds., John Wiley & Sons, New York, 1982, 325.

36. Radosevich, S. R. and Holt, J. S., Physiological responses and fitness of susceptible and resistant weed biotypes to triazine herbicides, in *Herbicide Resistance in Plants*, LeBaron, H. M. and Gressel, J., Eds., John Wiley & Sons, New York, 1982, 163.

37. Heap, I. M., International survey of herbicide-resistant weeds and state by state surveys of the U.S., WeedSmart, Corvallis, OR, 2000. http://www.weedscience.com

38. Pratley, J., Urwin, N., Stanton, R., Baines, P., Broster, J., Cullis, K., Schafer, D., Bohn, J., and Krueger, J., Resistance to glyphosate in *Lolium rigidum*. I. Bioevaluation, *Weed Sci.*, 47, 405, 1999.

39. Powles, S. B., Loraine-Colwill, D. F., Dellow, J. J., and Preston, C., Evolved resistance to glyphosate in rigid ryegrass (*Lolium rigidum*) in Australia, *Weed Sci.*, 46, 604, 1998.

40. Tran, M., Baerson, S., Brinker, R., Casagrande, L., Faletti, M., Feng, Y., Nemeth, M., Reynolds, T., Rodriguez, D., Schafer, D., Stalker, D., Taylor, N., Teng, Y., and Dill, G., Characterization of glyphosate resistant *Eleusine indica, Proc. 17th Asian-Pacific Weed Sci. Soc. Conf.*, Bangkok, Thailand, 1999, 527.

41. Lee, L. J. and Ngim, J., A first report of glyphosate resistant goosegrass (*Eleusine indica*) in Malaysia, *Pest Manage. Sci.,* 56, 336, 2000.

42. Leah, J. M., Caseley, C. J., Riches, C. R., and Valverde, B., Elevated activity of aryl-acylamidase is associated with propanil resistance in jungle rice *Echinochloa colona* (L.) Link., *Pestic. Sci.*, 42, 281, 1999.

43. Carey, V. F., III, Hoagland, R. E., and Talbert, R. E., Resistance mechanisms in propanil-resistant barnyardgrass. II. *In vivo* metabolism of the propanil molecule, *Pestic. Sci.*, 49, 333, 1997.

44. Huang, B.-Q. and Gressel, J., Barnyardgrass *(Echinochloa crus-galli)* resistance to both butachlor and thiobencarb in China, *Resistant Pest Manage.,* 9, 5, 1997.

45. Fischer, A. J., Areh, C. M., Bayer, D. E., and Hill, J. E., Herbicide-resistant *Echinochloa oryzoides* and *E. phyllopogon* in California *Oryza sativa* fields, *Weed Sci.,* 48, 225, 2000.

46. Zelaya, I. A. and Owen, M. D. K., Differential response of common waterhemp (*Amaranthus rudis*) to glyphosate in Iowa, *Weed Sci. Soc. Am. Abstr.,* 40, 151, 2000.

47. Holm, L. G., Plucknett, J. D., Pancho, L. V., and Herberger, J. P., *The World's Worst Weeds, Distribution and Biology,* University Press of Hawaii, Honolulu, 1977, 609.

48. Retzinger, E. J. and Mallory-Smith, C., Classification of herbicides by site of action for weed resistance management strategies, *Weed Technol.*, 11, 384, 1997.

49. Schmidt, R. R., HRAC classification of herbicides according to mode of action, *Proc. Brit. Crop Prot. Conf. — Weeds*, 1997, 1133.

2 Biochemical Mechanisms, Inheritance, and Molecular Genetics of Herbicide Resistance in Weeds

Christopher Preston and Carol A. Mallory-Smith

CONTENTS

0-8493-2219-7/01/$0.00+$.50
© 2001 by CRC Press LLC

2.1 INTRODUCTION

As discussed in Chapter 1, herbicide resistance has been reported to most herbicide chemical classes and has been documented in 147 different weed species.[1] A complete and updated list can be found at http://www.weedscience.com. Resistance occurs as a result of heritable changes to biochemical processes that enable plant survival when treated with a herbicide. This chapter deals with the specific changes to plant biochemistry that endow resistance to herbicides. Any biochemical change that allows a plant to survive herbicide application can be selected. This means that even for the same herbicide, resistance can be endowed by a number of different biochemical mechanisms. Resistance can result from changes to the herbicide target site such that binding of the herbicide is reduced, or over-expression of the target site may occur. Alternatively, there may be a reduction in the amount of herbicide that reaches the target enzyme through detoxification, sequestration, or reduced absorption of herbicide. Finally, the plant may survive through the ability to protect plant metabolism from toxic compounds produced as a consequence of herbicide action. This chapter considers major mechanisms of resistance to herbicides in weed species where resistance has evolved in the field.

2.2 TARGET-SITE-BASED RESISTANCE

All herbicides act by binding or otherwise interacting with one or more proteins with consequent negative effects on plant metabolism or growth. Plants can become resistant to the effects of herbicides through modifications to these proteins that reduce or eliminate the ability of the herbicide to bind or interact. In such cases, resistance is described as being target-site-based. An alternative type of target-site-based resistance is over-production of herbicide-binding proteins. Target-site-based resistance mechanisms have been extensively reviewed in the past few years.[2–4] Reviews of target-site resistance to the PS II-,[5] ALS-,[6,7] and ACCase-[8,9] inhibiting herbicides have been extensive and, therefore, need not be repeated here. This section will only briefly cover areas reviewed previously and will focus on consideration of new information.

2.2.1 Resistance to Photosystem II-Inhibiting Herbicides

Photosynthetic electron transport is inhibited by a number of different herbicide chemical classes, including the triazines, ureas, and nitriles that block electron transport on the reducing side of photosystem II (PS II).[5] This blockage of electron flow leads to production of excess singlet oxygen, which results in destruction of lipids and chlorophyll.

The first example of resistance to PS II-inhibiting herbicides was reported in 1970,[10] and many new examples of resistance occurred shortly thereafter. By 1990, more than 50 weed species had populations with resistance to PS II-inhibiting herbicides.[11] Triazine resistance has now been documented in 61 different weed species around the world.[1] With few exceptions, resistance to PS II-inhibiting herbicides is target-site-based and due to changes in the herbicide-binding domain on the D1 protein of PS II.[5] Cross-resistance patterns vary among the various chemical classes of PS II-inhibiting herbicides. Usually plants that are resistant to triazines are not cross-resistant to the substituted ureas or nitriles.[12] The molecular basis of most field-selected triazine-resistant weeds has been identified as a single amino acid substitution of Ser 264 to Gly in the D1 protein encoded by the chloroplastic *psbA* gene.[13] This amino acid substitution has been identified in resistant populations of numerous weed species and removes a hydrogen bond that is important for binding herbicides (Figure 2.1).

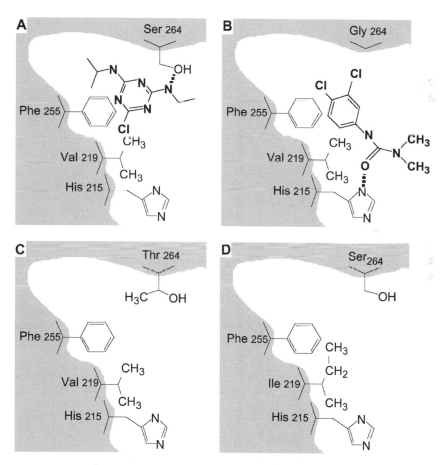

FIGURE 2.1 The effect on binding of herbicides of mutations within the D1 protein of PS II Wild type D1 protein (A), Gly 264 substitution (B), Thr 264 substitution (C), and Ile 219 substitution (D). Binding of atrazine (A) and diuron (B) are shown. Ser 264 provides a hydrogen bond to atrazine and the phenyl ring of Phe 255 is involved in stacking interactions with the triazine ring of atrazine. His 215 provides a hydrogen bond to diuron and the phenyl ring of Phe 255 is involved in stacking interactions with the phenyl ring of diuron. Val 219 is the amino acid sitting directly above His 215 in α-helix IV. The positions of amino acids and herbicides are based on the structures shown for the reaction center of *Rhodopseudomonus viridis* (as in [14]).

Apart from the Ser 264 to Gly change, there are only two other mutations in the D1 protein reported in weed populations resistant to PS II-inhibiting herbicides. A population of *Portulaca oleracea* resistant to linuron and atrazine has a single amino acid substitution at the Ser 264 resulting in a Thr substitution.[15,16] A Val 219 to Ile substitution confers diuron and metribuzin resistance in two different populations of *Poa annua* from fields in Oregon.[17] In the latter case, there was no change at Ser 264 or in any other positions in the herbicide-binding region. Previously, the Val 219 to Ile substitution had been reported in green algae but not in a higher plant species under field selection.[13]

The PS II reaction center proteins of higher plants are homologous to the reaction centers of photosynthetic bacteria such as *Rhodopseudomonas viridis*.[5,13] The reaction center of *R. viridis* has been crystallized, providing an insight into the likely structure of the plant PS II reaction center.[18] Crystals of the *R. viridis* reaction center containing bound herbicides[19–21] provide evidence for the role of important amino acids in herbicide binding (Figure 2.1). Ser 264 provides a hydrogen bond to triazine herbicides; however, substituted urea herbicides bind deeper into the Q_B binding site, hydrogen bonding with His 215. Phe 255 provides a hydrophobic interaction with both groups of herbicides. Thus the Ser 264 to Gly 264 change removes a hydrogen bond necessary for binding of triazine herbicides, but not substituted urea herbicides. Therefore, these mutants are highly resistant to triazine herbicides only.[13] The Ser 264 to Thr 264 change does not remove the hydrogen bond, but may distort its position. This change most likely causes interference with entry of herbicides to the Q_B site, or sterically interferes with herbicide interaction with Phe 255. In this case, resistance to both groups of herbicides is found.[15] The Val 219 to Ile 219 change is a conservative change. However, Val 219 is positioned immediately above His 215 in α-helix IV.[14] As Ile has a larger side chain than Val, this change may interfere with binding of herbicides to His 215. Therefore, the Val 219 to Ile 219 change would be predicted to result in substituted urea resistance. Triazine resistance also might be expected due to interference in the hydrophobic interactions between Phe 255 and the herbicide or changing the position of a tightly bound water molecule that provides an additional hydrogen bond to triazines.[21]

Because the mutations that confer resistance to PS II-inhibiting herbicides are encoded on plastid DNA, resistance is maternally inherited.[22] However, very occasional transmission of PS II resistance through pollen is reported, particularly where large numbers of crosses are made.[23] In addition, recent reports have indicated heterogeneity of chloroplasts within plants.[24] This raises the possibility that some plants resistant to PS II inhibitors may revert to the susceptible type with time through decreases in the proportion of plastids containing the mutation.

2.2.2 RESISTANCE TO MICROTUBULE ASSEMBLY-INHIBITING HERBICIDES

The dinitroaniline chemical family, which contains several herbicides including ethalfluralin, pendimethalin, prodiamide, oryzalin, and trifluralin, inhibits the assembly of microtubules. The microtubules are formed when heterodimer subunits of α- and β-tubulin polymerize. The dinitroaniline herbicides bind to tubulin, prevent polymerization, and so prevent cell division and cell elongation.[25]

Resistance to the dinitroanilines has been reported in seven species,[1] of which dinitroaniline-resistant *Eleusine indica* has been the most widely studied. *E. indica* populations with either high (R) or intermediate (I) dinitroaniline resistance have been identified. When dinitroaniline resistance was last reviewed,[26] the mechanism of resistance in *E. indica* was suggested to be a difference in tubulin structure based on differences in electrophoresis of tubulin subunits. Since then, the mechanism of resistance and the mutations endowing resistance have been identified in *E. indica*. Resistance has been established to result from single amino acid substitutions in α-tubulin.[27–29] When sequences of the cDNA of α-tubulin were compared, base changes were identified in the R and I types compared to α-tubulin cDNA from susceptible (S) populations. The R population contained two point mutations that resulted in amino acid changes from Thr 239 to Ile and Ala 340 to Thr in the TUA1 and TUA2 α-tubulin genes, respectively.[29] The authors speculated the change in hydrophobicity with these amino acid substitutions could alter the binding of herbicides to the protein. The I population had a different substitution, from Met 268 to Thr in TUA1, which also would decrease hydrophobicity.[29] The difference in amino acid substitutions between the R and I populations demonstrates that different mutations within α-tubulins differentially affect binding of dinitroaniline herbicides.

Four β-tubulin genes from the R and S populations were sequenced and no mutations that would result in amino acid replacements were found.[30] This provides further evidence that β-tubulin genes are not involved in dinitroaniline resistance in *E. indica*. Phylogenetic analysis of the β-tubulin genes from resistant and susceptible populations indicated that resistance in *E. indica* originated independently at multiple locations rather than spreading from one site.[30]

Definitive evidence that resistance to dinitroaniline herbicides is due to amino acid modifications in the α-tubulin gene was obtained when the gene containing the Thr 239 to Ile substitution was used to transform maize and tobacco. The transforming α-tubulin gene was shown to confer dinitroaniline resistance in maize cells and regenerated tobacco plants.[27,31] The substitution either changes the configuration to the herbicide-binding site or increases the stability of the interaction between α- and β-tubulin.[28]

With the dinitroaniline-resistant *E. indica*, reciprocal crosses between the R and S or I and S populations and subsequent analysis of the F_2 and F_3 generations established that dinitroaniline resistance is controlled by a single, recessive nuclear gene.[32] No plants with intermediate resistance were found when crosses between the R and S populations were made, which indicated that the I population is not a hybrid between the R and S populations. This was confirmed when the α-tubulin gene of the I population was sequenced showing an amino acid substitution different from the R population.[29] Analysis of crosses between the I and S populations showed that intermediate resistance also is controlled by a single, recessive nuclear gene.[33] Similarly, Jasieniuk et al.[34] reported that a single, recessive nuclear gene also controls trifluralin resistance in *Setaria viridis*. However, when Wang et al.[35] crossed the trifluralin-resistant *S. viridis* with *S. italica* (foxtail millet), trifluralin resistance in the F_2 populations derived from selfed F_1 plants was recessive, but apparently not monogenic. This indicates that minor genes play a part in trifluralin resistance in the interspecific hybrid. The authors suggested several possibilities for the difference in reports of the inheritance of trifluralin resistance. Segregation may have been affected by interspecific hybridization or the resistance gene is tightly linked to a detrimental factor that is conserved over backcross generations. They further suggested that there might be two linked loci, one more important for resistance than the other, with the additive action of both loci required for resistance to high doses of trifluralin. This study points out that the inheritance of resistance may be more complex than is sometimes reported.

2.2.3 RESISTANCE TO ACETOLACTATE SYNTHASE-INHIBITING HERBICIDES

At least five different chemical families of herbicides inhibit acetolactate synthase (ALS), also called acetohydroxyacid synthase. These are the sulfonylurea, imidazolinone, triazolopyrimidine, pyridinyloxybenzoate, and sulfonylaminocarbonyltriazolinone chemistries.[36] ALS is the first common enzyme in the biosynthetic pathway for the production of the branched-chain amino acids. The exact binding site for herbicides on the enzyme has not been determined but has been suggested to be a residual quinone-binding site based on homology with pyruvate oxidase.[37]

The ALS-inhibiting herbicides were introduced into world agriculture in the 1980s and have been used extensively worldwide[38] with the result that resistance has evolved in a large number of weed species.[1] The number of weed species with populations resistant to ALS-inhibiting herbicides has increased rapidly over the past decade and now numbers 58, second only to triazine resistance.[1] In all but a few instances, resistance to the ALS-inhibiting herbicides is due to a less sensitive ALS. In contrast to triazine resistance, target-site-based resistance to the ALS-inhibiting herbicides can be conferred by a number of different point mutations. Amino acid substitutions that have been reported to provide resistance to ALS-inhibiting herbicides in weeds are listed in Table 2.1. Differences occur in target-site cross-resistance among the different chemical classes of herbicides that inhibit ALS. The differences are related to particular amino acid substitutions that occur within the binding region. Saari et al.[6] suggested that target-site cross-resistance between the sulfonylureas and the triazolopyrimidines is common and related to mutations at Pro 197, while target-site cross-resistance between the sulfonylureas and imidazolinones is less predictable.[50]

Multiple point mutations at Pro 173 (equivalent to Pro 197 in *Arabidopsis thaliana*) in ALS in *Kochia scoparia* have been reported.[42,44] Indeed, six different substitutions of Ala, Arg, Glu, Leu, Ser, or Thr for Pro 173 have been observed in different *K. scoparia* populations. In addition to having different amino acid substitutions, the populations were from different geographic regions, indicating that there were multiple independent selections of ALS-resistant populations.

Analysis of three *Sisymbrium orientale* populations resistant to ALS inhibitors revealed differences in amino acid substitutions in ALS among populations.[41] Two of the populations, from different geographic regions, had the same Trp to Leu change, while the third population had a Pro to Ile substitution, the latter resulting

TABLE 2.1
Amino Acid Substitutions Endowing Resistance to ALS Inhibitors in Resistant Weed Populations

Species	Population	Amino Acid Substitution[a]	Ref.
Amaranthus sp.	Iowa	Trp 591 · Leu	39
A. rudis	Illinois	Trp 591 · Leu	40
Brassica tournefortii	WBT1	Pro 197 · Ala	41
Kochia scoparia	KS-R	Pro 197 · Thr	42
	ND-R	Pro 197 · Arg	42
	MAN-R	Pro 197 · Leu	42
	MT-R	Pro 197 · Gln	42
	ID#5-R, TX-R	Pro 197 · Ala	42
	SLV-R	Pro 197 · Ser	42
	Illinois	Trp 591 · Leu	43
Lactuca serriola	Idaho	Pro 197 · Thr	44
	SLS1	Pro 197 · His	45
Salsola iberica	Washington	Pro 197 · Leu	Cited in 3
Sisymbrium orientale	SS01	Pro 197 · Ile	41
	NS01, SS03	Trp 591 · Leu	41
Xanthium sp.	MO-XANST	Trp 591 · Leu	46
	MI-XANST	Ala 122 · Thr	47
		Ala 205 · Val	48

[a] To aid comparison, the position numbers of amino acids for the various species have been altered to those of *Arabidopsis thaliana*.[49]

from two nucleotide substitutions.[41] Whether the two mutations occurred at the same time or whether they were due to separate events cannot be determined. However, differences in the point mutations among the populations provide strong evidence that there are independent selections for ALS resistance. No differences in the substrate binding of the enzyme were measured, but two of the resistant populations had a higher maximum rate of reaction, which could indicate over-expression of ALS in these populations.

The different patterns of target-site cross-resistance among ALS-inhibiting herbicides suggest that the sulfonylurea and imidazolinone herbicides bind to different areas of the herbicide-binding pocket on ALS (Figure 2.2). Trp 591 most likely interacts via hydrophobic interactions with aromatic ring systems of the herbicides. Substitution of Leu for Trp removes this hydrophobic interaction. This mutation results in cross-resistance to sulfonylurea and imidazolinone herbicides (Table 2.1). Substitutions at Ala 122 result in resistance to imidazolinone herbicides only.[47] Clearly, this amino acid must be positioned in proximity to where imidazolinone herbicides bind. Substitutions at Pro 197 commonly result in strong resistance to sulfonylurea herbicides with little or no resistance to imidazolinone herbicides.[6] However, some mutations result in resistance to imidazolinone herbicides as well.[41]

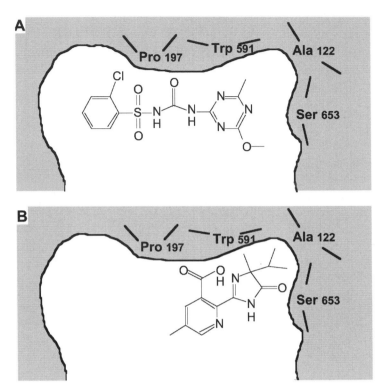

FIGURE 2.2 Proposed interaction of sulfonylurea (A) and imidazolinone (B) herbicides with amino acids of acetolactate synthase. This interpretation is based on cross-resistance patterns of ALS mutants.[3] The sulfonylurea herbicides interact with Trp 591, probably through hydrophobic interactions, and amino acids around Pro 197. The imidazolinone herbicides interact with Trp 591 and Ser 653; the latter amino acid may provide a hydrogen bond to the herbicide.

Substitutions at Pro 197 would change the tertiary structure of ALS and may displace an amino acid essential for binding of sulfonylurea herbicides. Depending on the amino acid substitution at Pro 197, interference with binding of imidazolinone herbicides might also occur.

Isolines of *Lactuca sativa* differing in resistance to ALS-inhibiting herbicides were produced by hybridizing *L. serriola* resistant to ALS inhibitors with herbicide-susceptible *L. sativa*, and then back-crossing to *L. sativa* for five generations.[51] The resistant isoline contained a Pro 197 to His substitution, previously identified in the resistant *L. serriola* parent, and a Ser 431 to Asn substitution. The amino acid substitutions in the resistant isoline had an adverse effect on the enzyme and the specific activity of ALS from the resistant isoline was reduced compared to the specific activity of the susceptible isoline. Feedback inhibition by valine, leucine, and isoleucine also was reduced in the resistant isoline, while the concentration of all three amino acids was higher in seed, and valine and leucine were higher in the leaves of the resistant isoline when compared to the susceptible isoline. Despite reduced specific activity of ALS, biomass production of the resistant isoline was faster than biomass production for the susceptible isoline, and seed size also was larger. These studies revealed differences in the two isolines that have not been shown to occur at the whole plant level in fitness and competition studies with ALS-resistant and -susceptible weeds.[52]

Resistance to the ALS-inhibiting herbicides in all species investigated is controlled by a single, nuclear-encoded gene that is either dominant or has incomplete dominance (also referred to as semidominant).[41,53-55] Heterozygous plants often show symptoms when treated with ALS-inhibiting herbicides but do not die.[53] The response of F_1 hybrids, produced by crossing ALS-inhibiting herbicide-resistant and -susceptible *S. orientale* populations, to chlorsulfuron was intermediate to that of the parents, indicating that the trait is nuclear encoded and not cytoplasmic.[41] The F_1 hybrid was more sensitive to chlorsulfuron than the resistant parent, which suggests the gene has incomplete rather than complete dominance. F_2 generations of several different species show segregation ratios of either 3:1 or 1:2:1 for the resistance trait, which indicates it is controlled by a single gene.[41,53-55] This pattern of inheritance of resistance to ALS-inhibiting herbicides appears independent of the specific amino acid modifications that endow resistance. This suggests that each of the known amino acid substitutions within ALS will result in dominant-resistance inheritance.

2.2.4 RESISTANCE TO ACETYL-COENZYME A CARBOXYLASE-INHIBITING HERBICIDES

Acetyl-coenzyme A carboxylase (ACCase) is the target site of the aryloxyphenoxy-propanoate (APP) and cyclohexanedione (CHD) herbicides.[56,57] ACCase is the first dedicated enzyme in the biosynthetic pathway for lipid synthesis.[8,9] Herbicides that inhibit ACCase are often termed graminicides as they only affect grass species with virtually no effect on a wide range of dicotyledonous species.[8,58] For this reason, ACCase-inhibiting herbicides are used for grass weed control in dicotyledenous crop species. Grasses contain two different ACCase enzymes with about 80% of the activity being associated with the plastid form of the enzyme. The plastidic enzyme of nongrasses is structurally different from that of grasses, being composed of different subunits as compared to a single protein of about 220 to 230 kDa in grasses.[59] All plant species have a separate cytoplasmic multidomain ACCase. Only the plastidic enzyme of grasses is sensitive to the APP and CHD herbicides.[60]

Resistance in grass weeds following widespread usage of ACCase-inhibiting herbicides has been widely reported,[1,8,58] with 19 different grass weeds with populations resistant to ACCase inhibitors.[1] With few exceptions, resistance to the ACCase-inhibiting herbicides has been attributed to an insensitive ACCase; however, the difficulty in extracting and purifying ACCase has raised questions about the conclusions reached in some studies.[9,61] The mutation(s) responsible for resistance to the ACCase-inhibiting herbicides have not been identified, unlike PS II and ALS resistances. Therefore, studies on the molecular genetics of target-site-based resistance to ACCase-inhibiting herbicides have not been conducted. However, it is clear from differences in target-site cross-resistance patterns of different resistant populations that multiple mutations within ACCase conferring resistance are possible.[3,62]

The plastidic ACCase from grass species is a large enzyme containing each of the four domains that in nongrasses are encoded on separate proteins.[59] These are the biotin carboxylase, biotin carboxy carrier protein (BCCP), and carboxyltransferase α and β (Figure 2.3). Therefore, the plastidic ACCase of grasses is similar in structure to the cytoplasmic ACCase of all species, including grasses. The only difference is that the grass plastidic ACCase is susceptible to herbicides whereas the cytoplasmic ACCase is tolerant. Elegant experiments by Nikolskaya et al.[63] constructed chimeric ACCase from wheat by combining parts of the herbicide-sensitive plastidic and herbicide-tolerant cytoplasmic proteins. These experiments localized herbicide susceptibility of the plastidic ACCase to a 411-amino acid region containing the carboxyltransferase β domain (Figure 2.3). Therefore, it seems likely that when amino acid modifications responsible for resistance to ACCase-inhibiting herbicides are discovered they will occur around the carboxyltransferase β domain.

A single dominant or partially dominant gene confers target-site resistance to ACCase-inhibiting herbicides in resistant populations of *Avena* spp.[66–68] and *Lolium* spp.[69,70] A single gene conferred resistance to both APP and CHD herbicides in an *A. fatua* population.[67] In a subsequent study, resistance in two different populations was controlled by different alleles of the same gene.[71]

2.2.5 RESISTANCE TO GLYPHOSATE

Glyphosate, the world's most widely used herbicide, inhibits 5-enolpyruvylshikimate-3-phosphate (EPSP) synthase, an enzyme in the biosynthetic pathway for the production of aromatic amino acids.[38,72] Glyphosate is a nonselective herbicide effective in controlling many plants. Given the importance of glyphosate, it is understandable that considerable effort was expended attempting to create herbicide-resistant forms of EPSP synthase, including site-directed mutation of the active site of the enzyme.[73] However, these experiments demonstrated the difficulty in achieving a glyphosate-resistant, yet catalytically active, EPSP synthase. The conclusion drawn from these studies was that target-site-based resistance to glyphosate would be very unlikely to evolve in the field.[72]

Resistance to glyphosate appeared in the field in populations of *Lolium rigidum* in Australia in 1995 following more than 15 years of use of the herbicide.[74,75] Since

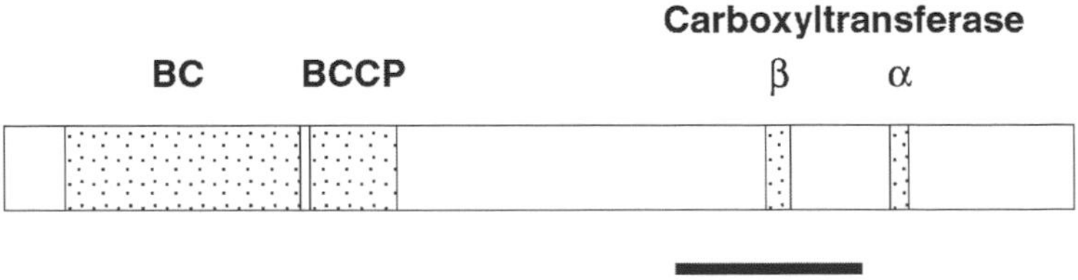

FIGURE 2.3 Linear structure of plastidic acetyl-coenzyme A carboxylase from wheat showing biotin carboxylase, biotin carboxyl carrier protein, and carboxyltransferase β and α domains. The solid line underneath indicates the position of the 411 amino acid region that endows herbicide sensitivity in the plastidic ACCase.[63] (From a comparison of the plastidic sequence[64] with the cytoplasmic sequence.[65])

then glyphosate resistance has occurred in populations of *Eleusine indica* in Malaysia.[76] Importantly, one glyphosate-resistant population of *E. indica* from Malaysia has been found to contain an EPSP synthase with reduced sensitivity to glyphosate.[77] The enzyme of the resistant plants is fivefold less sensitive to glyphosate while at the whole plant level there is threefold resistance.[77]

Two amino acid modifications were identified when the gene for EPSP synthase of the resistant population of *E. indica* was sequenced.[78] One of the amino acid substitutions, Ser substitution for Pro 106, occurred within the putative glyphosate-binding site.[73] This amino acid substitution in EPSP synthase is known to confer about tenfold resistance to glyphosate.[73]

2.2.6 CONCLUSIONS — TARGET-SITE-BASED RESISTANCE

The early identification of target-site-based resistance to the PS II-inhibiting herbicides and the identification of the single amino acid modification involved had tended to influence thinking about herbicide resistance. To some extent, a dogma of only one amino acid substitution endowing resistance to herbicides prevailed. Therefore, assumptions were made that the same mutation had occurred. For example, until the identification of the Val 219 to Ile substitution in *Poa annua*, only changes at Ser 264 had been identified in field-selected PS II-resistant weed species.[13,17] It may be that few mutations will confer resistance to the PS II inhibitors or it may be that researchers were too quick to accept that resistance would be conferred by the same mutation in all species. The actual mutation in many species has not been elucidated; therefore, there may be other mutations present in weed species that have resistance to the PS II inhibitors. In contrast, many different point mutations have been identified in weed populations with resistance to the ALS-inhibiting herbicides. These point mutations are related to patterns of target-site cross-resistance. Likewise, the different target-site cross-resistance patterns that have been identified for ACCase resistance suggest that different point mutations in the ACCase gene will account for the different patterns of cross-resistance to these herbicides. Thus far, only two mutations have been shown to be responsible for dinitroaniline resistance providing different levels of resistance and a single mutation identified in EPSP synthase. Based on experience with PS II, researchers should not reject the idea that multiple mutations that confer resistance to herbicides may be possible in other target enzymes.

2.3 HERBICIDE DETOXIFICATION-BASED RESISTANCE MECHANISMS

The vast majority of herbicides can be detoxified to some extent by plants. The exceptions are herbicides such as paraquat and glyphosate, which are poorly metabolized by plants[72,79] and are thus used as nonselective herbicides. Indeed, the very concept of selective herbicides, lethal to weed species but not to the crop, usually depends on more rapid metabolism of the herbicide by the crop species.[80] For example, chlorsulfuron and other wheat-selective herbicides do not kill wheat plants because wheat rapidly detoxifies them. Often the targeted weed species have some ability to metabolize the herbicide, but this is insufficient to stop the weed from being killed.[81] Generally, similar enzymatic systems are responsible for metabolism in crops and weeds, the difference being the rate and extent of metabolism are much greater in the crop.

Populations of 18 weed species have evolved resistance to herbicides through increased rates of herbicide detoxification (Table 2.2). Studies of herbicide resistance in weeds due to increased herbicide detoxification are extensive at the biochemical level; however, less information is available concerning the genetic control of resistance and none concerning the specific mutations endowing resistance. Three enzyme systems are thus far implicated in metabolism-based resistance to herbicides: the glutathione transferases, the aryl acylamidases, and the cytochrome P450 monooxygenases. The majority of examples of resistance due to increased herbicide metabolism are catalyzed by cytochrome P450 monooxygenases. These are also the most complex examples of resistance, frequently resulting in nontarget-site cross-resistance to unrelated herbicides.

TABLE 2.2
Weed Species with Populations Resistant to Herbicides as a Result of Enhanced Herbicide Detoxification

Species	Proposed Enzymatic System	Herbicides	Ref.
Alopecurus myosuroides	Cytochrome P450	Chlorotoluron Diclofop-methyl Propaquizafop Chlorsulfuron	82–89
	GST	Fenoxaprop-p-ethyl	90, 91
Abutilon theophrasti	GST	Atrazine	92, 93
Avena fatua	Cytochrome P450? (loss of activation)	Triallate	94
A. sterilis	Cytochrome P450	Diclofop-methyl	95
Digitaria sanguinalis	Unknown, possibly cytochrome P450	Fluazifop-p-butyl	96
Echinochloa colona	Aryl acylamidase	Propanil	97
E. crus-galli	Aryl acylamidase	Propanil	98
Hordeum leporinum	Unknown	Fluazifop-p-butyl	99
Lolium rigidum	Cytochrome P450	Diclofop-methyl Chlorsulfuron Chlorotoluron Metribuzin Simazine Chlorotoluron	100–110
	Unknown, possibly cytochrome P450	Tralkoxydim Fluazifop-p-butyl	108, 111
Phalaris minor	Cytochrome P450	Isoproturon	113
Sinapis arvensis	Cytochrome P450	Ethametsulfuron-methyl	114
Stellaria media	Cytochrome P450	Mecoprop	115

2.3.1   GLUTATHIONE TRANSFERASES

Glutathione transferases (historically abbreviated as GSTs) belong to a family of enzymes that attach the tripeptide glutathione through the cysteine residue to electrophilic, hydrophobic compounds.[116] Although the endogenous roles of GSTs are yet to be fully elucidated, they are involved in stress responses, often being induced following the onset of stress or pathogen attack.

Enzymes from this family have been well characterized because of their role in tolerance of maize and sorghum crops to triazine and chloroacetanilide herbicides.[117] GSTs function as dimers of subunits ranging from about 25 to 29 kDa. Different combinations of subunits, either as homodimers or heterodimers, may have different specificities.[116,118–120] The reaction mechanism involves a nucleophilic displacement of an electrophilic group, with glutathione binding to the substrate through the cysteine sulfur (Figure 2.4). With some herbicides a nucleophilic displacement ether cleavage occurs. Once glutathione becomes bound to the substrate, the glutathione conjugate is normally exported to the vacuole for further processing.[122]

Several populations of *Abutilon theophrasti* from the United States are resistant to the PS II-inhibiting triazine herbicides as a result of increased herbicide metabolism catalyzed by GSTs.[92,93] Further investigations of one resistant population have established that activity of two GST isoenzymes with atrazine as the substrate are highly elevated in the resistant individuals. These isoenzymes have specific activity for triazine herbicides and consequently the resistant populations are only resistant to triazine herbicides.[92] More recent studies with purified GST from the resistant

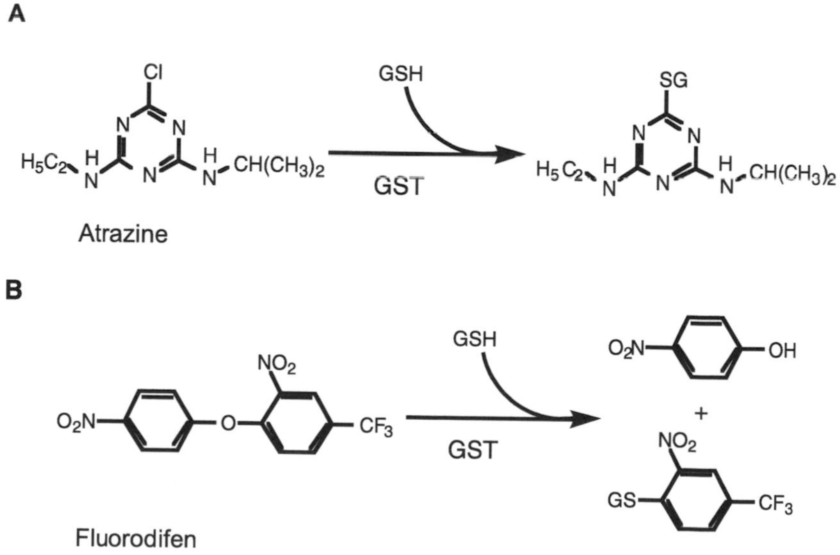

FIGURE 2.4   Glutathione transferase-mediated detoxification of herbicides by direct conjugation, atrazine (A), and by ether cleavage followed by conjugation, fluorodifen (B). (Reaction schemes after Reference 121.)

A. theophrasti indicate resistance is the result of increased catalytic activity rather than over-expression of GST.[123]

Cummins et al.[90] suggested that GSTs play a role in resistance to fenoxaprop-*p*-ethyl and other herbicides in some resistant populations of *Alopecurus myosuroides*. Indeed, these resistant populations have higher GST content and higher expression of certain isoenzymes; however, fenoxaprop-*p*-ethyl metabolism by GST *in vitro* is low.[90,91] One suggestion is that glutathione peroxidase activity associated with the GST is responsible for protection of the resistant plant from the effects of the herbicide.[124] However, this hypothesis is unable to account for the more rapid metabolism of fenoxaprop and diclofop by resistant *A. myosuroides*.[86] Clearly, further investigation is required to determine the role of glutathione peroxidase activity in herbicide resistance in *A. myosuroides*.

Genetic studies conducted with one atrazine-resistant population of *Abutilon theophrasti* indicate that GST-endowed resistance is inherited as a single nuclear-encoded gene with partial dominance.[125] Further studies demonstrated the level of atrazine metabolism in all F_1 progeny, regardless of maternal parent, was intermediate between that of the parental populations.[92]

2.3.2 ARYL ACYLAMIDASES

Aryl acylamidase is an enzyme that catalyzes the hydrolysis of certain acylamides.[126] The endogenous role of aryl acylamidase is not known, but it may be involved in nitrogen metabolism in plants.[127] This enzyme is also responsible for the metabolism of propanil in rice, thereby providing selectivity to this herbicide.[128]

Several populations of *Echinochloa colona* and *E. crus-galli* have evolved resistance to propanil in rice culture at various locations around the world (see Chapter 6 and [129,130]). In both *E. colona* and *E. crus-galli*, propanil resistance is the result of increased rates of propanil detoxification.[97,98] In one population of *E. colona*, elevated aryl acylamidase activity against propanil was demonstrated.[97] In crude enzyme extracts, the resistant population had a threefold enhancement of propanil metabolism by aryl acylamidase. The K_m for propanil of the resistant enzyme is unchanged, indicating resistance is probably the result of a threefold over-expression of the enzyme rather than a more efficient enzyme.[97] The resistant population also had increased aryl acylamidase activity against a range of other chemically related compounds.

Aryl acylamidase activity in propanil-resistant *E. colona* can be inhibited by a number of carbamate and organophosphate pesticides.[97] As described in Chapter 6, this has led to management of propanil-resistant *E. colona* populations in the field by a mixture of propanil and piperophos herbicides.[130,131]

2.3.3 CYTOCHROME P450 MONOOXYGENASES

The cytochrome P450 monooxygenases are a large family of enzymes responsible for the addition of a single oxygen atom to hydrophobic substrates.[132] In plants, these enzymes are involved in a myriad of biosynthetic pathways including the synthesis of heme, allelochemicals, phytoalexins, and suberins.[133,134] These are membrane-bound heme-containing proteins and require the addition of two electrons from NADPH reductase, but occasionally from other electron donors.[134] The reaction sequence catalyzed by cytochrome P450 monooxygenases involves the concerted splitting of molecular oxygen and addition of a single oxygen atom to the substrate (Figure 2.5). This reaction occurs in four steps. First, the substrate is bound to the enzyme, displacing water as a ligand to the heme iron. Next, an electron is added from NADPH, then O_2 is bound. Addition of another electron causes splitting of the oxygen molecule with a hydroxyl group inserted into the substrate and the other oxygen atom removed as water. The hydroxylated substrate then leaves the enzyme. During this process, the oxidation state of the heme iron cycles between Fe^{2+} and Fe^{3+}.[136]

Cytochrome P450 monooxygenases are important for selectivity of a large number of herbicides within certain crops that are able to rapidly metabolize herbicides to inactive compounds.[80] The reactions catalyzed by cytochrome P450 monooxygenases are diverse; however, herbicides are typically hydroxylated or dealkylated by this enzyme system. Hydroxylation is generally followed by conjugation to sugars, often glucose.[80] These conjugates are subsequently exported to the vacuole and/or incorporated in the cell wall.[137]

Cytochrome P450 monooxygenases are the major enzymatic system implicated in detoxification-based herbicide resistance in grasses. Table 2.2 details examples where cytochrome P450-dependent monooxygenases are deduced to endow herbicide resistance. Cytochrome P450-dependent herbicide metabolism is generally difficult to study because of problems in purifying active enzyme and the low rate of metabolism of herbicides *in vitro*.[138] Therefore, these studies have typically used inhibitors of cytochrome P450 monooxygenases, such as 1-amino benzotriazole (ABT) or piperonyl butoxide (PBO), or identification of signature hydroxylated or dealkylated products to demonstrate the involvement of cytochrome P450 monooxygenases. To date, microsomes isolated from these resistant weed populations have not convincingly demonstrated herbicide metabolism *in vitro*.

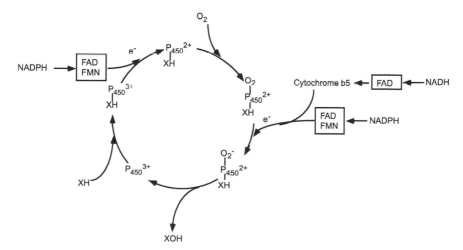

FIGURE 2.5   Reaction mechanism for cytochrome P450 monooxygenases. (From Davies, J. and Caseley, M., Herbicide safenes: a review, *Pestic. Sci.*, 55, 1043, 1999. © Society of Chemical Industry. With permission.)

TABLE 2.3
Cytochrome P450 Monooxygenase Inhibitors Differentially Inhibit Herbicide Metabolism in Multiple-Resistant *Lolium rigidum*

	Herbicide Remaining Unmetabolized (%)				
Inhibitor	Diclofop-methyl + diclofop acid	Tralkoxydim	Simazine	Chlorotoluron	Chlorsulfuron
None	38	23	35	45	78
1-Amino-benzotriazole	52*	22[ns]	61*	64*	84[ns]
Malathion	38[ns]	26[ns]	48[ns]	54[ns]	91*
Piperonyl butoxide	44[ns]	17[ns]	63*	63*	78[ns]
Tetcyclasis	45[ns]	18[ns]	89**	82**	83[ns]

Note: Values are herbicide remaining 24 h (diclofop-methyl, simazine, and chlorotoluron) or 6 h (tralkoxydim and chlorsulfuron) after treatment.

ns = not significantly different to metabolism in the absence of inhibitor.
* = significant decrease in herbicide metabolism ($P < 0.05$).
** = very significant decrease in herbicide metabolism ($P < 0.01$).

Source: Collated from Reference 108.

Over 180 genes or gene sequences belonging to the cytochrome P450 family have been identified in *Arabidopsis thaliana*.[139] This huge diversity of genes raises the possibility that individual herbicides are likely to be metabolized by specific enzymes. This has been established for one population of *Lolium rigidum* from Australia that is resistant to a large number of herbicides.[108] In this population, metabolism of five herbicides from different chemistries was, based on inhibition profiles, due to at least four different cytochrome P450 enzymes (Table 2.3).

This diversity of genes encoding cytochrome P450 monooxygenases also means that in different weed species, different enzymes may be recruited to metabolize a single herbicide. For example, isoproturon is metabolized by ring-hydroxylation in resistant *Phalaris minor* from India;[112,113] however, this population is not resistant to chlorotoluron,[140] a very similar chemical (Figure 2.6). In contrast, a population of *L. rigidum* from Australia resistant to chlorotoluron, mainly due to enhanced ring-hydroxylation,[100,109] also is resistant to isoproturon,[109] as is a population of *A. myosuroides* from the U.K.[82] Clearly, cytochrome P450 monooxygenases of different substrate specificity have been recruited to metabolize substituted urea herbicides in these weed species. One consequence of cytochrome P450 monooxygenase diversity in plants is that there is probably an enzyme available in all plants capable of metabolizing any herbicide that cytochrome P450 monooxygenases can attack.

Some *A. fatua* populations from Canada resistant to triallate demonstrate a different example of metabolism-based herbicide resistance. Rather than enhanced

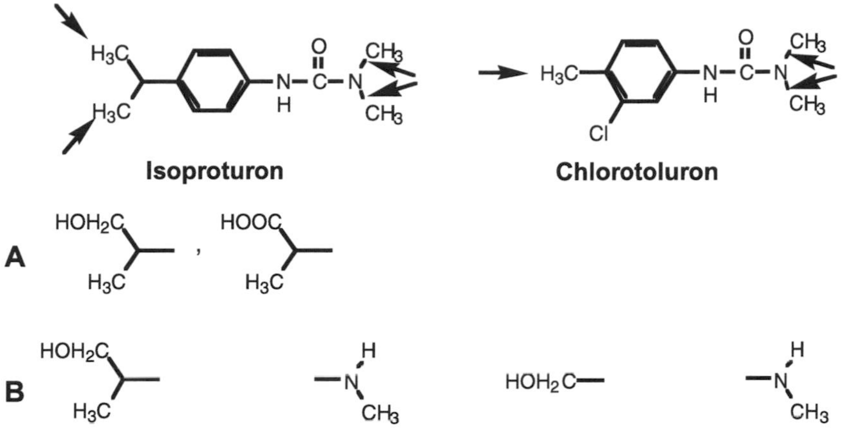

FIGURE 2.6 Cytochrome P450 monooxygenase-dependent metabolism of isoproturon and chlorotoluron. Arrows indicate points of attack. Cytochrome P450 monooxygenase reactions that are elevated in resistant populations compared to susceptible populations of *Alopecurus myosuroides* (A) or *Phalaris minor* (B) are indicated. (Collated from References 82 and 113.)

rates of metabolism, these populations show reduced metabolism of triallate compared to susceptible types.[94] Triallate requires activation to the sulfone for activity and this activation step is probably catalyzed by cytochrome P450 monooxygenases.[141] In the case of triallate-resistant *A. fatua*, reduced activation of triallate is suggested to be the resistance mechanism.[94]

The genetic inheritance of enhanced metabolism resistance mechanisms is poorly understood. Studies by Chauvez (cited in [142]) on *A. myosuroides* resistant to chlorotoluron suggested that resistance alleles at two or more genes contributed to resistance. Studies with *L. rigidum* from Australia resistant to a wide range of herbicides show that resistance to both simazine and chlorotoluron is nuclearly inherited with partial dominance (Figure 2.7). The F_1 plants show intermediate resistance compared to the susceptible and resistant parents. Further studies with one population are consistent with a major gene involvement in resistance for each herbicide, but do not rule out some contribution from minor genes.[143]

2.4 MECHANISMS OF HERBICIDE RESISTANCE OTHER THAN TARGET-SITE MODIFICATIONS OR HERBICIDE METABOLISM

As stated in the introduction to this chapter, herbicide resistance may result from one or more of a variety of biochemical mechanisms. Resistance to several groups of herbicides has been identified with resistance mechanisms that do not involve reduced target-site sensitivity or increased herbicide detoxification. In many instances, the actual mechanism is still unknown largely due to the difficulties in identifying alternative mechanisms of herbicide resistance.

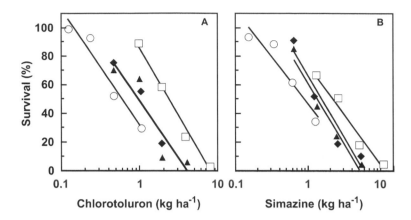

FIGURE 2.7 Dose response to chlorotoluron (A) and simazine (B) of susceptible VLR 1 (○), resistant VLR 69 (□), and reciprocal crosses (▲,♦) between susceptible and resistant populations of *Lolium rigidum*.[143]

2.4.1 Resistance to Bipyridyl Herbicides

The bipyridyl herbicides, paraquat and diquat, are nonselective and active on a wide range of annual plants. These herbicides interact with the photosynthetic electron transport chain accepting electrons from the reducing side of PS I and are thereby reduced. The resulting anion radical reacts with oxygen and forms the highly reactive superoxide anion and hydroxyl radical. These radicals in turn cause lipid peroxidation and the destruction of cell membranes.[144] The bipyridyl herbicides act rapidly in strong light, producing symptoms within hours of application.

Paraquat resistance has evolved in populations of 26 weed species,[1] mainly in perennial cropping systems, including tree crops, vine crops, and alfalfa,[144] but also in broad area annual cropping.[145] The mechanism of resistance to paraquat has been studied extensively in a few resistant populations. Despite this effort, no conclusive mechanisms have been determined for resistance to the bipyridyl herbicides.

Decreased translocation and decreased penetration to the active site were proposed as the mechanism of resistance to the bipyridyl herbicides in *Hordeum glaucum* and *H. leporinum*.[146,147] In other studies with the same population of *H. glaucum*, increased sequestration in the vacuole of the resistant plants was suggested as the mechanism of resistance.[148] This hypothesis is not in conflict with the conclusion that resistance is due to decreased penetration to the active site. Paraquat-resistant populations of *H. glaucum* and *H. leporinum* displayed unusual temperature sensitivity, being highly resistant in winter but weakly resistant in summer (Figure 2.8). It was established that at 30°C the resistance mechanism breaks down such that the resistant population is not able to survive high rates of paraquat.[149] Under high temperatures, there was increased translocation of paraquat and more of the herbicide reached the active site in the resistant populations of the *Hordeum* spp.[149] This

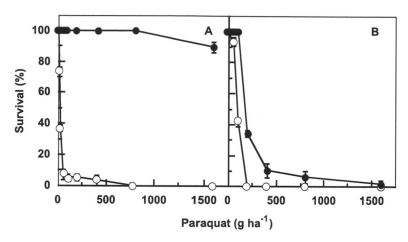

FIGURE 2.8　Dose response of susceptible (○) and resistant (●) populations of *Hordeum glaucum* to paraquat under cool (A) or warm (B) conditions showing the dramatic reduction in resistance under warm conditions. (From Purba, E., Preston, C., and Powles, S. B., The mechanism of resistance to paraquat is strongly temperature dependent in resistant *Hordeum leporinum* Link and *H. glaucum* Steud., *Planta*, 196, 1995. © Springer-Verlag. With permission.)

response supports the argument that resistance is related to decreased amounts of the herbicide reaching the active site.

In *Arctotheca calendula*, resistance did not appear to be related to decreased translocation of diquat but reduced herbicide penetration to the active site was proposed as the mechanism.[150] Further experiments with *A. calendula* have demonstrated that long-distance translocation of paraquat is dependent on the amount of leaf damage occurring. The resistant population has less leaf damage following paraquat application; therefore, paraquat translocation does not occur as rapidly.[151] These experiments support the hypothesis that paraquat is kept from the active site in the resistant population.

Paraquat sequestration has also been proposed as a resistance mechanism in *Conyza bonariensis*.[152] In these studies, there was much reduced lateral movement of paraquat in the resistant population compared to the susceptible population. Results of previous studies showed that paraquat was not metabolized in either the resistant or susceptible population.[153] Paraquat-resistant *Erigeron philadelphicus* and *E. canadensis* also showed restricted movement of the paraquat in leaves compared to susceptible individuals of these species.[154]

In *H. glaucum*, *H. leporinum*, and *A. calendula*, a single, incompletely dominant gene confers paraquat resistance.[155,156] However, the degree of dominance varies. Dominance is low for the grass species, heterozygotes are 2- to 8-fold resistant compared to 250-fold resistant for homozygotes, but higher for *A. calendula*.[156] In contrast to the *Hordeum* spp., paraquat resistance in *E. canadensis* is inherited as a single gene with a high level of dominance.[157] These differences in dominance of the resistant allele suggest that the biochemical basis of paraquat resistance might be different in *A. calendula* and *E. canadensis* compared to *H. glaucum* and

H. leporinum. With the exception of *E. canadensis*,[157] the populations segregated in a 1:2:1 ratio in the F_2 generation, with the intermediate plants presumed to be heterozygous. In *H. glaucum*, the F_3 generation progeny from the intermediate F_2 plants segregated again in a 1:2:1 ratio, while plants produced from the resistant plants were all resistant.[155] In *A. calendula*, back-crosses from the F_1 generation segregated in a 1:1 ratio, confirming the single gene hypothesis.[156] In *E. canadensis*, the F_2 generation segregated in a 1:3 ratio, indicating a single dominant gene was responsible for resistance.[157]

An alternative mechanism proposed to explain paraquat resistance is the elevation of oxygen radical detoxifying enzymes in the chloroplasts of resistant plants. Elevated activities of such enzymes have been measured in resistant populations of *Conyza bonariensis*[158,159] and *E. philadelphicus*,[160] and inferred for *C. canadensis*.[161] In *C. bonariensis*, elevated activities of a whole host of enzymes including superoxide dismutase, ascorbate peroxidase, glutathione reductase, and glutathione peroxidase were recorded.[158,159,162–164] In addition, the resistant population has elevated polyamine contents and increased activity of ornithine decarboxylase, an enzyme important to polyamine synthesis.[165] Activity of oxygen radical detoxifying enzymes was induced dramatically by paraquat application to resistant plants and to a lesser extent in susceptible plants.[166] Paraquat resistance in *C. bonariensis* is proposed due to a single, partially dominant gene co-segregating with increased activity of detoxification enzymes.[167] Other studies on resistant populations of *C. bonariensis* and *C. canadensis* have disputed a major role for oxygen radical detoxifying enzymes in resistance.[153,168,169] This dispute has been considered in detail elsewhere,[144,170] with the conclusion that without some way of removing or immobilizing paraquat the oxygen radical detoxification mechanism could only provide a modest degree of resistance.

2.4.2 RESISTANCE TO GLYPHOSATE

As described in Section 2.2.5, glyphosate resistance has evolved in several populations of both *Lolium rigidum* and *Eleusine indica*. While the mechanism of glyphosate resistance in one population of *E. indica* has been established as a target-site modification (see Section 2.2.5), glyphosate resistance in *L. rigidum* is not due to reduced sensitivity of EPSP synthase to glyphosate.[171,172] The resistance mechanism remains to be determined. Studies with two different glyphosate-resistant *L. rigidum* populations have found no differences in glyphosate absorption, translocation, or metabolism compared to susceptible populations.[171,173] One resistant *L. rigidum* population (Echuca) had a higher level of EPSP synthase activity than a susceptible population;[172] however, the other glyphosate-resistant population (Orange) did not have increased EPSP activity.[171] Reduced movement of glyphosate to its site of action in the plastid was proposed as a possible mechanism of resistance in the Orange population.[171]

A single nuclear gene with incomplete dominance controls resistance in the Orange population.[174] In contrast, Pratley et al.[75] reported increased levels of resistance in subsequent generations of the Echuca population that had been selected with glyphosate and suggested the involvement of more than one gene. However, the number of genes involved in the resistance has not been determined.[75]

Several populations of *Convolvulus arvensis* with differences in sensitivity to glyphosate were identified in 1984.[175] The authors did not refer to these populations as resistant; however, the differences in sensitivity to glyphosate are of the same magnitude reported for resistant *L. rigidum* and *E. indica* populations and for the rest of this discussion will be referred to as resistant. In this case, resistance has not evolved in the populations in response to herbicide selection pressure, but elucidation of the biochemical mechanisms of resistance may prove illuminating. Increased activity of the shikimate pathway has been suggested as the mechanism that provides glyphosate resistance in one *C. arvensis* population.[176] The resistant population contains a higher level of 3-deoxy-D-*arabino*-heptulosonate-7-phosphate synthase (DAHP synthase), the first enzyme in the shikimate pathway, than the susceptible population. The authors suggest there is higher activity of the shikimate pathway and that greater carbon flow through the pathway along with a greater phenylalanine pool in the resistant population allows this population to be less affected by glyphosate. The authors also hypothesize that multiple mechanisms were responsible for the resistance. Five populations with differing sensitivities to glyphosate were used as parents in a diallel cross experiment. Genetic analysis indicated that multiple genes were involved in the resistance and that maternal influences were important.[177]

2.4.3 RESISTANCE TO AUXIN MIMICS

The auxin mimic herbicides, which include the phenoxy acetic acid, benzoic, and picolinic herbicide families, have been widely used in agriculture since the 1940s.[178] They are termed auxin mimics because they mimic the action of the natural plant hormone indol acetic acid (IAA) to excess. Despite the extensive use of these herbicides, resistance to the auxin mimic herbicides has been identified in 19 weed species only.[1] Even though these herbicides have been used since 1945, the exact details on how the herbicides act are still unknown. Dicotyledonous plants treated with auxin mimic herbicides display multiple symptoms including epinasty, abnormal cell elongation, and adventitious root initiation. Resistance to auxin mimic herbicides was extensively reviewed by Coupland,[178] who identified that in many cases the basis of the resistance had not been fully elucidated. With one exception,[115] differences in absorption, translocation, or metabolism of herbicides were not responsible for resistance.[178] Since that time, resistance in two species, *Centaurea solstitialis* and *Sinapsis arvensis* (sometimes identified as *Brassica kaber*), has been extensively researched.

The resistant population of *C. solstitialis* had similar amounts of foliar absorption, translocation, metabolism, and cellular absorption of picloram compared to a susceptible population.[179] The picloram-resistant *C. solstitialis* population is cross-resistant to clopyralid. Foliar absorption, translocation, metabolism, and cellular absorption of clopyralid are also similar between resistant and susceptible populations.[180] In addition, there were no differences in ethylene production between resistant and susceptible plants.[180,181] Although studies were conducted to examine differences in auxin/picloram binding sites between the populations, auxin/picloram binding could not be consistently detected and so it is still unknown whether such differences in picloram recognition might be involved with the resistance mechanism. Following picloram application, several genes are expressed differentially in resistant plants compared to susceptible plants, but these have not yet been characterized.[182]

Resistance in *S. arvensis/B. kaber* was not found to be due to differences in herbicide absorption, translocation, or metabolism.[183] After application of picloram, the susceptible plants did produce more ethylene.[184] This difference was attributed to differences in pathway regulation. Differences in ATP-dependent activity were measured and the authors attribute these differences to modulation of calcium ion channels. When auxin-binding proteins were compared, it was found that in the susceptible population the auxin-binding protein was more sensitive to picloram and dicamba than that in the resistant population.[185] This was confirmed in other experiments, which demonstrated differential effects of picloram on auxin-induced activities in resistant plants.[186] In addition, cytokinin levels were higher in the resistant population than in the susceptible population, which may mean that there are differences in hormonal regulation between the populations.[187]

There have been limited studies of inheritance of resistance to auxin mimic herbicides. However, it is known that a single, recessive nuclear gene controls picloram resistance in *C. solstitialis*,[188] whereas a single, dominant nuclear gene[189,190] controls dicamba resistance in *S. arvensis/B. kaber*. These differences in inheritance indicate there may be more than one mechanism responsible for auxin mimic herbicide resistance.

2.5 MULTIPLE-RESISTANCE MECHANISMS

Multiple-resistance is defined as resistance due to more than one resistance mechanism.[191] In the worst cases, this may result in weed populations with simultaneous resistance to many different herbicides. These populations can be particularly difficult to control as there may be few herbicide options remaining. Understanding the biochemistry and genetics of multiple-resistance is considerably more challenging as a result of having to separate the resistance mechanisms. Therefore, while a number of cases of multiple-resistance have been suspected, few have followed up with detailed biochemical studies to determine the mechanisms of resistance (Table 2.4), and even fewer genetic studies have occurred. The examples in Table 2.4 show that multiple-resistance can occur via combinations of target-site-based mechanisms, target-site- and metabolism-based mechanisms, and other combinations of mechanisms.

2.5.1 EVOLUTION AND BIOCHEMISTRY OF MULTIPLE-RESISTANCE

Multiple-resistance can arise within a weed population in several ways: through a change in selection history; through selection of multiple mechanisms by a single herbicide; or through crossing of individuals containing different resistance mechanisms. Examples of all three processes are known.

TABLE 2.4
Populations of Weed Species Showing Examples
of Multiple Herbicide Resistance

Species and Population	Herbicides	Mechanisms	Ref.
Alopecurus myosuroides Lincs E1	ACCase inhibitors PS II inhibitors	Target site Metabolism	86, 89, 192
Amaranthus rudis Illinois	ALS inhibitors PS II inhibitors	Target site Target site	40
Avena fatua	Triallate Difenzoquat	Reduced activation Increased binding?	94, 193
A. sterilis NAS 4	ACCase inhibitors	Target site Metabolism	95
Brachypodium distachyon	PS II inhibitors	Target site Metabolism	194
Conyza canadensis	PS II inhibitors Paraquat	Target site Unknown	195
Echinochloa crus-galli	Atrazine Quinclorac	Target site Unknown	196
Galium spurium	ALS inhibitors Quinclorac	Target site Unknown	197
Hordeum leporinum	ACCase inhibitors	Target site Metabolism	99
Kochia scoparia	ALS inhibitors PS II inhibitors	Target site Target site	43
Lolium rigidum SLR 31	ACCase inhibitors ALS inhibitors Tubulin elongation inhibitors	Target site Metabolism Others?	102, 104, 105, 198
L. rigidum VLR 69	ACCase inhibitors ALS inhibitors PS II inhibitors	Target site Metabolism	100, 101, 108, 199
L. rigidum WLR 1	ALS inhibitors	Target site Metabolism	200
L. rigidum WLR 2	PS II inhibitors Amitrole	Metabolism Unknown	100, 101, 106
Setaria faberi	PS II inhibitors	Target site Metabolism	201
S. viridis	PS II inhibitors	Target site Metabolism	201

2.5.1.1 Multiple-Resistance by Sequential Herbicide Selection

There are several examples of multiple-resistance evolving sequentially as a result
of a change in selection history (Table 2.4). The example of triazine- and sulfonylurea-resistant *Kochia scoparia* will be used to illustrate this process.[43] Triazine
herbicides were used extensively to control weeds, including the widespread weed

K. scoparia across the northern United States and Canada. Triazine resistance subsequently evolved in *K. scoparia* populations.[202] Following the introduction of the ALS-inhibiting sulfonylurea herbicides, these were used extensively to control *K. scoparia* and other weeds. Inevitably, ALS-inhibiting herbicides were used against populations that had previously evolved resistance to the triazine herbicides. At least one population of triazine-resistant *K. scoparia* subsequently evolved resistance to ALS-inhibiting herbicides.[43] As expected, these multiple-resistant plants contain two distinct resistance mechanisms. The resistant population has a D1 protein insensitive to triazine herbicides due to a substitution of Gly for Ser at position 264. In addition, a substitution of Leu for Trp at position 570 in ALS resulted in a modified ALS enzyme insensitive to sulfonylurea and imidazolinone herbicides.[43] A similar case has been reported for triazine and ALS-inhibiting herbicide-resistant *Amaranthus rudis*.[40] Given the past and continuing widespread use of the PS II- and ALS-inhibiting herbicides in world agriculture, it is likely that numerous more examples of multiple-resistance to these two modes of action will occur.

Selection for multiple-resistance does not have to be sequential. In principle, similar results could occur from rotating herbicide modes of action. While there is anecdotal evidence for resistance occurring in this fashion,[38,203] there are few documented examples.

2.5.1.2 Multiple-Resistance by Selection with a Single Herbicide

Selection of multiple mechanisms of resistance by a single herbicide also can occur. This type of multiple-resistance is less commonly reported, probably because many researchers examine for a single major resistance mechanism. Two such examples are a *Brachypodium distachion* population resistant to triazine herbicides[194] and a *Lolium rigidum* population resistant to sulfonylurea herbicides.[200] In both cases there is an increased rate of herbicide metabolism; however, a strongly insensitive target enzyme dominates resistance. A more clear-cut example of multiple-resistance to a single herbicide is diclofop-methyl resistance in an *Avena sterilis* population (NAS 4). This population is strongly resistant to diclofop-methyl; however, it has an ACCase enzyme with only tenfold resistance to diclofop acid.[95] This population also demonstrated more rapid metabolism of diclofop acid. In this population at least two genes must encode herbicide resistance, one endowing resistance due to a modified ACCase and at least one endowing resistance due to increased detoxification of diclofop acid. As *A. sterilis* is a self-pollinated species, these genes most likely have been selected together.

Cross-pollinating weed species can also accumulate resistance mechanisms through gene flow. This may occur through selection with a single herbicide where two individuals, each containing a different resistance mechanism, cross. Some of the resulting progeny will inherit both resistance mechanisms.[62,108,203] However, where there is also a complex herbicide selection history or gene flow from other fields with different herbicide use patterns, particularly complicated patterns of multiple resistance can occur. This happens with *L. rigidum* populations in Australia.[62,203] Such populations may evolve resistance to a wide range of herbicides and be particularly difficult to control. *L. rigidum* population VLR 69 is a particularly good example of multiple-resistance.[204] This population contains an ACCase resistant to aryloxy-phenoxypropanoate herbicides and enhanced metabolism of triazine, substituted urea, triazinone, aryloxyphenoxypropanoate, cyclohexanedione, and sulfonylurea herbicides.[108,203] In addition, about 5% of the population also contains a herbicide-resistant ALS.[199] As described above, at least four different cytochrome P450 mono-oxygenases contribute to herbicide resistance in this population.[108]

2.5.2   GENETICS OF MULTIPLE-RESISTANCE

There is a direct relationship between genes and resistance mechanisms such that a single gene will contribute to a single mechanism. Therefore, all cases of multiple-resistance, by definition, must be encoded by more than one gene. However, it is also theoretically possible for more than one gene to contribute to the same resistance mechanism. Therefore, sorting out the genetics of multiple-resistance can be a formidable task.

The example of a *K. scoparia* population resistant to both triazine and ALS inhibitors described above[43] is relatively simple, as only two genes, one encoding a modification to PS II and the second a modification to ALS, contribute to resistance. However, where enhanced metabolism of herbicides is one of the mechanisms of resistance, the genetics can be considerably more complicated. This is because a single gene may control increased expression of several genes through coordinate expression. Alternatively, some herbicides, such as atrazine, can be metabolized by more than one enzyme system.[80] In this way, more than one gene could contribute to metabolism-based resistance.

An illustration of the complexity of genetics of multiple-resistance comes from studies with *L. rigidum* population VLR 69. This population would be expected to have different genes endowing target-site-based resistance at ACCase and ALS, as well as an additional gene(s) endowing metabolism-based resistance.[203] Studies were conducted to determine the linkage between the metabolism-based resistances in *L. rigidum* population VLR 69. To do this, segregating F_2 families were selected with high rates of several herbicides. The survivors were crossed within treatments, and the progeny analyzed for cross-resistance. This analysis demonstrated distinct linkages of resistance (Figure 2.9). For example, selection of the F_2 with chloro-toluron resulted in more progeny with resistance to simazine and chlorotoluron than with resistance to tralkoxydim or chlorsulfuron. Similar results were obtained with selection of the F_2 with simazine.[143] This demonstrates that resistance mechanisms for simazine and chlorotoluron are closely linked and probably controlled by a single gene. In all, at least five different genes appear to contribute to multiple-resistance in *L. rigidum* population VLR 69.[143]

2.6   CONCLUSIONS AND PROSPECTS

The intensive use of herbicides to control weeds inevitably has resulted in the evolution of herbicide-resistant weed populations. On surveying the large body of

FIGURE 2.9 Cross-resistance and lack of cross-resistance to (left to right) chlorsulfuron, chlorotoluron, tralkoxydim, or simazine in progeny of simazine-selected F_2 families created from a cross between *Lolium rigidum* populations VLR 69 and VLR 1.[143]

work on resistance mechanisms, it is clear that the majority of studied examples relate to changes in the target enzyme. This may reflect biological reality but a note of caution, in that it may be biased by the relative ease with which target-site changes can be detected. Despite this, it is clear that single amino acid changes to target enzymes can be easily selected by herbicides in the field. However, it must be remembered that target-site changes are not the only mechanism that can endow resistance to herbicides. Any biochemical change that allows a plant to survive application of herbicide can be selected. Other resistance mechanisms detected include increased herbicide detoxification and reduced penetration of herbicide to its active site. It is also possible for a single herbicide to select different resistance mechanisms in different populations, or indeed the same population, of a species.

No herbicide should be considered resistance proof. The ability of weeds to evolve resistance to paraquat, where functional target-site modifications are very unlikely and herbicide metabolism is nonexistent,[144] demonstrates that resistance is possible through other mechanisms. However, there are herbicides that are much more prone to resistance evolution than others. The rarity of resistance to the widely used auxin mimics is testament to the difficulty of evolving herbicide resistance to some modes of action.[178]

Overwhelmingly, the genetic basis of resistance can be attributed to a single gene, usually a single base pair change in DNA resulting in a single amino acid change in a protein. In cases of multiple-resistance it appears that several genes, each endowing resistance to non-overlapping groups of herbicides, are involved. Even in these cases, the direct linkage of one gene to one resistance mechanism appears to hold. As herbicides provide an enormous selection pressure, it is to be expected that single gene mechanisms endowing a large change in phenotype will be selected.[205,206] The few examples where more than one gene contributes to a single resistance mechanism are worthy of note. As yet, such examples are poorly understood; however, full elucidation of the genetic and biochemical behavior of these populations will be illuminating.

The evolution of herbicide resistance in weed populations has not led to less herbicide application or a dramatic increase in nonchemical control methods. Instead, alternative herbicides have been used to manage resistance weeds. This has resulted in, and will continue to result in, ever more complicated patterns of multiple herbicide resistance. The use of other herbicide strategies such as mixtures and variable dose rates to delay the onset of resistance has been promoted[207,208] but not widely adopted. An understanding of the mechanisms and genetics of resistance has helped elucidate why these strategies will not be widely effective. As resistance is for the most part endowed by single, mostly dominant genes of large effect, such strategies are unlikely to greatly delay the inevitable.[206] The only certain way to delay the evolution of herbicide resistance is to use each herbicide less often and to introduce significant nonchemical control methods.

REFERENCES

1. Heap, I., International survey of herbicide resistant weeds: lessons and limitations, *Proc. 1999 Brighton Conf. – Weeds*, 1999, 769.
2. Powles, S. B. and Holtum, J. A. M., Eds., *Herbicide Resistance in Plants: Biology and Biochemistry*, Lewis Publishers, Boca Raton, FL, 1994, 353 pp.
3. Devine, M. D. and Eberlein, C. V., Physiological, biochemical and molecular aspects of herbicide resistance based on altered target sites, in *Herbicide Activity: Toxicology, Biochemistry and Molecular Biology*, Roe, R. M., Burton, J. D., and Kuhr, R. J., Eds., IOS Press, Amsterdam, 1997, 159.
4. Smeda, R. J. and Vaughn, K. C., Mechanisms of resistance to herbicides, in *Molecular Mechanisms of Resistance to Agrochemicals*, 13, Sjut, V., Ed., Springer-Verlag, Berlin, 1997, 79.
5. Gronwald, J. W., Resistance to phytosystem II inhibiting herbicides, in *Herbicide Resistance in Plants: Biology and Biochemistry*, Powles, S. B. and Holtum, J. A. M., Eds., Lewis Publishers, Boca Raton, FL, 1994, 27.
6. Saari, L. L., Cotterman, J. C., and Thill, D. C., Resistance to acetolactate synthase inhibiting herbicides, in *Herbicide Resistance in Plants: Biology and Biochemistry*, Powles, S. B. and Holtum, J. A. M., Eds., Lewis Publishers, Boca Raton, FL, 1994, 83.
7. Saari, L. L. and Maxwell, C. A., Target-site resistance for acetolactate synthase inhibitor herbicides, in *Weed and Crop Resistance to Herbicides*, De Prado, R., Jorrín, J., and García-Torres, L., Eds., Kluwer Academic Publishers, Dordrecht, the Netherlands, 1997, 81.
8. Devine, M. D., Mechanisms of resistance to acetyl-coenzyme A carboxylase inhibitors: a review, *Pestic. Sci.*, 51, 259, 1997.
9. Incledon, B. J. and Hall, J. C., Acetyl-coenzyme A carboxylase: quanternary structure and inhibition by graminicidal herbicides, *Pestic. Biochem. Physiol.*, 57, 255, 1997.
10. Ryan, G. F., Resistance of common groundsel to simazine and atrazine, *Weed Sci.*, 18, 614, 1970.
11. Holt, J. S. and LeBaron, H. M., Significance and distribution of herbicide resistance, *Weed Technol.*, 4, 141, 1990.
12. Fuerst, E. P., Arntzen, C. J., Pfister, K., and Penner, D., Herbicide cross-resistance in triazine-resistant biotypes of four species, *Weed Sci.*, 34, 344, 1986.
13. Trebst, A., The molecular basis of plant resistance to photosystem II herbicides, in *Molecular Genetics and Evolution of Pesticide Resistance*, ACS Symp. Ser. 645, Brown, T. M., Ed., American Chemical Society, Washington, D.C., 1996, 5.
14. Sinning, I., Herbicide binding in the bacterial photosynthetic reaction center, *Trends Biochem. Sci.*, 17, 150, 1992.

15. Masbani, J. G. and Zandstra, B. H., Discovery of a common purslane (*Portulaca oleracea*) biotype resistant to linuron, *Weed Technol.*, 13, 599, 1999.

16. Masabni, J. G. and Zandstra, B. H., A serine-to-threonine mutation in linuron-resistant *Portulaca oleracea*, *Weed Sci.*, 47, 393, 1999.

17. Mengistu, L. W., Mueller-Warrant, G. W., Liston, A., and Barker, R. E., *psbA* mutation (valine219 to isoleucine) in *Poa annua* resistant to metribuzin and diuron, *Pest Manage. Sci.*, 56, 209, 2000.

18. Deisenhofer, J., Epp, O., Miki, K., Huber, R., and Michel, H., X-ray structure analysis of a membrane protein complex, electron density map at 3Å resolution and a model of the chromophores of the photosynthetic reaction center from *Rhodopseudomonas viridis*, *J. Mol. Biol.*, 180, 385, 1984.

19. Michel, H., Epp, O., and Deisenhofer, J., Pigment-protein interactions in the photosynthetic reaction center from *Rhodopseudomonas viridis*, *EMBO J.*, 5, 2445, 1986.

20. Sinning, I., Koepke, J., and Michel, H., Recent advances in the structure analysis of *Rhodopseudomonas viridis* mutants resistant to the herbicide terbutryn, in *Reaction Centers of Photosynthetic Bacteria*, Michel-Beyerle, M. E., Ed., Springer-Verlag, Berlin, 1990, 199.

21. Lancaster, C. R. D. and Michel, H., Refined crystal structures of reaction centers from *Rhodopseudomonas viridis* in complexes with the herbicide atrazine and two chiral atrazine derivatives also lead to a new model of the bound carotenoid, *J. Mol. Biol.*, 286, 883, 1999.

22. Darmency, H., Genetics of herbicide resistance in weeds and crops, in *Herbicide Resistance in Plants: Biology and Biochemistry*, Powles, S. B. and Holtum, J. A. M., Eds., Lewis Publishers, Boca Raton, FL, 1994, 263.

23. Darmency, H. and Gasquez, J., Inheritance of triazine resistance in *Poa annua*: consequences for population dynamics, *New Phytol.*, 89, 487, 1981.

24. Frey, J. E., Müller-Schärer, H., Frey, B., and Frei, D., Complex relationship between triazine-susceptible phenotype and genotype in the weed *Senecio vulgaris* may be caused by chloroplast DNA polymorphism, *Theor. Appl. Genet.*, 99, 578, 1999.

25. Strachan, S. D. and Hess, F. D., The biochemical mechanism of action of the dinitroaniline herbicide oryzalin, *Pestic. Biochem. Physiol.*, 20, 141, 1983.

26. Smeda, R. J. and Vaughn, K. C., Resistance to dinitroaniline herbicides, in *Herbicide Resistance in Plants: Biology and Biochemistry*, Powles, S. B. and Holtum, J. A. M., Eds., Lewis Publishers, Boca Raton, FL, 1994, 215.

27. Anthony, R. G., Waldin, T. R., Ray, J. A., Bright, S. W. J., and Hussey, P. J., Herbicide resistance caused by spontaneous mutation of the cytoskeletal protein tubulin, *Nature*, 393, 260, 1998.

28. Anthony, R. G. and Hussey, P. J., Dinitroaniline herbicide resistance and the microtubule cytoskeleton, *Trends Plant Sci.*, 4, 112, 1999.

29. Yamamoto, E., Zeng, L., and Baird, W. V., α-Tubulin missense mutations correlate with antimicrotubule drug resistance in *Eleusine indica*, *Plant Cell*, 10, 297, 1998.

30. Yamamoto, E., Zeng, L., and Baird, W. V., Molecular characterization of four β-tubulin genes from dinitroaniline susceptible and resistant biotypes of *Eleusine indica*, *Plant Mol. Biol.*, 39, 45, 1999.

31. Anthony, R. G., Reichelt, S., and Hussey, P. J., Dinitroaniline herbicide-resistant transgenic tobacco plants generated by co-overexpression of a mutant α-tubulin and a β-tubulin, *Nature Biotechnol.*, 17, 712, 1999.

32. Zeng, L. and Baird, W. V., Genetic basis of dinitroaniline herbicide resistance in a highly resistant biotype of goosegrass (*Eleusine indica*), *J. Heredity*, 88, 427, 1997.

33. Zeng, L. and Baird, W. V., Inheritance of resistance to anti-microtubule dinitroaniline herbicides in an "intermediate" resistant biotype of *Eleusine indica* (Poaceae), *Am. J. Bot.*, 86, 940, 1999.

34. Jasieniuk, M., Brûlé-Babel, A., and Morrison, I. N., Inheritance of trifluralin resistance in green foxtail (*Setaria viridis*), *Weed Sci.*, 42, 123, 1994.

35. Wang, T., Fleury, A., Ma, J., and Darmency, H., Genetic control of dinitroaniline resistance in foxtail millet (*Setaria italica*), *J. Heredity*, 87, 423, 1996.

36. Singh, B. K. and Shaner, D. L., Biosynthesis of branched chain amino acids: from test tube to field, *Plant Cell*, 7, 935, 1995.

37. Schloss, J. V., Ciskanik, L. M., and Van Dyk, D. E., Origin of the herbicide binding site of acetolactate synthase, *Nature*, 331, 330, 1988.

38. Powles, S. B., Preston, C., Bryan, I. B., and Jutsum, A. R., Herbicide resistance: impact and management, *Adv. Agron.*, 58, 57, 1997.

39. Woodworth, A. R., Rosen, B. A., and Bernasconi, P., Broad range resistance to herbicides targeting acetolactate synthase (ALS) in a field isolate of *Amaranthus* sp. is conferred by a Trp to Leu mutation in the ALS gene (Accession No. U55852) (PGR96-051), *Plant Physiol.*, 111, 1353, 1996.

40. Foes, M. J., Liu, L., Tranel, P. J., Wax, L. M., and Stoller, E. W., A biotype of common waterhemp (*Amaranthus rudis*) resistant to triazine and ALS herbicides, *Weed Sci.*, 46, 514, 1988.

41. Boutsalis, P., Karotam, J., and Powles, S. B., Molecular basis of resistance to acetolactate synthase-inhibiting herbicides in *Sisymbrium orientale* and *Brassica tournefortii*, *Pestic. Sci.*, 55, 507, 1999.

42. Guttieri, M. J., Eberlein, C. V., Mallory-Smith, C. A., Thill, D. C., and Hoffman, D. L., DNA sequence variation in domain A of the acetolactate synthase genes of herbicide-resistant and -susceptible weed biotypes, *Weed Sci.*, 40, 670, 1992.

43. Foes, M. J., Liu, L., Vigue, G., Stoller, E. W., Wax, L. M., and Tranel, P. J., A kochia (*Kochia scoparia*) biotype resistant to triazine and ALS-inhibiting herbicides, *Weed Sci.*, 47, 20, 1999.

44. Guttieri, M. J., Eberlein, C. V., and Thill, D. C., Diverse mutations in the acetolactate synthase gene confer chlorsulfuron resistance in kochia (*Kochia scoparia*) biotypes, *Weed Sci.*, 43, 175, 1995.

45. Stone, L., Rieger, M. A., and Preston, C., unpublished data, 1997.

46. Bernasconi, P., Woodworth, A. R., Rosen, B. A., Subramanian, M. V., and Siehl, D. L., A naturally occurring point mutation confers broad range tolerance to herbicides that target acetolactate synthase, *J. Biol. Chem.*, 270, 17381, 1995.

47. Bernasconi, P., Woodworth, A. R., Rosen, B. A., Subramanian, M. V., and Siehl, D. L., A naturally occurring point mutation confers broad range tolerance to herbicides that target acetolactate synthase — correction, *J. Biol. Chem.*, 271, 13925, 1996.

48. Woodworth, A. R., Bernasconi, P., Subramanian, M. V., and Rosen, B. A., A second naturally occurring point mutation confers broad based tolerance to acetolactate synthase inhibitors, *Plant Physiol.*, 111, 105, 1996.

49. Mazur, B. J., Chui, C.-F., and Smith, J. K., Isolation and characterization of plant genes coding for acetolactate synthase, the target enzyme for two classes of herbicides, *Plant Physiol.*, 85, 1110, 1987.

50. Guttieri, M. J., Eberlein, C. V., Mallory-Smith, C. A., and Thill, D. C., Molecular genetics of target-site resistance to acetolactate synthase inhibiting herbicides, in *Molecular Genetics and Evolution of Pesticide Resistance*, ACS Symp. Ser. 645, Brown, T. M., Ed., American Chemical Society, Washington, D.C., 1996, 10.

51. Eberlein, C. V., Guttieri, M. J., Berger, P. H., Fellman, J. K., Mallory-Smith, C. A., Thill, D. C., Baerg, R. J., and Belknap, W. R., Physiological consequences of mutation for ALS-inhibitor resistance, *Weed Sci.*, 47, 383, 1999.

52. Alcocer-Ruthling, M., Thill, D. C., and Shafii, B., Differential competitiveness of sulfonylurea-resistant and -susceptible prickly lettuce (*Lactuca serriola*), *Weed Technol.*, 6, 303, 1992.

53. Mallory-Smith, C. A., Thill, D. C., Dial, M. J., and Zemetra, R. S., Inheritance of sulfonylurea herbicide resistance in *Lactuca* spp., *Weed Technol.*, 4, 787, 1990.

54. Thompson, C. R., Thill, D. C., Mallory-Smith, C. A., and Shafii, B., Characterization of chlorsulfuron resistant and susceptible kochia (*Kochia scoparia*), *Weed Technol.*, 8, 470, 1994.

55. Boutsalis, P. and Powles, S. B., Inheritance and mechanism of resistance to herbicides inhibiting acetolactate synthase in *Sonchus oleraceus* L., *Theor. Appl. Genet.*, 91, 242, 1995.

56. Burton, J., Gronwald, J., Somers, D., Connelly, J., Gengenbach, B., and Wyse, D., Inhibition of plant acetyl-CoA carboxylase by the herbicides sethoxydim and haloxyfop, *Biochem. Biophys. Res. Commun.*, 148, 1039, 1987.

57. Rendina, A. R. and Felts, J. M., Cyclohexanedione herbicides are selective and potent inhibitors of acetyl-CoA carboxylase from grasses, *Plant Physiol.*, 86, 983, 1988.

58. Devine, M. D. and Shimabukuro, R. H., Resistance to acetyl coenzyme A carboxylase inhibiting herbicides, in *Herbicide Resistance in Plants: Biology and Biochemistry*, Powles, S. B. and Holtum, J. A. M., Eds., Lewis Publishers, Boca Raton, FL, 1994, 141.

59. Herbert, D., Walker, K. A., Price, L. J., Cole, D. J., Pallett, K. E., Ridley, S. M., and Harwood, J. L., Acetyl-CoA carboxylase – a graminicide target site, *Pestic. Sci.*, 50, 67, 1997.

60. Konishi, T., Shinohara, K., Yamada, K., and Sasaki, Y., Acetyl-CoA carboxylase in higher plants: most plants other than gramineae have both the prokaryotic and eukaryotic forms of this enzyme, *Plant Cell Physiol.*, 37, 117, 1996.

61. Shukla, A., Dupont, S., and Devine, M. D., Resistance to ACCase-inhibitor herbicides in wild oat: evidence for target site-based resistance in two biotypes from Canada, *Pestic. Biochem. Physiol.*, 57, 147, 1997.

62. Tardif, F. J., Preston, C., and Powles, S. B., Mechanisms of herbicide multiple resistance in *Lolium rigidum*, in *Weed and Crop Resistance to Herbicides*, De Prado, R., Jorrín, J., and García-Torres, L., Eds., Kluwer Academic Publishers, Dordrecht, the Netherlands, 1997, 117.

63. Nikolskaya, T., Zagnitko, O., Tevzadze, G., Haselkorn., and Gornicki, P., Herbicide sensitivity determinant of wheat plastid acetyl-CoA carboxylase is located in a 400-amino acid fragment of the carboxyltransferase domain, *Proc. Natl. Acad. Sci. U.S.A.*, 96, 14647, 1999.

64. Gornicki, P., Faris, J., King, I., Podowinski, J., Gill, B., and Haselkorn, R., Plastid-localized acetyl-CoA carboxylase of bread wheat is encoded by a single gene on each of the three ancestral chromosome sets, *Proc. Natl. Acad. Sci. U.S.A.*, 94, 14179, 1997.

65. Gornicki, P., Podowinski, J., Scappino, L. A., DiMaio, J., Ward, E., and Haselkorn, R., Wheat acetyl-coenzyme A carboxylase: cDNA and protein structure, *Proc. Natl. Acad. Sci. U.S.A.*, 91, 6860, 1994.

66. Barr, A. R., Mansooji, A. M., Holtum, J. A. M., and Powles, S. B., The inheritance of herbicide resistance in *Avena sterilis* ssp. *ludoviciana*, biotype SAS, *Proc. 1st Int. Weed Cont. Congr.*, 2, 70, 1992.

67. Murray, B. G., Morrison, I. N., and Brûlé-Babel, A., Inheritance of acetyl-CoA carboxylase inhibitor resistance in wild oat (*Avena fatua*), *Weed Sci.*, 43, 233, 1995.
68. Seefeldt, S. S., Hoffman, D. L., Gealy, D. R., and Fuerst, E. P., Inheritance of diclofop resistance in wild oat (*Avena fatua* L.) biotypes from the Willamette Valley of Oregon, *Weed Sci.*, 46, 170, 1998.
69. Betts, K. J., Ehlke, N. J., Wyse, D. L., Gronwald, J. W., and Somers, D. A., Mechanism of inheritance of diclofop resistance in Italian ryegrass (*Lolium multiflorum*), *Weed Sci.*, 40, 184, 1992.
70. Tardif, F. J., Preston, C., Holtum, J. A. M., and Powles, S. B., Resistance to acetyl-coenzyme A carboxylase-inhibiting herbicides endowed by a single major gene encoding a resistant target site in a biotype of *Lolium rigidum*, *Aust. J. Plant Physiol.*, 23, 15, 1996.
71. Murray, B. G., Brûlé-Babel, A., and Morrison, I. N., Two distinct alleles encode for Acetyl-CoA carboxylase inhibitor resistance in wild oat (*Avena fatua*), *Weed Sci.*, 44, 476, 1996.
72. Bradshaw, L. D., Padgette, S. R., Kimball, S. L., and Wells, B. H., Perspectives on glyphosate resistance, *Weed Technol.*, 11, 189, 1997.
73. Padgette, S. R., Re, D. B., Gasser, C. S., Eichholtz, D. A., Frazier, R. B., Hironaka, C. M., Levime, E. B., Shah, D. M., Fraley, R. T., and Kishore, G. M., Site-directed mutagenesis of a conserved region of the 5-enolpyruvylshikimate 3 phosphate synthase active site, *J. Biol. Chem.*, 266, 22364, 1991.
74. Powles, S. B., Lorraine-Colwill, D. F., Dellow, J. J., and Preston, C., Evolved resistance to glyphosate in rigid ryegrass (*Lolium rigidum*) in Australia, *Weed Sci.*, 46, 604, 1998.
75. Pratley, J., Urwin, N., Stanton, R., Baines, P., Broster, J., Cullis, K., Schafer, D., Bohn, J., and Krueger, R., Resistance to glyphosate in *Lolium rigidum*. I. Bioevaluation, *Weed Sci.*, 47, 405, 1999.
76. Lee, L. J. and Ngim, J., A first report of glyphosate-resistant goosegrass (*Eleusine indica* (L) Gaertn) in Malaysia, *Pest Manage. Sci.*, 56, 336, 2000.
77. Tran, M., Bearson, S., Brinker, R., Casagrande, L., Faletti, M., Feng, Y., Nemeth, M., Reynolds, T., Rodriguez, D., Schafer, D., Stalker, D., Taylor, N., Teng, Y., and Dill, G., Characterization of glyphosate resistant *Eleusine indica* biotypes from Malaysia, *Proc. 17th Asian-Pacific Weed Sci. Congr.*, 1999, 527.
78. Dill, G., personal communication, 2000.
79. Calderbank, A. and Slade, P., Diquat and paraquat, in *Herbicides: Chemistry, Degradation and Mode of Action*, Vol. 2, 2nd ed., Kearney, P. C. and Kaufman, D. D., Eds., Marcel Dekker, New York, 1976, 501.
80. Hatzios, K. K., Biotransformations of herbicides in higher plants, in *Environmental Chemistry of Herbicides*, Vol. II, Grover, R. and Cessna, A. J., Eds., CRC Press, Boca Raton, FL, 1991, 142.
81. Cole, D. J., Detoxification and activation of herbicides by plants, *Pestic. Sci.*, 42, 209, 1994.
82. Kemp, M. S., Moss, S. R., and Tomas, T. H., Herbicide resistance in *Alopecurus myosuroides*, in *Managing Resistance to Agrochemicals: From Fundamental Research to Practical Strategies*, Green, M. B., LeBaron, H. M., and Moberg, W. K., Eds., American Chemical Society, Washington, D.C., 1990, 376.
83. Mendenez, J. M., De Prado, R., Jorrin, J., and Taberner, A., Penetration, translocation and metabolization of diclofop-methyl in chlorotoluron-resistant and -susceptible biotypes of *Alopecurus myosuroides*, *Proc. Brighton Crop Prot. Conf. — Weeds*, 1993, 213.

84. Mendenez, J. M., De Prado, R., and Devine, M. D., Chlorsulfuron cross-resistance in a chlorotoluron-resistant biotype of *Alopecurus myosuroides*, *Proc. Brighton Crop Prot. Conf. — Weeds*, 1997, 319.
85. Hall, L. M., Moss, S. R., and Powles, S. B., Mechanism of resistance to chlorotoluron in two biotypes of the grass weed *Alopecurus myosuroides*, *Pestic. Biochem. Physiol.*, 53, 180, 1995.
86. Hall, L. M., Moss, S. R., and Powles, S. B., Mechanism of resistance to aryloxyphenoxypropionate herbicides in two resistant biotypes of *Alopecurus myosuroides* (blackgrass): herbicide metabolism as a cross-resistance mechanism, *Pestic. Biochem. Physiol.*, 57, 87, 1997.
87. Mendenez, J. and De Prado, R., Metabolism of propaquizafop in chlorotoluron-resistant and -susceptible biotypes of *Alopecurus myosuroides*, in *Proc. 2nd Int. Weed Cont. Congr.*, Brown, H., Cussans, G. W., Devine, M. D., Duke, S. O., Fernandez-Quintanilla, C., Helweg, A., Labrada, R. E., Landes, M., Kudsk, P., and Streibig, J. C., Eds., Department of Weed Control and Pesticide Ecology, Slagelse, Denmark, 1996, 517.
88. Mendenez, J. and De Prado, R., Diclofop-methyl cross-resistance in a chlorotoluron-resistant biotype of *Alopecurus myosuroides*, *Pestic. Biochem. Physiol.*, 56, 123, 1996.
89. Cocker, K. M., Moss, S. R., and Coleman, J. O. D., Multiple mechanisms of resistance to fenoxaprop-P-ethyl in United Kingdom and other European populations of herbicide-resistant *Alopecurus myosuroides* (black-grass), *Pestic. Biochem. Physiol.*, 65, 189, 1999.
90. Cummins, I., Moss, S., Cole, D. J., and Edwards, R., Glutathione transferases in herbicide-resistant and herbicide-susceptible black-grass (*Alopecurus myosuroides*), *Pestic. Sci.*, 51, 1997, 244.
91. Reade, J. P. H., Hull, M. R., and Cobb, A. H., A role for glutathione-*S*-transferases in herbicide resistance in black-grass (*Alopecurus myosuroides*), *Proc. Brighton Crop Prot. Conf. — Weeds*, 777, 1997.
92. Anderson, M. P. and Gronwald, J. W., Atrazine resistance in a velvetleaf (*Abutilon theophrasti*) biotype due to enhanced glutathione-*S*-transferase activity, *Plant Physiol.*, 96, 104, 1991.
93. Gray, J. A., Balke, N. E., and Stoltenberg, D. E., Increased glutathione conjugation of atrazine confers resistance in a Wisconsin velvetleaf (*Abutilon theophrasti*) biotype, *Pestic. Biochem. Physiol.*, 55, 157, 1996.
94. Kern, A. J., Peterson, D. M., Miller, E. K., Colliver, C. C., and Dyer, W. E., Triallate resistance in *Avena fatua* L. is due to reduced herbicide activation, *Pestic. Biochem. Physiol.*, 56, 163, 1996.
95. Maneechote, C., Preston, C., and Powles, S. B., A diclofop-methyl-resistant *Avena sterilis* biotype with a herbicide-resistant acetyl-coenzyme A carboxylase and enhanced metabolism of diclofop-methyl, *Pestic. Sci.*, 49, 105, 1995.
96. Hidayat, I. and Preston, C., Enhanced metabolism of fluazifop acid in a biotype of *Digitaria sanguinalis* resistant to the herbicide fluazifop-P-butyl, *Pestic. Biochem. Physiol.*, 57, 137, 1997.
97. Leah, J. M., Caseley, J. C., Riches, C. R., and Valverde, B., Association between elevated activity of aryl acylamidase and propanil resistance in jungle-rice, *Echinochloa colona*, *Pestic. Sci.*, 42, 281, 1994.
98. Carey, V. F., III, Hoagland, R. E., and Talbert, R. E., Resistance mechanism of propanil-resistant barnyardgrass. II. *In vivo* metabolism of the propanil molecule, *Pestic. Sci.*, 49, 333, 1997.

99. Matthews, N., Powles, S. B., and Preston, C., Mechanisms of resistance to acetyl-coenzyme A carboxylase inhibiting herbicides in a *Hordeum leporinum* population, *Pest Manage. Sci.*, 56, 441, 2000.

100. Burnet, M. W. M., Loveys, B. R., Holtum, J. A. M., and Powles, S. B., A mechanism of chlorotoluron resistance in *Lolium rigidum*, *Planta*, 190, 182, 1993.

101. Burnet, M. W. M., Loveys, B. R., Holtum, J. A. M., and Powles, S. B., Increased detoxification is a mechanism of simazine resistance in *Lolium rigidum*, *Pestic. Biochem. Physiol.*, 46, 207, 1993.

102. Christopher, J. T., Powles, S. B., Liljegren, D. R., and Holtum, J. A. M., Cross-resistance to herbicides in annual ryegrass (*Lolium rigidum*). II. Chlorsulfuron resistance involves a wheat-like detoxification system, *Plant Physiol.*, 95, 1036, 1991.

103. Cotterman, J. C. and Saari, L. L., Rapid metabolic inactivation is the basis for cross-resistance to chlorsulfuron in a diclofop-methyl-resistant rigid ryegrass (*Lolium rigidum*) biotype SR4/84, *Pestic. Biochem. Physiol.*, 36, 61, 1992.

104. Holtum, J. A. M., Matthews, J. M., Häusler, R. E., Liljegren, D. R., and Powles, S. B., Cross-resistance to herbicides in annual ryegrass (*Lolium rigidum*). III. On the mechanism of resistance to diclofop-methyl, *Plant Physiol.*, 97, 1026, 1991.

105. Preston, C. and Powles, S. B., Amitrole inhibits diclofop metabolism and synergises diclofop-methyl in a diclofop-methyl-resistant biotype of *Lolium rigidum*, *Pestic. Biochem. Physiol.*, 62, 179, 1998.

106. Burnet, M. W. M., Mechanisms of Herbicide Resistance in *Lolium rigidum*, Ph.D. thesis, University of Adelaide, Adelaide, Australia, 1992.

107. Christopher, J. T., Preston, C., and Powles, S. B., Malathion antagonizes metabolism-based chlorsulfuron resistance in *Lolium rigidum*, *Pestic. Biochem. Physiol.*, 49, 172, 1994.

108. Preston, C., Tardif, F. J., Christopher, J. T., and Powles, S. B., Multiple resistance to dissimilar herbicide chemistries in a biotype of *Lolium rigidum* due to enhanced activity of several herbicide degrading enzymes, *Pestic. Biochem. Physiol.*, 54, 123, 1996.

109. Preston, C. and Powles, S. B., Light-dependent enhanced metabolism of chlorotoluron in a substituted urea herbicide-resistant biotype of *Lolium rigidum* Gaud., *Planta*, 201, 202, 1997.

110. De Prado, R., De Prado, J. L., and Menendez, J., Resistance to substituted urea herbicides in *Lolium rigidum* biotypes, *Pestic. Biochem. Physiol.*, 57, 126, 1997.

111. Preston, C., unpublished data, 1997.

112. Singh, S., Kirkwood, R. C., and Marshall, G., Effect of ABT on the activity and rate of degradation of isoproturon in susceptible and resistant biotypes of *Phalaris minor* and in wheat, *Pestic. Sci.*, 53, 123, 1998.

113. Singh, S., Kirkwood, R. C., and Marshall, G., Effect of monooxygenase inhibitor piperonyl butoxide on the herbicidal activity and metabolism of isoproturon in herbicide resistant and susceptible biotypes of *Phalaris minor* and wheat, *Pestic. Biochem. Physiol.*, 59, 143, 1998.

114. Veldhuis, L. J., Hall, L. M., O'Donovan, J. T., Dyer, W., and Hall, J. C., Metabolism-based resistance of a wild mustard (*Sinapis arvensis* L.) biotype to ethametsulfuron-methyl, *J. Agric. Food Chem.*, 48, 2986, 2000.

115. Coupland, D., Lutman, P. J. W., and Heath, C. R., Uptake, translocation and metabolism of mecoprop in a sensitive and a resistant biotype of *Stellaria media*, *Pestic. Biochem. Physiol.*, 36, 61, 1990.

116. Marrs, K. A., The functions and regulation of glutathione-*S*-transferases in plants, *Annu. Rev. Plant Physiol. Plant Mol. Biol.*, 47, 127, 1996.

117. Farago, S., Brunhold, C., and Kreuz, K., Herbicide safeners and glutathione metabolism, *Physiol. Plant.*, 91, 537, 1994.

118. Cummins, I., Cole, D. J., and Edwards, R., Purification of multiple glutathione transferases involved in herbicide detoxification from wheat (*Triticum aestivum* L.) treated with the safener fenchlorazole-ethyl, *Pestic. Biochem. Physiol.*, 59, 35, 1997.

119. Dixon, D. P., Cole, D. J., and Edwards, R., Characterisation of multiple glutathione transferases containing the GST I subunit with activities toward herbicide substrates in maize (*Zea mays*), *Pestic. Sci.*, 50, 72, 1997.

120. Jepson, I., Lay, V. J., Holt, D. C., Bright, S. W. J., and Greenland, A. J., Cloning and characterization of maize herbicide safener-induced cDNAs encoding subunits of glutathione *S*-transferase isoforms I, II and IV, *Plant Mol. Biol.*, 26, 1855, 1994.

121. Lamoureux, G. L. and Rusness, D. G., The role of glutathione-*S*-transferases in pesticide metabolism, selectivity, and mode of action in plants and insects, in *Glutathione: Chemical, Biochemical, and Medical Aspects, Part B*, Dolphin, D., Poulson, R., and Avramovic, O., Eds., John Wiley & Sons, New York, 1989, 153.

122. Rea, P. A., Li, Z. S., Lu, Y. P., Drozdowicz, Y. M., and Martinoia, E., From vacuolar GS-X pumps to multispecific ABC transporters, *Ann. Rev. Plant Physiol. Plant Mol. Biol.*, 49, 727, 1998.

123. Plaisance, K. L. and Gronwald, J. W., Enhanced catalytic constant for glutathione *S*-transferase (atrazine) activity in an atrazine-resistant *Abutilon theophrasti* biotype, *Pestic. Biochem. Physiol.*, 63, 34, 1999.

124. Cummins, I., Cole, D. J., and Edwards, R., A role for glutathione transferases functioning as glutathione peroxidases in resistance to multiple herbicides in black-grass, *Plant J.*, 18, 285, 1999.

125. Anderson, R. N. and Gronwald, J. W., Noncytoplasmic inheritance of atrazine tolerance in velvetleaf (*Abutilon theophrasti*), *Weed Sci.*, 35, 496, 1987.

126. Frear, D. S. and Still, G. G., The metabolism of 3,4-dichloropropionanilide in plants. Partial purification and properties of an aryl acylamidase from rice, *Phytochemistry*, 7, 913, 1968.

127. Hirase, K. and Matsunaka, S., Physiological role of the propanil hydrolyzing enzyme (aryl acylamidase I) in rice plants, *Pestic. Biochem. Physiol.*, 41, 82, 1991.

128. Yih, R. Y., McRae, H., and Wilson, H. F., Mechanism of selective action of 3,4-dichloropropionanilide, *Plant Physiol.*, 43, 1291, 1968.

129. Carey, V. F., III, Hoagland, R. E., and Talbert, R. E., Verification and distribution of propanil-resistant barnyardgrass in Arkansas, *Weed Technol.*, 9, 366, 1995.

130. Valverde, B. E., Management of herbicide resistant weeds in Latin America: the case of propanil-resistant *Echinochloa colona* in rice, in *Proc. 2nd Int. Weed Cont. Congr.*, Brown, H., Cussans, G. W., Devine, M. D., Duke, S. O., Fernandez-Quintanilla, C., Helweg, A., Labrada, R. E., Landes, M., Kudsk, P., and Streibig, J. C., Eds., Department of Weed Control and Pesticide Ecology, Slagelse, Denmark, 1996, 415.

131. Daou, H. and Talbert, R.E., Control of propanil-resistant barnyardgrass (*Echinochloa crus-galli*) in rice (*Oryza sativa*) with carbaryl/propanil mixtures, *Weed Technol.*, 13, 65, 1999.

132. Bollwell, G. P., Bozak, K., and Zimmerlin, A., Plant cytochrome P450, *Phytochemistry*, 37, 1491, 1994.

133. Schuler, M. A., Plant cytochrome P450 monooxygenases, *Crit. Rev. Plant Sci.*, 15, 235, 1996.

134. Halkier, B. A., Catalytic reactivities and structure/function relationships of cytochrome P450 enzymes, *Phytochemistry*, 43, 1, 1996.

135. Davies, J. and Caseley, J. C., Herbicide safeners: a review, *Pestic. Sci.*, 55, 1043, 1999.

136. Porter, T. D. and Coon, M. J., Cytochrome P450. Multiplicity of isoforms, substrates, and catalytic and regulatory mechanisms, *J. Biol. Chem.*, 266, 13469, 1991.
137. Sandermann, H., Jr., Plant metabolism of xenobiotics, *Trends Biochem. Sci.*, 17, 82, 1992.
138. Barrett, M., Metabolism of herbicides by cytochrome P450 in corn, *Drug Metab. Drug Interact.*, 12, 299, 1995.
139. Nelson, D. R., Cytochrome P450 and the individuality of species, *Arch. Biochem. Biophys.*, 369, 1, 1999.
140. Singh, S., Kirkwood, R. C., and Marshall, G., Management approaches for isoproturon-resistant *Phalaris minor* in India, *Proc. Brighton Crop Prot. Conf. — Weeds*, 1997, 357.
141. Lay, M.-M. and Casida, J. E., Dichloroacetamide antidotes enhance thiocarbamate sulfoxide detoxification by elevating corn root glutathione content and glutathione *S*-transferase activity, *Pestic. Biochem. Physiol.*, 6, 442, 1976.
142. Chauvel, B. and Gasquez, J., Relationships between genetic polymorphism and herbicide resistance within *Alopecurus myosuroides* Huds., *Heredity*, 72, 336, 1994.
143. Preston, C., unpublished data, 1999.
144. Preston, C., Resistance to photosystem I disrupting herbicides, in *Herbicide Resistance in Plants: Biology and Biochemistry*, Powles, S. B. and Holtum, J. A. M., Eds., Lewis Publishers, Boca Raton, FL, 1994, 61.
145. Alizadeh, H. A., Preston, C., and Powles, S. B., Paraquat-resistant biotypes of *Hordeum glaucum* from zero-tillage wheat, *Weed Res.*, 38, 139, 1998.
146. Bishop, T., Powles, S. B., and Cornic, G., Mechanism of paraquat resistance in *Hordeum glaucum*. II. Paraquat uptake and translocation, *Aust. J. Plant Physiol.*, 14, 539, 1987.
147. Preston, C., Holtum, J. A. M., and Powles, S. B., On the mechanism of resistance to paraquat in *Hordeum glaucum* and *H. leporinum*, *Plant Physiol.*, 100, 630, 1992.
148. Lasat, M. M., DiTomaso, J. M., Hart, J. J., and Kochian, L. V., Evidence for vacuolar sequestration of paraquat in roots of a paraquat-resistant *Hordeum glaucum* biotype, *Physiol. Plant.*, 99, 255, 1997.
149. Purba, E., Preston, C., and Powles, S. B., The mechanism of resistance to paraquat is strongly temperature dependent in resistant *Hordeum leporinum* Link and *H. glaucum* Steud., *Planta*, 196, 1995.
150. Preston, C., Balachandran, S., and Powles, S. B., Investigations of mechanisms of resistance to bipyridyl herbicides in *Arctotheca calendula* (L.) Levyns, *Plant Cell Environ.*, 17, 1113, 1994.
151. Soar, C., Karotam, J., Powles, S. B., and Preston, C., unpublished data, 1999.
152. Norman, M. A., Smeda, R. J., Vaughn, K. C., and Fuerst, E. P., Differential movement of paraquat in resistant and sensitive biotypes of *Conyza*, *Pestic. Biochem. Physiol.*, 50, 31, 1994.
153. Norman, M. A., Fuerst, E. P., Smeda, R. J., and Vaughn, K. C., Evaluation of paraquat resistance mechanisms in *Conyza*, *Pestic. Biochem. Physiol.*, 46, 236, 1993.
154. Tanaka, Y., Chisaka, H., and Saka, H., Movement of paraquat in resistant and susceptible biotypes of *Erigeron philadelphicus* and *E. canadensis*, *Physiol. Plant.*, 66, 605, 1986.
155. Islam, A. K. M. R. and Powles, S. B., Inheritance of resistance to paraquat in barley grass *Hordeum glaucum* Steud., *Weed Res.*, 28, 393, 1988.
156. Purba, E., Preston, C., and Powles, S. B., Inheritance of bipyridyl herbicide resistance in *Arctotheca calendula* and *Hordeum leporinum*, *Theor. Appl. Genet.*, 87, 598, 1993.

157. Yamasue, Y., Kamiyama, K., Hanioka, Y., and Kusanagi, T., Paraquat resistance and its inheritance in seed germination of the foliar-resistant biotypes of *Erigeron canadensis* L. and *E. sumatrensis* Retz., *Pestic. Biochem. Physiol.*, 44, 21, 1992.

158. Shaaltiel, Y. and Gressel, J., Multienzyme oxygen radical detoxifying system correlated with paraquat resistance in *Conyza bonariensis, Pestic. Biochem. Physiol.*, 26, 22, 1986.

159. Shaaltiel, Y., Glazer, A., Bocion, T. F., and Gressel, J., Cross-tolerance to herbicides and environmental oxidants of plant biotypes tolerant to paraquat, sulfur dioxide, and ozone, *Pestic. Biochem. Physiol.*, 31, 12, 1988.

160. Matsunaka, S. and Ito, K., Paraquat resistance in Japan, in *Herbicide Resistance in Weeds and Crops*, Caseley, J. C., Cussans, G. W., and Atkin, R. K., Eds., Butterworth-Heinemann, Oxford, 1991, 77.

161. Lehoczki, E., Laskay, G., Gaál, I., and Szigeti, Z., Mode of action of paraquat in leaves of paraquat-resistant *Conyza canadensis* (L.) Cronq., *Plant Cell Environ.*, 15, 531, 1992.

162. Amsellem, Z., Jansen, M. A. K., Driesenaar, A. R. J., and Gressel, J., Developmental variability of photoxidative stress tolerance in paraquat-resistant *Conyza, Plant Physiol.*, 103, 1097, 1993.

163. Ye, B. and Gressel, J., Constitutive variation of ascorbate peroxidase activity during development parallels that of superoxide dismutase and glutathione reductase in paraquat-resistant *Conyza, Plant Sci.*, 102, 147, 1994.

164. Ye, B., Faltin, Z., Ben-Hayyim, G., Eshdat, Y., and Gressel, J., Correlation of glutathione peroxidase to paraquat/oxidative stress resistance in *Conyza* determined by direct fluorometric assay, *Pestic. Biochem. Physiol.*, 66, 182, 2000.

165. Ye, B., Müller, H. H., Zhang, J., and Gressel, J., Constitutively elevated levels of putrescine and putrescine-generating enzymes correlated with oxidant stress resistance in *Conyza bonariensis* and wheat, *Plant Physiol.*, 115, 1443, 1997.

166. Ye, B. and Gressel, J., Transient, oxidant-induced antioxidant transcript and enzyme levels correlate with greater oxidant-resistance in paraquat-resistant *Conyza bonariensis, Planta*, 211, 50, 2000.

167. Shaaltiel, Y., Chua, N.-H., Gepstein, S., and Gressel, J., Dominant pleiotrophy controls enzymes co-segregating with paraquat resistance in *Conyza bonariensis, Theor. Appl. Genet.*, 75, 850, 1988.

168. Vaughn, K. C. and Fuerst, E. P., Structural and physiological studies of paraquat-resistant *Conyza, Pestic. Biochem. Physiol.*, 24, 86, 1985.

169. Turcsányi, E., Darkó, É., Borbély, G., and Lehoczki, E., The activity of oxyradical-detoxifying enzymes is not correlated with paraquat resistance in *Conyza canadensis* (L.) Cronq., *Pestic. Biochem. Physiol.*, 60, 1, 1998.

170. Hart, J. J. and DiTomaso, J. M., Sequestration and oxygen radical detoxification as mechanisms of paraquat resistance, *Weed Sci.*, 42, 277, 1994.

171. Lorraine-Colwill, D. F., Hawkes, T. R., Williams, P. H., Warner, S. A. J., Sutton, P. B., Powles, S. B., and Preston, C., Resistance to glyphosate in *Lolium rigidum, Pestic. Sci.*, 55, 486, 1999.

172. Gruys, K. J., Biest-Taylor, N. A., Feng, P. C. C., Baerson, S. R., Rodriguez, D. J., You, J., Tran, M., Feng, Y., Krueger, R. W., Pratley, J. E., Urwin, N. A., and Stanton, R. A., Resistance of glyphosate in annual ryegrass (*Lolium rigidum*). II. Biochemical and molecular analyses, *Weed Sci. Soc. Am. Abstr.*, 39, 82, 1999.

173. Feng, P. C. C., Pratley, J. E., and Bohn, J. A., Resistance to glyphosate in *Lolium rigidum*. II. Uptake, translocation, and metabolism, *Weed Sci.*, 47, 412, 1999.

174. Lorraine-Colwill, D. F., Powles, S. B., Hawkes, T. R., and Preston, C., Inheritance of glyphosate resistance in *Lolium rigidum* Gaud., *Theor. Appl. Genet.*, in press.
175. DeGennaro, F. P. and Weller, S. C., Differential susceptibility of field bindweed (*Convolvulus arvensis*) biotypes to glyphosate, *Weed Sci.*, 32, 472, 1984.
176. Westwood, J. H. and Weller, S. C., Cellular mechanisms influence differential glyphosate sensitivity in field bindweed (*Convolvulus arvensis*) biotypes, *Weed Sci.*, 45, 2, 1997.
177. Duncan, C. N. and Weller, S. C., Heritability of glyphosate susceptibility among biotypes of field bindweed, *J. Hered.*, 78, 257, 1987.
178. Coupland, D., Resistance to the auxin analog herbicides, in *Herbicide Resistance in Plants: Biology and Biochemistry*, Powles, S. B. and Holtum, J. A. M., Eds., Lewis Publishers, Boca Raton, FL, 1994, 171.
179. Fuerst, E. P., Sterling, T. M., Norman, M. A., Prather, T. S., Irzyk, G. P., Wu, Y., Lownds, N. K., and Callihan, R. H., Physiological characterization of picloram resistance in yellow starthistle, *Pestic. Biochem. Physiol.*, 56, 149, 1996.
180. Valenzuela-Valenzuela, J., Mechanisms of Cross-Resistance to Clopyralid in Picloram-Resistant Yellow Starthistle, Ph.D. thesis, New Mexico State University, Las Cruces, 1998.
181. Sabba, R. P., Sterling, T. M., and Lownds, N. K., Effect of picloram on resistant and susceptible yellow starthistle: the role of ethylene, *Weed Sci.*, 46, 297, 1998.
182. Sterling, T. M., personal communication, 1999.
183. Peniuk, M. G., Romano, M. L., and Hall, J. C., Physiological investigations into the resistance of a wild mustard (*Sinapsis arvensis* L.) biotype of auxinic herbicides, *Weed Res.*, 33, 431, 1993.
184. Hall, J. C., Alam, S. M. M., and Murr, D. P., Ethylene biosynthesis following foliar application of picloram to biotypes of wild mustard (*Sinapis arvensis* L.) susceptible or resistant to auxinic herbicides, *Pestic. Biochem. Physiol.*, 47, 36, 1993.
185. Webb, S. R. and Hall, C. J., Auxinic herbicide-resistant and -susceptible wild mustard (*Sinapis arvensis* L.) biotypes: effect of auxinic herbicides on seedling growth and auxin-binding activity, *Pestic. Biochem. Physiol.*, 52, 137, 1995.
186. Deshpande, S. and Hall, J. C., Auxinic herbicide resistance may be modulated at the auxin-binding site in wild mustard (*Sinapis arvensis* L.): a light scattering study, *Pestic. Biochem. Physiol.*, 66, 41, 2000.
187. Hall, J. C. and Romano, M. L., Morphological and physiological differences between the auxinic herbicide-susceptible (S) and -resistant (R) wild mustard (*Sinapis arvensis* L.) biotypes, *Pestic. Biochem. Physiol.*, 52, 149, 1995.
188. Sabba, R. P., Sterling, T. M., and Lownds, N. K., Complex genetics of yellow starthistle (*Centaurea solstitialis* L.) resistance to the auxinic herbicides picloram and clopyralid, *Weed Sci. Soc. Am. Abstr.*, 36, 4, 1997.
189. Sabba, R. P., Sterling, T. M., and Lownds, N. K., Inheritance of yellow starthistle resistance to the auxinic herbicides picloram and clopyralid, *Plant Physiol.*, 111S, 17, 1996.
190. Jasieniuk, M., Morrison, I. N., and Brûlé-Babel, A., Inheritance of dicamba resistance in wild mustard (*Brassica kaber*), *Weed Sci.*, 43, 192, 1995.
191. Hall, L. M., Holtum, J. A. M., and Powles, S. B., Mechanisms responsible for cross resistance and multiple resistance, in *Herbicide Resistance in Plants: Biology and Biochemistry*, Powles, S. B. and Holtum, J. A. M., Eds., Lewis Publishers, Boca Raton, FL, 1994, 243.

192. Hall, L. M., Tardif, F. J., and Powles, S. B., Mechanisms of cross and multiple resistance in *Alopecurus myosuroides* and *Lolium rigidum*, *Phytoprotection*, 75, S17, 1994.

193. Kern, A. J. and Dyer, W. E., Compartmental analysis of herbicide efflux in susceptible and difenzoquat-resistant *Avena fatua* L. suspension cells, *Pestic. Biochem. Physiol.*, 61, 27, 1998.

194. Gressel, J., Regev, S., Malkin, S., and Kleifeld, Y., Characterization of an *s*-triazine-resistant biotype of *Brachypodium distachyon*, *Weed Sci.*, 31, 450, 1983.

195. Pölös, E., Mikulàs, J., Szigeti, Z., Matkovics, B., Hai, D. Q., Pàrducz, À., and Lehoczki, E., Paraquat and atrazine co-resistance in *Conyza canadensis* (L.) Cronq., *Pestic. Biochem. Physiol.*, 30, 142, 1988.

196. Lopez-Martinez, N., Marshall, G., and De Prado, R., Resistance of barnyardgrass (*Echinochloa crus-galli*) to atrazine and quinclorac, *Pestic. Sci.*, 51, 171, 1997.

197. Hall, L. M., Stromme, K. M., Horsman, G. P., and Devine, M. D., Resistance to acetolactate synthase inhibitors and quinclorac in a biotype of false cleavers (*Galium spurium*), *Weed Sci.*, 46, 390, 1998.

198. McAlister, F. M., Holtum, J. A. M., and Powles, S. B., Dinitroaniline herbicide resistance in rigid ryegrass (*Lolium rigidum*), *Weed Sci.*, 43, 55.

199. Burnet, M. W. M., Christopher, J. T., Holtum, J. A. M., and Powles, S. B., Identification of two mechanisms of sulfonylurea resistance within one population of rigid ryegrass (*Lolium rigidum*) using a selective germination medium, *Weed Sci.*, 42, 468, 1994.

200. Christopher, J. T., Powles, S. B., and Holtum, J. A. M., Resistance to acetolactate synthase-inhibiting herbicides in annual ryegrass (*Lolium rigidum*) involves at least two mechanisms, *Plant Physiol.*, 100, 1909, 1992.

201. De Prado, R., Lopez-Martinez, N., and Gonzalez-Gutierrez, J., Identification of two mechanisms of atrazine resistance in *Setaria faberi* and *Setaria viridis* biotypes, *Pestic. Biochem. Physiol.*, 67, 114, 2000.

202. Salhoff, C. R. and Martin, A. R., *Kochia scoparia* growth response to triazine herbicides, *Weed Sci.*, 34, 40, 1986.

203. Preston, C., Tardif, F. J., and Powles, S. B., Multiple mechanisms and multiple herbicide resistance in *Lolium rigidum*, in *Molecular Genetics and Evolution of Pesticide Resistance*, Brown, T. M., Ed., American Chemical Society, Washington, D.C., 1996, 117.

204. Burnet, M. W. M., Hart, Q., Holtum, J. A. M., and Powles, S. B., Resistance to nine herbicide classes in a biotype of rigid ryegrass (*Lolium rigidum*), *Weed Sci.*, 42, 369, 1994.

205. Jasieniuk, M., Brûlé-Babel, A., and Morrison, I. N., The evolution and genetics of herbicide resistance in weeds, *Weed Sci.*, 44, 176, 1996.

206. Preston, C. and Roush, R.T., Variation in herbicide dose rates: risks associated with herbicide resistance, in *Precision Weed Management*, Medd, R. W. and Pratley, J. E., Eds., CRC for Weed Management Systems, Adelaide, Australia, 1999, 128.

207. Wrubel, R. P. and Gressel, J., Are herbicide mixtures useful for delaying the rapid evolution of resistance? A case study, *Weed Technol.*, 8, 635, 1994.

208. Gardner, S. N., Gressel, J., and Mangel, M., A revolving dose strategy to delay the evolution of both quantitative vs. major monogene resistances to pesticides and drugs, *Int. J. Pest Manage.*, 44, 161, 1998.

3 The Population Dynamics and Genetics of Herbicide Resistance — A Modeling Approach

Art J. Diggle and Paul Neve

CONTENTS

0-8493-2219-7/01/$0.00+$.50
© 2001 by CRC Press LLC

61

3.1   INTRODUCTION

The development and spread of herbicide resistance in weed populations is an evolutionary phenomenon. Evolution occurs when populations respond to selection and it depends on the existence within these populations of genetic variability. In plant populations from natural ecosystems, genetic variability is maintained by environmental heterogeneity. Agriculture, on the other hand, seeks to create homogenous environments that maximize crop production. Weed species have co-evolved with modern agricultural practices and, as a result, generally exhibit lower levels of genetic variability than natural plant populations.[1] This is not to say that potential for evolution does not exist in weed populations; indeed, the reverse is true, with the typically strong selection pressures imposed by agricultural weed control resulting in rapid evolution. Herbicides are very intense selective agents and where genetic variability for herbicide response exists in weed populations, evolution of herbicide resistance is the inevitable result.

The probability and rate of herbicide-resistance evolution depend on the interplay between the population dynamics and population genetics of weed populations.[2–5] Important evolutionary factors include:

- The intensity of selection
- The frequency of resistance alleles in natural (unselected) populations
- The mode of inheritance of resistance

- The relative fitness of susceptible and resistant biotypes in the presence and absence of herbicide
- Gene flow within and between populations

The intrinsic population dynamics of weed populations may also be important, especially in the area of seed bank dynamics where it is recognized that a persistent seed bank can act as a buffer to evolution.[5] Agronomic practices (e.g., tillage, weed seed capture at harvest, crop competition) clearly affect weed densities and seed production, but will only influence rates of resistance evolution where these act as selective agents for or against resistant biotypes. This will only be the case when differences exist in the phenology or population dynamics of resistant and susceptible biotypes.

Several simulation models of the population genetics and dynamics of herbicide resistance in weed populations have been developed.[5–10] This chapter discusses and reviews previous models and provides novel approaches to the development of simulation models of herbicide resistance. The final section provides examples of the use of these models as tools for long-term resistance management.

3.1.1 THE GENETIC BASIS OF HERBICIDE RESISTANCE

Herbicide resistance may be conferred on individuals within a population by any of a number of mechanisms, each of which is under genetic control. These include conformational changes to the target site for herbicide action, enhanced metabolism and degradation, reduced uptake and translocation, repair of herbicide-induced damage, and sequestration of the herbicide within the plant cell.[11] Variation in response to herbicides may be present in unselected weed populations due to the existence, at low frequencies of major resistance-conferring genes, which have mutated from the wild type. Subsequent selection with herbicides increases the frequency of these genes in weed populations. Alternatively, quantitative (continuous) variation in herbicide response may exist. In this instance, recurrent selection with herbicides will result in a progressive increase in the resistance status of the population as genes at many loci are selected.[2]

3.1.2 WHY A MODELING APPROACH?

Several factors govern the rate and probability of herbicide-resistance evolution in weed populations. The development of modeling approaches that simulate the population processes resulting in resistance evolution has many applications. To the plant population ecologist interested in evolutionary processes they enable, through sensitivity analyses, an assessment of the relative importance of each of these factors to the development and spread of resistance. While progress continues to be made with basic research, models can help to direct this research and provide focus for the key driving forces in resistance evolution. Perhaps more importantly, once models have been developed they can be used to explore management options that will delay, or perhaps even prevent, the evolution of resistance over the long term. One of the major constraints to the adoption of proactive, integrated management strategies to prevent or delay resistance is the perception among farmers that this will cost them money (see Chapter 5). Many are not convinced of the logic of dealing with a problem before it arises, preferring instead to adopt a more reactive approach. Simulation models together with bioeconomic analysis can be used to demonstrate the benefits of far-sighted management in terms of weed densities and ultimately economic returns. In this way, models can be used as decision-support tools for long-term resistance management.

3.1.3   EXISTING HERBICIDE-RESISTANCE MODELS

A number of population models have been developed to simulate the evolution and subsequent dynamics of herbicide resistance in weed populations. The first of these, developed in 1978 by Gressel and Segel,[6] sought to explore sensitivities among the major factors that regulate the appearance and spread of herbicide resistance. In particular, fitness, selection pressure, and soil seed bank dynamics were implicated. A greater understanding of these processes and their relative importance enables prescriptive management recommendations to prevent or reduce the risk of resistance evolution and spread. A model developed by Putwain and Mortimer[7] uses actual field data for the dynamics and relative fitness of triazine-resistant and -susceptible biotypes of *Senecio vulgaris* to predict the influence of herbicide use strategies and selection pressure on the relative abundance of these biotypes. Perhaps the most comprehensive and widely cited of the herbicide-resistance models published to date is that of Maxwell et al.[8] This model combines a detailed model of the demographics of an annual weed with an inheritance submodel based on the Hardy-Weinberg concept of gene segregation. A novel approach with this model was to include processes of seed and pollen immigration (gene flow) and these processes, together with those that determine the ecological fitness of susceptible and resistant biotypes, were demonstrated as key driving forces in the evolution and spread of resistance.

Further attempts at herbicide-resistance modeling have addressed particular issues and management options. In a modified version of their earlier model, Gressel and Segel[9] addressed the question of the effectiveness of herbicide rotations and mixtures as strategies to delay or prevent resistance evolution. Mortimer et al.[5] studied the influence of ecological fitness on the spread of herbicide resistance. More recently, Gardner et al.[10] developed a modeling framework that simulates the concomitant evolution of quantitative and major monogene resistances and used this to advocate a strategy of revolving doses to delay the evolution of both types of resistance. The most recent innovation in herbicide-resistance modeling has been the introduction of spatial dynamics in a model that combines population dynamics, population genetics, and spatial spread of herbicide resistance.[12] Such approaches may have considerable application in the future for modeling processes of pollen and seed dispersal within and between populations and for predicting spatial trends in resistance evolution and competitive processes in aggregated or patchy weed populations.

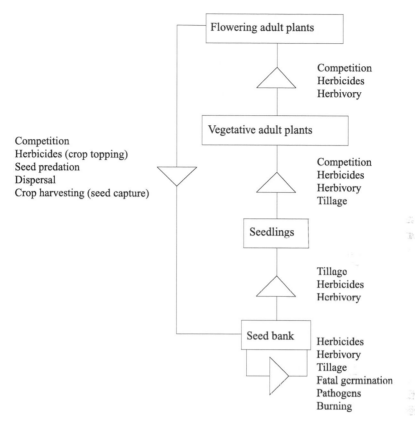

FIGURE 3.1   A simplified life history diagram for a single aged cohort of an annual plant species. The major causes of losses of plants from each life history stage are shown. (Adapted from Reference 13.)

3.2   THE POPULATION DYNAMICS OF ANNUAL WEED SPECIES

The basic structure for any model of the population dynamics of a weed species is determined by its life cycle. The flow diagram in Figure 3.1 shows a simplified life history for a typical annual plant. Plants exist in a number of states, or life history stages, and population dynamics is concerned with fluxes between these states, which are mediated by density-dependent and -independent processes. The reference point for most studies of the population dynamics of annual weedy species is the seed bank. Individual seeds may reside in the seed bank for varying lengths of time depending on the interplay between various biotic (dormancy, predation) and environmental (soil moisture, temperature) factors, which regulate the survival and germination of individual seeds. The study of fluxes into and out of the seed bank and of cycling between various dormancy states is commonly known as seed bank dynamics.

Once a seed germinates it typically becomes established as a seedling and then as a mature adult vegetative plant that flowers and sets seeds that are returned to the soil seed bank. Associated with the transition of individuals within a population between each of these life history stages is a probability. For example, as illustrated in Figure 3.1, the probability that an established seedling will become an adult plant depends upon biotic factors such as density-dependent mortality and herbivory and also on extrinsic factors, which in the case of weeds of agricultural systems are usually manifested by agronomic disturbances such as tillage and/or herbicides. Figure 3.1 summarizes the range of factors that influence fluxes between the various life history stages.

3.3 THE POPULATION GENETICS OF HERBICIDE RESISTANCE

3.3.1 ORIGINS OF GENETIC VARIABILITY FOR HERBICIDE SUSCEPTIBILITY IN PLANTS

We have already established that evolution of resistance will occur only in plant populations where variation for herbicide response exists. In unselected plant populations this variation may be pre-existing or it may arise *de novo* as a result of mutation or from gene flow from elsewhere. The level of pre-existing variation and the rates of spontaneous mutation and gene flow ultimately determine gene frequencies for resistance in unselected populations.

3.3.1.1 Genetic Mutation to Resistance

Spontaneous mutations are generated at gene loci throughout the genome at characteristic rates from generation to generation. While these rates may vary between gene loci, species, and between populations of the same species, typical rates have been reported as between 1×10^{-5} to 1×10^{-6} gametes per locus per generation.[14] Where weeds exhibit high levels of resistance this has been shown, in most cases, to be caused by a single mutation at the gene coding for the herbicide's target enzyme.[2,15,16] In these instances, frequencies of resistance will be directly related to mutation rates. Actual mutation rates at loci conferring herbicide resistance are not known in any weed species, although these have been estimated to be approximately 1×10^{-9} for ALS mutations in *Arabidopsis thaliana*.[17]

There has been some suggestion that mutation rates may be regulated by environmental factors and or management practices. These theories of directed[18,19] or hypermutation[20] have been controversial, stating that increased mutation rates may be an evolved response enabling organisms to increase their genetic variation when they are subjected to environmental stresses. Clearly, herbicides would constitute such a stress. Indeed, Bettini et al.[21] reported that exposing susceptible populations of *Chenopodium album* to sublethal doses of the triazine herbicides resulted in progeny with some similar characteristics to resistant biotypes. In this instance, resistance apparently is induced by exposure to the herbicide. These phenomena remain controversial and in the absence of further evidence it is accepted that spontaneous mutations generate the genetic variability upon which subsequent herbicide selection acts.

3.3.1.2 Initial Gene Frequencies for Herbicide Resistance

Simulation models of herbicide-resistance evolution require an estimate of the frequency of resistance alleles in unselected weed populations. In practice, these are difficult to obtain. Population genetics models (see Jasieniuk et al.),[15] which themselves rely on assumptions about mutation rates to resistance and the selective disadvantage of these mutations in the absence of selection with herbicides, suggest that these may be in the order of 1×10^{-5} to 1×10^{-6}. Confirmation of these figures requires that millions of plants be screened for resistance and, in most cases, this has been impractical. Some attempts have been made, however, with surprising results. Matthews and Powles[22] collected a number of previously unsprayed *Lolium rigidum* populations from agricultural land across southern Australia and obtained a gene frequency of 0.02 for resistance to diclofop-methyl. Darmency and Gasquez[23] measured the frequency of triazine-resistant mutants in *Chenopodium album* collected from private gardens in France; these ranged from 1×10^{-4} to 1×10^{-3}. However, we can use a simulation model to examine the influence of initial gene frequencies on predicted rates of resistance evolution. Results of a sensitivity analysis that used a one-gene explicit genotype model of herbicide resistance (see Sections 3.4.2.3 and 3.4.2.4) show the effect of initial gene frequency on rates of resistance evolution. As we might expect, increasing initial gene frequencies results in a more rapid evolution of resistance (Figure 3.2).

Each simulation presented in Figure 3.2 was run with the assumption of a constant weed density (population size) such that the effect of initial gene frequency alone was being compared. In reality, it is the absolute number of resistant individuals within a population, or the chance of a resistant individual occurring within that population, that drives the rate and probability of resistance evolution. In simple terms (gene flow from and to adjacent populations is zero) where a constant weed density is assumed, the probability of a resistant individual occurring in a population is a function of population size (weed density × population area) and the initial gene frequency. This relationship is illustrated in Table 3.1.

While mutation rates and hence initial gene frequencies are beyond the control of management, weed population densities are not. Weed control practices, other than herbicides, which maintain low weed densities can considerably decrease the chance of resistance evolution by reducing the number of resistance alleles in a population.[24] In Australia, the resistance-prone grass weed, *L. rigidum*, was once widely established and nurtured as a desirable component of pastures. It is now, as a result, the most widespread weed of broadacre cropping, and the sheer density of this species plays a large role in its propensity for resistance evolution.[25]

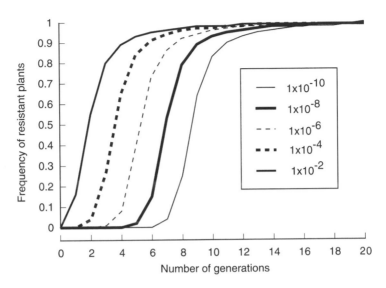

FIGURE 3.2 Predicted rates of herbicide-resistance evolution with varying initial gene frequencies. Herbicide is applied once every generation and the resistance trait is assumed to be completely dominant. Population size is constant in all simulations.

3.3.2 SELECTION PRESSURE

The existence within weed populations of genetic variation for herbicide response coupled with intense selection by herbicides provides the driving force for rapid evolution of resistance. The overall intensity of selection (selection pressure) is mediated by a number of factors. The selection coefficient is a measure of the extent to which the selective agent (herbicide) differentiates between resistant and susceptible individuals within the population.[2] This in turn is influenced by the residual activity of the herbicide. Nonpersistent herbicides will typically exert less intense selection than those that ensure season-long control of germinating weeds. This, of course, depends on the timing of herbicide application and the germination characteristics of the species and population under study. The frequency with which a selective agent is applied, either during a single season or over consecutive seasons, also influences the rate of evolution, as does the specificity of the herbicide's mode of action; those herbicides with a single target site are more prone to resistance evolution.

In the field, most herbicides are applied at rates that result in mortality of between 90 and 99% of susceptible individuals and, as such, they exert intense selection pressures. Actual quantitative measures of the selection coefficients (s) of herbicides in the field have been rare.[15] In population genetics theory, the selection coefficient is defined as the differential survival of alleles after the action of a selective agent.[2,15] In practice this has often been expressed as a measure of the differential mortality of resistant and susceptible phenotypes. However, a more accurate measure is

TABLE 3.1
Total Numbers of Resistant Individuals in Weed Populations of Varying
Size and Initial Gene Frequencies

Weed Density (Seeds m^{-2})	Population Size (m^2)	Initial Gene Frequency	Number of Resistant Individuals
10	10,000	1×10^{-8}	0.0001
		1×10^{-6}	0.01
	1,000,000	1×10^{-8}	0.1
		1×10^{-6}	10
100	10,000	1×10^{-8}	0.01
		1×10^{-6}	1
	1,000,000	1×10^{-8}	1
		1×10^{-6}	100
1000	10,000	1×10^{-8}	0.1
		1×10^{-6}	10
	1,000,000	1×10^{-8}	10
		1×10^{-6}	1000

obtained when differences in production of resistant and susceptible seed are considered.[2,5,6,15] This takes into account other factors that may result in differential seed production by resistant and susceptible survivors and allows for intrinsic density-dependent processes.

Figure 3.3 compares results from a simulation model where the selection pressure for resistance is varied by altering levels of control of susceptible individuals within the population (herbicide efficacy).

The results in Figure 3.3 demonstrate the effects of reduced selection pressure on rates of resistance evolution. These results should not, however, be interpreted without consideration of the effects of reducing herbicide efficacy on overall weed densities. Where, and if, reduced selection from herbicides is advocated as a strategy for reducing the risk of resistance evolution, then alternative weed control strategies must be employed to keep population densities in check. This issue will be explored in greater depth in Section 3.5.

3.3.3 INHERITANCE OF HERBICIDE RESISTANCE

Simulation models of herbicide resistance have, without exception, assumed diploidy such that an individual has two sets of chromosomes and two alleles for any gene locus. At a gene locus where allelic mutations may confer herbicide resistance the wild type is homozygous susceptible (SS). The degree of phenotypic resistance of heterozygotes (RS) depends on the relative dominance of the resistance allele. In the great majority of cases where the genetic basis of resistance has been clearly established, mutations at single, nuclear-encoded gene loci have been responsible.[15,16] However, this is not always true and exceptions will be discussed below. Before doing so, the implications of dominance and recessivity of resistance traits will be considered and explored using a modeling approach.

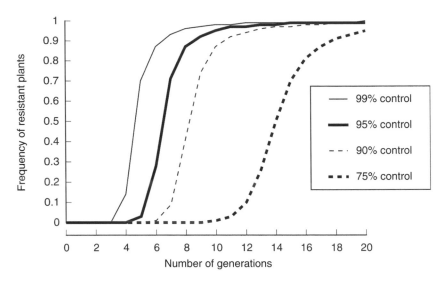

FIGURE 3.3   The influence of selection pressure on predicted rates of herbicide resistance evolution in *Lolium rigidum*. Initial gene frequency is 1×10^{-8}. Herbicide is applied once every generation and the resistance trait is assumed to be completely dominant. Population size is constant in all simulations.

3.3.3.1   Dominance

Herbicide-resistant alleles are typically dominant or semidominant.[15] The level of dominance of the resistance allele determines the phenotypic resistance status of heterozygous (RS) individuals. The frequency of heterozygotes within a population is determined largely by the breeding system of the species concerned. Breeding systems are considered in greater detail in Section 3.3.4.

Where resistance is completely recessive, heterozygotes will be phenotypically susceptible and hence neutral to selection from herbicides. In the case of a completely dominant trait, heterozygotes will be phenotypically resistant and indistinguishable from homozygous-resistant (RR) individuals. For many resistance traits the mutant allele may be incompletely dominant (e.g., paraquat resistance in *Hordeum glaucum*).[26] Where this is the case the observed phenotype depends on the rate at which the herbicide is applied. At lower doses the advantage conferred by a single copy of the mutant allele may be sufficient to ensure survival of the heterozygotes. At higher rates, however, only homozygous-resistant (RR) individuals are able to survive and heterozygotes may be effectively recessive. Strategies for the management of insecticide resistance have sought to exploit these differences by establishing doses that control heterozygotes.[27] In the early stages of resistance development, when gene frequencies are low, the overwhelming majority of resistance alleles will occur in heterozygous individuals (where initial gene frequency is 10^{-6}, RR individuals will occur at a frequency of 10^{-12} within a population). If a pesticide rate can be used which controls heterozygotes, selection for resistance is considerably reduced. For a number of reasons these strategies have not been pursued for herbicide-resistance management. In most weed species, mutant alleles, while not completely dominant, display a high degree of dominance and hence commercial doses do not discriminate between heterozygotes and homozygous-resistant individuals. To spray at sufficiently high rates to control heterozygotes would not be commercially viable and could create problems with crop phytotoxicity.

Figure 3.4 uses a simulation model (based on *L. rigidum*) to compare predicted rates of resistance evolution for a dominant, semidominant, and recessive trait in an outcrossing species. The herbicide is assumed to control 0, 50, and 100% of heterozygotes for the dominant, semidominant, and recessive traits, respectively.

3.3.3.2 Cytoplasmic Inheritance

To date, with the notable exception of the triazines, resistance to all herbicide classes is determined by nuclear-inherited genes. The gene conferring triazine resistance is, however, located in the chloroplast genome[28] and is therefore under somewhat different genetic control. While movement of nuclear encoded genes may take place via ovules or pollen, the chloroplast genome is maternally inherited, although low levels of transmission have been noted in pollen for some species.[29]

3.3.3.3 Quantitative Inheritance

In some instances, resistance to herbicides and other xenobiotics may be conferred by quantitative (additive) genetic variation at a number of genes of minor effect rather than by single major genes.[2,10] Circumstantial evidence for quantitatively controlled herbicide resistance has been presented by Gardner et al.[10] Holliday and Putwain[30] also demonstrated quantitative inheritance of mild resistance to triazine herbicides in *Senecio vulgaris*. This phenomenon has also been implicated in cases of insecticide resistance,[31] fungicide resistance,[32] and resistance to anticancer drugs.[33]

In accordance with published studies of the inheritance of herbicide resistance which demonstrate single-gene control of resistance (reviewed in [15] and [16]) most models of herbicide resistance have assumed resistance is conferred by single, major nuclear genes.[5-9] Gardner et al.[10] have developed a model able to simulate the simultaneous evolution of quantitatively controlled and major gene resistance and Via[31] has considered frameworks for models of quantitative resistance to pesticides.

3.3.4 Mating Systems

A major assumption of many models of herbicide resistance is that mating between plants in a population will be completely random.[5,6,34] There are a number of instances where this may not be the case. Random mating assumes self-incompatibility and 100% outcrossing between individuals within the population. In practice this rarely happens, with most weed species exhibiting at least low levels of selfing and a number that are highly self-fertilizing.[35] Mating systems may have a significant effect on the rate and probability of spread of resistance genes in weed populations.[15] The spread of resistance will be more rapid in an outcrossing species when resistance is conferred by a single dominant gene,[2,15] where in the early stages of resistance

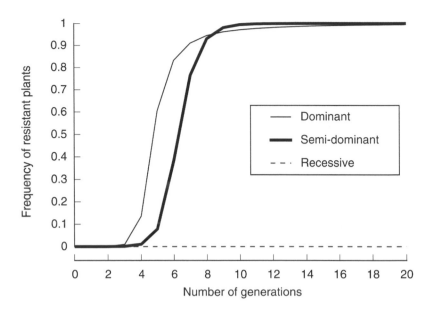

FIGURE 3.4 The influence of dominance on predicted rates of herbicide-resistance evolution in *Lolium rigidum*. Initial gene frequency is 1×10^{-6}. Herbicide is applied once every generation. Population size is constant in all simulations.

evolution, rare heterozygotes survive herbicide application and cross with susceptible survivors (escapes), a quarter of whose progeny will be heterozygous resistant. Selfing considerably reduces levels of heterozygosity in plant populations, slowing the spread of rare dominant mutations. The spread of recessive alleles, on the other hand, is favored by self-fertilization. In this case, rare heterozygotes (that are phenotypically susceptible) which survive herbicide application produce progeny that are one quarter homozygous resistant.

The assumption that mating within a population will be completely random is also somewhat questionable. In any plant population the likelihood is greater that mating will occur between near neighbors.[36] This issue is explored from a modeling perspective in Section 3.4.2.8, which considers the need and potential for spatially explicit models of herbicide resistance.

3.3.5 GENE FLOW

Gene flow between and within plant populations occurs through two primary agencies, pollen dispersal and seed movement.[2,15,37] Other factors such as introgression[38–40] between related species and the possible transfer of genes from genetically modified crop plants[37,41,42] also may be important. Flow of resistance alleles into weed populations provides an initial source of resistance upon which selection can act in a similar manner to mutations. Rates of gene flow are thought to be greater than mutation rates, but this relationship has not been well established for herbicide resistance. In general, rates of gene flow are thought to be less than 1% for populations separated by a few hundred meters and two magnitudes of order smaller for populations separated by 1.5 km or more.[43,44] Resistance genes most likely arise in an area through mutation and are spread among individuals within that area by gene flow.[15]

3.3.5.1 Pollen Dispersal

Movement of resistance genes in pollen has been measured for two herbicide-resistant grass species. In sulfonylurea-resistant *Kochia scoparia*, outcrossing of resistant pollen to susceptible plants was 1.4% at a distance of 29 m,[45] and in diclofop-methyl-resistant *Lolium multiflorum* this figure was 1% at 7 m.[46] Pollen dispersal is known to occur over greater distances than these, with a number of studies showing pollen dispersal follows a leptokurtic distribution with high levels near the source and rapidly declining levels at greater distances.[37] Of course, pollen traveling considerable distance is often unable to achieve fertilization due to loss of viability. Pollen dispersal may be greater in insect-pollinated species. For example, a study on *Raphanus sativus* has shown that approximately 6 to 18% of seeds had been fertilized by pollen from at least 100 m distance.[47] Regardless, seed movement is probably responsible for the majority of gene flow in weed populations.

3.3.5.2 Seed Movement

Levels of weed seed movement from one field to another by agricultural implements, in particular, harvesting equipment, may be considerable. McCanny and Cavers[48] showed that combine harvesters moved approximately 3% of seed of *Panicum miliaceum* from one field to a second field. Commercial grain stocks contaminated with herbicide-resistant weed seed may also be an important source of seed movement. In a survey conducted in Western Australia (Cawthray and Powles, unpublished), 66% of seed wheat stocks were contaminated with weed seeds. While the number of weed seeds was relatively low, this represents considerable potential for the transfer of resistant weed seed into new areas. Enormous quantities of weed seeds also are moved in hay and grain during times of drought.[49]

3.3.6 FITNESS

The concept of fitness in population genetics describes the relative survival, growth, and reproductive success of one genotype compared to another as a result of natural selection.[2] In other words, fitness is a measure of the relative contribution of a genotype to subsequent generations. Relative fitness is not, however, an absolute value and varies according to genotype by environment interactions. This may be particularly important when considering herbicide resistance where fitness may vary considerably depending on whether it is measured in the presence or absence of herbicide. Clearly, the possession of resistance alleles considerably increases the fitness of an individual in the presence of herbicide. However, the reverse may be true when individuals are compared in the absence of herbicides, as many novel mutations, especially at major gene loci, are associated with pleiotropic deleterious effects which reduce fitness.[50] Fitness penalties in the absence of selection by herbicides have been well documented for triazine-resistant mutants. More recent research to establish if fitness penalties are associated with ACCase- and ALS-resistance mutations have shown little or no differences in comparative growth and germination characteristics.[51–54] Fitness penalties in herbicide-resistant weed biotypes have been reviewed by other authors.[55,56] Here we are interested in how these penalties can be incorporated into models of resistance and how they can be exploited, where present, through management to slow the rate of herbicide-resistance evolution.

Most previous work on determination of fitness differentials between R (resistant) and S (susceptible) biotypes has attempted to quantify differences in growth and competitiveness.[51,53,54,57–59] In their model of herbicide resistance, Maxwell et al.[8] adapted a competition model developed by Firbank and Watkinson[60] to account for potential differences in the growth and seed yield of R and S biotypes. Mortimer et al.[5] used experimental data to develop a modeling framework, which illustrated that fitness, measured as per capita seed production, was both frequency and density dependent and that the interaction of density-dependent and -independent regulation at the population level was critical to the spread of resistance.

Fitness, however, should be measured at all stages of the life cycle if a complete understanding of the factors that determine the relative contribution of different genotypes to subsequent generations is to be achieved. More recent research has examined differences in the germination and dormancy characteristics,[52,61,62] emergence characteristics,[63] and seed longevity[64] of R and S biotypes. Each of the life cycle characteristics considered above, together with others such as differences in the phenology or pollen viability and competition of R and S biotypes, can be considered as a "component" of overall fitness. Where these differences exist, they can be incorporated into population dynamics models, such that each potential genotype has discrete characteristics in terms of life history traits. These differences become particularly important when they can be manipulated by nonherbicide weed control strategies. For example, an observation that resistant individuals germinated earlier in the season would suggest that delayed seeding may be a good option for increasing control of these individuals with a nonselective agent (preseeding tillage).

3.4 APPROACHES TO MODELING HERBICIDE RESISTANCE

3.4.1 POPULATION DYNAMICS

3.4.1.1 The Life Cycle

The life cycle is central to most models of the population dynamics of weeds (Figure 3.1). The reference point within the cycle that is commonly used for accounting is the number of seeds per unit area in the seed pool prior to germination. This number ($P_{istg-yj}$, see Tables A3.1 and A3.2) is the sum of seed that is added prior to germination, for example, as a contaminant of crop seed (P_{isd}), and seed that has survived in the soil up to that time (Eq. 3.1). Seed that has survived is the seed that was present in the soil after the previous seed-set multiplied by the fraction of those seeds that survived the time between seed-set and germination ($F_{istp \to stg\text{-}}$). The seed that was present at seed-set is made up of the ungerminated seed that had survived to that point (the product of the seed that was present before germination in the previous generation ($P_{istg\text{-}yj\text{-}1}$)), the fraction of that seed that did not germinate ($1 - F_{istg\text{-} \to g}$), and the fraction of the ungerminated seed that survived in the ground until seed-set ($F_{istg+ \to stp}$)) plus the newly produced seed that enters the seed pool (the total seed produced (P_{isp}) times the fraction of that seed that reaches the pool ($F_{isp \to stp}$)).

$$P_{istg\text{-}yj}\ P_{isd} + (P_{istg\text{-}yj\text{-}1}\ (1 - F_{istg\text{-} \to g})\ F_{istg+ \to stp} + P_{isp}\ F_{isp \to stp})\ F_{istp \to stg\text{-}} \qquad (3.1)$$

The number of mature plants (P_{ip}) is the product of number of seeds that are present in the ground before germination ($P_{istg\text{-}}$), the germination fraction ($F_{istg\text{-} \to g}$), the fraction of germinated seeds that become established seedlings ($F_{ig \to e}$), and the fraction of established seedlings that survive to maturity ($F_{ie \to p}$)(Eq. 3.2).

$$P_{ip} = P_{istg\text{-}}\ F_{ig \to g}\ F_{ig \to e}\ F_{ie \to p} \qquad (3.2)$$

Between them, Eqs. 3.1 and 3.2 define the entire life cycle with the exception of reproductive output (seed production). These equations allow simulation of the range of phenomena that influence survival of individuals. Longevity of seed is determined by the fraction of ungerminated seed that survives in the seed bank during the growing season ($F_{stg+ \to stp}$) and the fraction that survives between seed-set (harvest) and the beginning of the following season ($F_{istp \to stg\text{-}}$). Seed remaining viable but dormant during the growing season is calculated as $1 - F_{istg\text{-} \to g}$ (the germination fraction). Cultural weed control practices such as capture of weed seed at harvest, or burning, grazing, or otherwise removing stubble can be modeled by reducing the fraction of newly produced seed that reaches the seed bank ($F_{isp \to stp}$). Late applications of herbicide aimed at reducing production of weed seed could also be taken into account by reducing this fraction. Nonselective herbicides applied preseeding can be thought of as reducing the fraction of germinated seeds that establish as seedlings ($F_{ig \to e}$), while selective postemergent herbicides would reduce the fraction of seedlings that survive to maturity ($F_{ie \to p}$). Tillage affects the fraction of germinated seeds that establish and may also impact on the fraction of seeds that germinate.

In the simplest case, the life cycle can be modeled for a single seed pool. Where a population of seed can be divided into categories (based, for example, on herbicide susceptibility, age, or depth of burial) with different transition fractions between life history stages, separate pools can be used for each category, with appropriate transition functions between categories. For example, where germination rates change with age of seed, as may happen with survival of seed in the pool or the fraction of seed that germinates, it is possible to create separate pools for seed of different ages. Freshly matured seed enters a pool of new seed and ungerminated seed that survives in the soil until seed-set is transferred to a pool of seed that is one generation older. Where the germination fraction or survival fractions vary with depth of burial, separate pools can be modeled for different depth classes and transition between depths can occur in response to tillage or other physical or biological processes.

3.4.1.2   Competition

The part of the life cycle that remains to be calculated is seed production (P_{isp}). Seed production is regulated by the interaction between plants and their environment throughout the growing season. A major factor in this process is competition between plants for resources. For many purposes it is sufficient to employ a function that estimates seed production in one generation directly from the numbers of competing plants. For other purposes it is necessary to simulate one or more processes in greater detail using a shorter time-step.

3.4.1.2.1   Competition equations

Seed production can be estimated using the sorts of equations that have commonly been used to describe the results of competition experiments. A large number of such equations have been developed and several of them were compared by Cousens.[65] Many of these are variants on hyperbolic equations and are based on work by DeWit.[66] A simple example of this sort of equation can be derived from a hyperbolic plant density-seed production curve for a monoculture by assuming that competitors are identical to plants from the monoculture except that they do not produce seed (Eq. 3.3). A restatement of this assumption is that the competing plant types use an identical set of resources in an equivalent way. For a unit area, seed produced by plant type 1 (P_{1sp}) is the product of the maximum seed production for that type of plant (M_1), the density of that type of plant (P_{1p}), and a type-specific scaling factor (K_1), all divided by 1 + the total scaled plant density. The type-specific scaling factor is the inverse of the density of plants of that type grown in a mono-culture that will produce half of the maximum seed production. The total scaled population is the sum of $K_i P_{ip}$ for types 1, 2, ..., n where n is the number of competing types. In cases where the plant types do not compete equally for the same resources additional interaction terms of the form $K_{ij} P_{ip} P_{jp}$ can be added to the denominator.

$$P_{1sp} = \frac{M_1 K_1 P_{1p}}{1 + K_1 P_{1p} + K_2 P_{1p} + \ldots + K_n K_n P_{np}} \tag{3.3}$$

This equation can be used to calculate seed production for any of the competing plant types by substituting the plant number, maximum production, and scaling factor for that type in the denominator. Other types of production, such as weight of seed (i.e., yield of the crop) or biomass, can be estimated by using the appropriate maximum production value in the same units as the product (for example, tonnes/hectare).

As with Eqs. 3.1 and 3.2, Eq. 3.3 is the simplest case, where the maximum production and scaling factor is uniform for the entire population of a particular plant type. Where these factors vary between identifiable subpopulations, as happens, for example, when germination occurs in successive waves or cohorts rather than as a single event or when different genotypes have differential competitive abilities, the subpopulations can be represented as separate plant types in Eq. 3.3. The seed produced by the total population is then the sum of that produced by the subpopulations.

Under field conditions, production of seed would be expected to vary from generation to generation. This variability presumably affects the maximum seed production in particular, but it may also affect the other parameters of Eqs. 3.1 to 3.3. If seed bank dynamics are being simulated for a sequence of generations, mean values of the parameters are often used. However, if the variability in maximum seed production or any of the other parameters is known or can be estimated it is possible to choose random values for those parameters for each generation.

3.4.1.2.2   Simulation of competition

The competition equations thus far described operate on a time-step of one generation. In fact, competition depends on many variables that can continually change throughout the life of the plants. It is possible to simulate such processes using crop growth models.[67] Models of this sort are computing intensive, and they can be unwieldy because development of resistance often depends on several interacting populations of weeds over many generations. However, such models may be necessary in order to simulate any interactions that occur between selection pressure and variables that fluctuate during plant growth.

3.4.2   POPULATION GENETICS AND THE EVOLUTION OF HERBICIDE RESISTANCE

3.4.2.1   Introduction

From a modeling perspective, herbicides affect the probabilities of survival of weed populations at one or more points in their life cycle. In simple terms, herbicide resistance is a reduction of herbicide efficacy through time. Such a reduction can be represented in models using a number of methods. These methods vary widely in complexity and are useful for different purposes.

3.4.2.2   Simple Cases

3.4.2.2.1   Herbicide applications ('shots')

The simplest method for modeling development of herbicide resistance is to assume that a particular herbicide is effective for a set number of applications (or 'shots') after which it has no further effect. This method relies on a constant amount of selection pressure being applied per application of herbicide and assumes minimal interaction between selection pressure and other variables, which in many situations is close to correct. The use of shots is efficient in terms of computing resources and for many purposes, such as economic models or farm management decision aids, it is appropriate.[68]

3.4.2.2.2   Independent populations

Development of herbicide resistance can be modeled by assuming that susceptible and resistant plants form completely genetically separate but competing populations which produce progeny that are identical to the parents. The resistant population is initially small but increases relative to the susceptible population due to higher survival fractions during the part of the life cycle where the herbicide is applied.

An assumption of independent populations approximates reality in self-pollinated plants and cases where resistance genes are cytoplasmic.

3.4.2.3   Major Genes

Where cross pollination is occurring and herbicide resistance is conferred by nuclear genes, the dynamics are governed by the rules of genetic recombination. These rules are well understood and can be employed at a range of levels of complexity depending on the number of genes involved and the nature of the intercrossing that occurs.

In the majority of cases where the genetic basis of herbicide resistance has been determined it is conferred by a mutation at a single gene locus conferring major resistance (Section 3.1.1). Many common weeds of crops are diploid and hence have two copies of each of their genes. Where two alleles exist for a particular gene there are three possible genotypes. The alleles can be denoted using the upper and lower case of the same letter, i.e., 'A' and 'a'. We will use the upper case to denote the initially more common allele, which typically codes for susceptibility to herbicides. For a herbicide-resistance gene, the three genotypes would be homozygous susceptible (AA) where the plant has two copies of the allele for susceptibility, heterozygous (Aa) where there is one copy of each allele, and homozygous resistant (aa) where both alleles code for resistance.

During meiosis the alleles segregate into gametes, referred to as ovules and pollen, and are subsequently recombined to form the next generation of seeds. The ratio of alleles in the ovules and in the pollen produced by a plant is the same as that of the parent. The fraction of the resistance allele in the pollen that fertilizes the ovules of a particular plant ($F_{qai.}$) is the average of the fraction of the resistance allele in the gametes from that plant ($F_{qai \rightarrow}$) and the pollen from the population as a whole (F_{qa}) weighted by the self-pollination fraction ($F_{qi \rightarrow i}$), as shown in Eq. 3.4.

$$F_{qai \leftarrow} = F_{qai \rightarrow} F_{qi \rightarrow i} + F_{qa} (1 - F_{qi \rightarrow i}) \qquad (3.4)$$

The fractions of each genotype in the seed which is produced on a particular plant (Fs_{AApi}, F_{sAapi}, F_{saapi}) are functions of the fraction of the resistance allele in that plant ($F_{qai \rightarrow}$) and the fraction in the pollen that fertilizes the plant ($F_{qai \leftarrow}$), as shown in Eqs. 3.5 to 3. 7.

$$Fs_{AApi} = (1 - F_{qai \rightarrow}) (1 - F_{qai \leftarrow}) \qquad (3.5)$$

$$F_{sAapi} = F_{qai \rightarrow} (1 - F_{qai \leftarrow}) + F_{qai \leftarrow} (1 - F_{qai \rightarrow}) \qquad (3.6)$$

$$F_{saapi} = F_{qai \rightarrow} F_{qai \leftarrow} \qquad (3.7)$$

3.4.2.4   Explicit Genotypes

A model of development of herbicide resistance controlled by a single gene can be produced by keeping separate track of the population of each genotype. Survival of plants through the life cycle and the effects of management practices are calculated using Eqs. 3.1 and 3.2, and the crossing and proportion of seed of each genotype produced is calculated using Eqs. 3.4 to 3.7. We will refer to models of this type as explicit genotype models. Where numbers of seeds produced in each generation are calculated using a competition equation such as Eq. 3.3, the model can calculate population dynamics. If calculations of absolute populations are not required, the simpler alternative of normalizing the population in each generation will calculate changing ratios of alleles and genotypes.

3.4.2.5 Gene Ratios

For any population where mating is occurring completely at random the Hardy-Weinberg ratios[69] (Eqs. 3.8 to 3.10) can be used to calculate genotype frequencies (F_{sAA}, F_{sAa}, and F_{saa}) for a single gene where allele frequencies are known (F_{qa} is the frequency of allele a). In such cases the genotypes will achieve equilibrium after every mating cycle, regardless of the initial ratios of the genotypes. For many weed species, even where this condition is not completely met, Hardy-Weinberg equilibrium is an adequate approximation and it has been widely employed.[8]

$$F_{sAA} = (1 - F_{qa})^2 \tag{3.8}$$

$$F_{sAa} = 2 F_{qa} (1 - F_{qa}) \tag{3.9}$$

$$F_{saa} = F_{qa}^2 \tag{3.10}$$

The Hardy-Weinberg equations can replace Eqs. 3.4 to 3.7 in models of population dynamics or gene ratios. Use of the Hardy-Weinberg equations removes the requirement to keep separate track of the populations of each genotype. For resistance controlled by a single gene it is only necessary to carry forward the frequency of the allele for resistance. We have compared the explicit genotype and Hardy-Weinberg variants of the gene frequency model for a range of situations. Unless otherwise stated the parameters used in the following examples are those shown in Table 3.2.

Where mating is random, the two models produce identical results (Figure 3.5). Conditions that violate the assumption of random mating are common. They include less-than-complete germination, immigration and emigration of seed and pollen, staggered flowering time, and any bias toward self-pollination.

The predictions of the explicit genotype and Hardy-Weinberg variants of the model have been compared for conditions of incomplete germination. The fraction of seed that germinates ($F_{istg \to g}$) was set to 0.3, and all other parameters were the defaults (Table 3.2). In this case, the Hardy-Weinberg equations predict the same initial rate of evolution of resistance as the explicit model, but as the frequency of the resistance gene increases, the rate of selection is under-predicted (Figure 3.6). This effect occurs because susceptible seed (AA) that is ungerminated remains susceptible in the explicit genotype model. In the Hardy-Weinberg approximation, this ungerminated seed is effectively crossed with the more resistant new seed that enters the pool. For this reason, the amount of susceptible seed is reduced and the amount of heterozygous, phenotypically resistant seed is increased. In this example

TABLE 3.2
Default Values for Parameters Used in the Simulations Described in Section 3.4

Parameter	Symbol	Default Value
Fraction of seed from the seed pool that germinates	$F_{istg-\rightarrow g}$	1
Fraction of germinated seed that establishes as seedlings	$F_{ig\rightarrow e}$	1
Fraction of phenotypically susceptible seedlings (AA) that survive to maturity	$F_{AAe\rightarrow p}$	0.05
Fraction of phenotypically resistant seedlings (Aa and aa) that survive to maturity	$F_{Aae\rightarrow p}$ $F_{aae\rightarrow p}$	1
Amount of added seed (as a fraction of the seed in the pool before addition)	P_{isd}	0
Self-pollination fraction	$F_{qi\rightarrow i}$	0
The fraction of the gene pool of the initial seed that is allele a (resistant)	$F_{qastg-y0}$	10^{-6}
The fraction of the gene pool of the added seed that is allele a (resistant)	F_{qasd}	10^{-6}

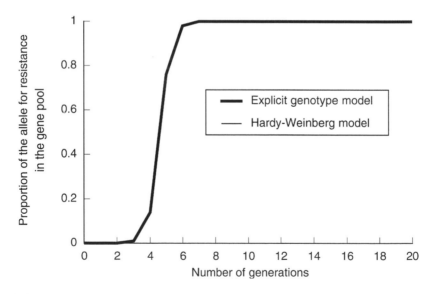

FIGURE 3.5 Simulated proportion of the gene pool that is allele a (which codes for resistance) through time for explicit genotype and Hardy-Weinberg variants of a one-gene model of proportional development of herbicide resistance, using the default settings.

in the tenth year of the simulation the explicit genotype model calculated 0.36, 0.38, and 0.26 as the respective fractions of the AA, Aa, and aa genotypes in the seed pool, giving a frequency of the allele for resistance of 0.45. If this population were in Hardy-Weinberg equilibrium this would equate to genotype frequencies of 0.30, 0.50, and 0.20, respectively.

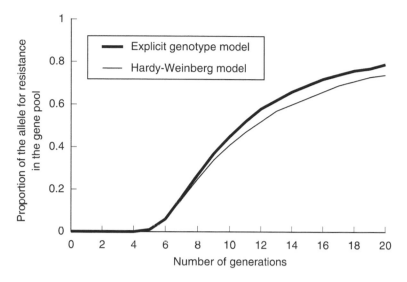

FIGURE 3.6 Simulated proportion of the gene pool that is allele a (which codes for resistance) through time for explicit genotype and Hardy-Weinberg variants of a one-gene model of proportional development of herbicide resistance. Germination fraction is 0.3.

A similar effect occurs when resistant seed is added to the seed pool. Figure 3.7 illustrates the results of a simulation using the default parameters with an addition of 5% homozygous resistant seed in the third year of the simulation. The Hardy-Weinberg model under-predicts the rate of evolution of resistance. This under-prediction occurs because the Hardy-Weinberg equilibrium after addition of the resistant seed includes a substantial fraction of heterozygous-resistant individuals and a commensurate reduced fraction of homozygous-susceptible individuals. The explicit genotype model maintains the added seed in the homozygous-resistant pool and as a consequence a larger fraction of the a alleles are exposed to the herbicide in the homozygous-resistant form.

Variants on the Hardy-Weinberg equations exist which calculate the genotype ratios for the situation where some degree of self-pollination occurs but where mating is otherwise random. Maxwell et al.[8] use Eqs. 3.11 to 3.13 (attributed to Wright[70]) for this purpose. The variables used in these equations are the same as those in Eqs. 3.8 to 3.10 with the addition of the self-pollination fraction ($F_{qi \to i}$). These equations calculate the genotype frequencies for self-pollination fractions between 0 and 1 as the arithmetic average of the frequencies with no self-pollination and with complete self-pollination, weighted by the self-pollination fraction.

$$F_{sAA} = (1 - F_{qa})\, F_{qi \to i} + (1 - F_{qa})^2\, (1 - F_{qi \to i}) \qquad (3.11)$$

$$F_{sAa} = 2\, F_{qa}\, (1 - F_{qa})\, (1 - F_{qi \to i}) \qquad (3.12)$$

$$F_{saa} = F_{qa}\, F_{qi \to i} + F_{qa}^2\, (1 - F_{qi \to i}) \qquad (3.13)$$

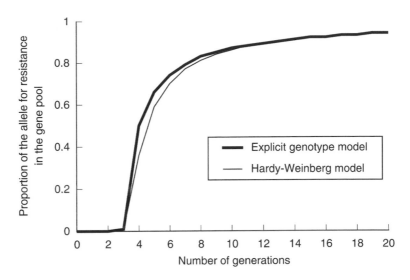

FIGURE 3.7 Simulated proportion of the gene pool that is allele a (which codes for resistance) through time for explicit genotype and Hardy-Weinberg variants of a one-gene model of proportional development of herbicide resistance. 5% resistant seed is added to the seed pool before germination in year 3.

An alternative set of equations that predict the equilibrium state of the explicit genotype model is presented in Eqs. 3.14 to 3.16 (Figure 3.9). With complete self-pollination both the Wright equations and the alternative equations agree with the standard Hardy-Weinberg equations and hence produce identical results to those shown in Figures 3.5 to 3.7. With complete self-pollination and with no self-pollination the alternative equations agree with the Wright equations (Figure 3.8). With intermediate levels of self-pollination, the Wright equations under-predict the equilibrium fraction of the heterozygotes and correspondingly over-predict the fraction of homozygotes relative to the explicit genotype model and the alternative equations.

$$F_{sAA} = \frac{F_{qi \to i}F_{qa}(1 - F_{qa})}{(2 - F_{qi \to i})} + (1 - F_{qa})^2 \qquad (3.14)$$

$$F_{sAa} = \frac{4F_{qa}(1 - F_{qa})(1 - F_{qi \to i})}{(2 - F_{qi \to i})} \qquad (3.15)$$

$$F_{saa} = \frac{F_{qi \to i}F_{qa}(1 - F_{qa})}{(2 - F_{qi \to i})} + F_{qa}^2 \qquad (3.16)$$

In the case of partial self-pollination, when the genotype ratios differ from the equilibrium state, the equilibrium is not re-established in a single generation (Figure 3.10). This lag occurs even where mating otherwise occurs at random. The Hardy-Weinberg equations presume instantaneous equilibration and hence do not account for this lag.

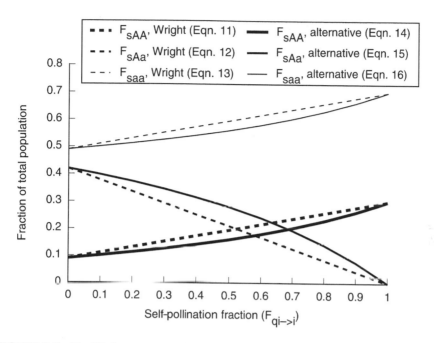

FIGURE 3.8 Equilibrium genotype fractions vs. self-pollination fraction as predicted by the Wright and alternative variants on the Hardy-Weinberg equations for 0.7 of the gene pool being allele a.

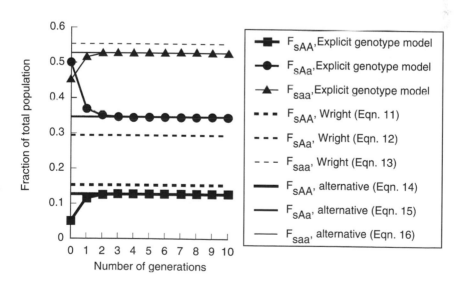

FIGURE 3.9 Genotype fractions through successive generations as predicted by the explicit genotype model and the Wright and alternative variants of the Hardy-Weinberg equations for 0.7 of the gene pool being allele a, self-pollination fraction of 0.3, and an initial frequency of heterozygotes of 0.5 (0.83 of the maximum possible for that gene fraction).

The lag in obtaining equilibrium can impact on the rate of evolution of resistance. Figure 3.9 illustrates a comparison of the explicit genotype model and models using the Wright and alternative variants on the Hardy-Weinberg equations. The three models agree on the rate of evolution of resistance in the early generations, but when change occurs most rapidly, the Hardy-Weinberg models over-predict this rate. This phenomenon occurs because homozygous-susceptible plants are killed by the herbicide and heterozygous plants are not. The frequency of heterozygotes in the plant population is consequently higher than the Hardy-Weinberg equilibrium. The equilibrium is fully regained in the seed in the Hardy-Weinberg models but only partially regained in the explicit genotype model. Hence the proportion of the susceptible allele that is exposed to the herbicide in homozygous plants is too large in the Hardy-Weinberg models and selection occurs more rapidly. The discrepancy is larger in the case of the Wright equations because the Wright equations underestimate the equilibrium proportion of heterozygotes relative to the other methods of calculation.

3.4.2.6 Two Genes

The single-gene explicit genotype model described above can be extended to cases where two genes are involved. Such cases include development of resistance to two herbicide modes of action where each is controlled by a single gene, or development of resistance to a single herbicide where two resistance mechanisms are possible.

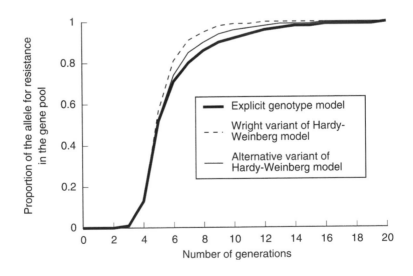

FIGURE 3.10 Simulated proportion of the gene pool that is allele a (which codes for resistance) through time for explicit genotype and the Wright and alternative variants on the Hardy-Weinberg single-gene model of proportional development of herbicide resistance. The self-pollination fraction is 0.5.

As for the one-gene case the Hardy-Weinberg approximations can be employed for each gene individually. This approach requires an assumption about how the genotypes for one gene are distributed across the genotypes for the other gene. In the equilibrium state in a randomly mating population these distributions would not be expected to be correlated. However, as in the case of partial self-pollination, the equilibrium would not be attained in a single generation.

Where there are two genes (designated A and B), each of which controls resistance to a type of herbicide (also designated A and B), and where the allele for both genes that confers resistance (a and b) is dominant and initially rare in relation to the alleles that confer susceptibility (A and B), the equilibrium frequency of the double heterozygous-resistant genotype (AaBb) would be extremely low (the product of the frequencies of Aa and Bb in the population). The frequency of the double homozygous-resistant genotype (aabb) would be much lower again.

Should an individual of type AaBb arise, however, it would cross almost exclusively with the common type (AABB). The seed produced on the AaBb plant would be equal proportions AABB, AABb, AaBB, and AaBb. Assuming equal levels of fecundity all of the genotypes in the population, an equal number of individuals of each of these genotypes would be produced on AABB plants pollinated by the double-resistant plant. Consequently, the number of individuals of type AaBb in the progeny would be 0.5 of the seed produced by the AaBb plant. This proportion is far higher than the equilibrium proportion that would be achieved after many generations of intercrossing, which is the proportion that would be calculated by a two-gene Hardy-Weinberg model.

Modeling a two-gene system by simulating the three genotypes for each gene explicitly is also possible and, like the two-gene Hardy-Weinberg model, assume that the genotypes for each of the genes are evenly distributed across the genotypes for the other gene. Such a model requires that the total population plus two values, the number of individuals for two of the three genotypes, be carried forward between generations for each gene. The population of the third genotype for each gene then can be calculated by difference. The two-gene Hardy-Weinberg model, in contrast, requires only that the total population and one additional value, the gene ratio, be carried forward for each gene modeled.

A third method for modeling a two-gene system is to account separately for the populations and seed production of each of the genotypes and to calculate the genotypes of the progeny using the standard rules of genetic recombination, which are illustrated for the one-gene case in Eqs. 3.4 to 3.7. For two genes there are ten genotypes that are distinguishable on the basis of the genotypes of the parental gametes. These ten genotypes and the details of the progeny that they produce are described in Table 3.3.

Two of the ten genotypes, AaBb and AabB, are identical in terms of frequencies of their alleles, but they differ in the genotypes of their parental gametes. The AaBb individuals get both of their resistance genes from the same parent and the AabB individuals get one from each. Where the genes are unlinked the gametes produced by these two types are identical but where linkage exists, the parental gamete genotypes will be favored in proportion to the tightness of the linkage.

TABLE 3.3
Parent Alleles, Alleles Produced, and the Genotypes and Frequencies of Progeny Assuming That Crossing Occurs with a Population That Produces Equal Quantities of All Alleles and Assuming That the Genes Are Not Linked

Genotype	Parent Alleles	Alleles Produced	Genotype and Frequency of Progeny When Crossed with a Population with an Equal Frequency of All Alleles
AABB	AB	AB	AABB(1/4), AABb(1/4), AaBB(1/4), AaBb(1/4)
AABb	AB, Ab	AB, Ab	AABB(1/8), AABb(1/4), AAbb(1/8), AaBB(1/8), AaBb(1/8), AabB(1/8), Aabb(1/8)
AAbb	Ab	Ab	AABb(1/4), AAbb(1/4), AabB(1/4), Aabb(1/4)
AaBB	AB, aB	AB, aB	AABB(1/8), AABb(1/8), AaBB(1/4), AaBb(1/8), AabB(1/8), aaBB(1/8), aaBb(1/8)
AaBb	AB, ab	AB, Ab, aB, ab	AABB(1/16), AABb(1/8), AAbb(1/16), AaBB(1/8), AaBb(1/8), AabB(1/8), Aabb(1/8), aaBB(1/16), aaBb(1/8), aabb(1/16)
AabB	Ab, aB	AB, Ab, aB, ab	AABB(1/16), AABb(1/8), AAbb(1/16), AaBB(1/8), AaBb(1/8), AabB(1/8), Aabb(1/8), aaBB(1/16), aaBb(1/8), aabb(1/16)
Aabb	Ab, ab	Ab, ab	AABb(1/8), AAbb(1/8), AaBb(1/8), AabB(1/8), Aabb(1/4), aaBb(1/8), aabb(1/8)
aaBB	aB	aB	AaBB(1/4), AabB(1/4), aaBB(1/4), aaBb(1/4)
aaBb	aB, ab	aB, ab	AaBB(1/8), AabB(1/8), AabB(1/8), Aabb(1/8), aaBB(1/8), aaBb(1/4), aabb(1/8)
aabb	ab	ab	AaBb(1/4), Aabb(1/4), aaBb(1/4), aabb(1.4)

A model which accounts for all ten genotypes of two genes (the two-gene explicit genotype model) has been compared with a model which accounts the genotypes of the two genes separately and assumes an equilibrium distribution of the genotypes across each other (the two-gene equilibrium model). Both models predict weed population through time in competition with a crop. All parameters are the defaults from Table 3.2 except as indicated in Table 3.4.

The two models have been run for the situation where both herbicide A and herbicide B have been applied in each year. Simulations were run with no added weed seed and with an addition of 20 weed seeds per square meter per generation

TABLE 3.4
Default Values for Parameters Used in the Two-Gene Simulations Described in Section 3.4.2.6

Parameter	Symbol	Default Value
Fraction of germinated weed seed that establishes as seedlings	$F_{ig \to e}$	0.5
The amount of weed seed that was present at the start (seeds/m²)	$P_{wstg-y0}$	50
Fraction of phenotypically double-susceptible weed seedlings (AABB) that survive to maturity	$F_{ie \to p}$	0.0025
Fraction of phenotypically herbicide A resistant, B susceptible weed seedlings (AaBB and aaBB) that survive to maturity	$F_{ie \to p}$	0.05
Fraction of phenotypically herbicide B resistant, A susceptible weed seedlings (AABb and AAbb) that survive to maturity	$F_{ie \to p}$	0.05
Fraction of phenotypically double-resistant weed seedlings (AaBb, AabB, Aabb, aaBb, and aabb) that survive to maturity	$F_{ie \to p}$	1
Amount of added weed seed (seeds/m²)	P_{wsd}	0 or 20
The fraction of the gene pool of the initial and added weed seed that is allele a (resistant to herbicide A)	$F_{qastg-y0}$ F_{qasd}	10^{-6}
The fraction of the gene pool of the initial and added weed seed that is allele b (resistant to herbicide B)	$F_{qbstg-y0}$ F_{qbsd}	10^{-6}
Maximum weed seed production (seeds/m²)	M_w	30,000
Weed scaling factor	K_w	1/25
Crop scaling factor	K_c	3/25
Crop density (plants/m²)	P_{cp}	100

with the same low level of resistance as in the initial seed pool. The two-gene explicit genotype model always predicts more rapid evolution of resistance. Where there is no seed immigration the difference is minor. With immigration of susceptible seed, however, the explicit genotype model calculates a 1-year delay in evolution of resistance while the equilibrium model predicts a much longer delay (Figure 3.11). These differences result from the fact already outlined that double-resistant individuals (AaBb and AabB) do not reach equilibrium with a susceptible population (AABB) in a single year of crossing. Rather, the number of phenotypically double-resistant progeny is equal to one half of the seed produced on the double-resistant plants. Hence, the addition of both herbicides controls a maximum of one half of the progeny of the double-resistant plants crossed with susceptible plants in the explicit genotype model, but it controls most of the progeny from the same cross in the equilibrium model. The effect is accentuated by addition of susceptible seed because the double-susceptible type dominates the population for longer.

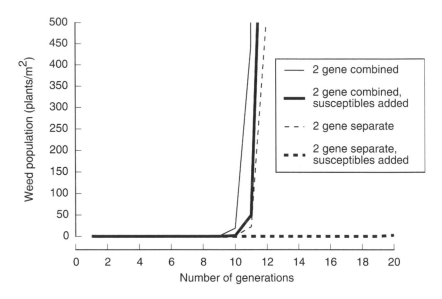

FIGURE 3.11 Simulated population of weeds through time with and without the addition of susceptible seed each year for a model that calculates all of the combined genotypes for two genes, and a model that calculates the genotypes for the individual genes separately. Herbicides A and B have been applied each year.

3.4.2.7 Polygenic Resistance

It is possible to create models that explicitly simulate the combined genotypes for more than two genes, but the number of genotypes that must be taken into account increases exponentially with the number of genes being considered. The total number of genotypes with distinct parental alleles for n genes (G_n) is given by the sum of integers from 1 to 2n (Eq. 3.17). For three genes this value is 36 and for four genes it is 136.

$$G_n = \sum_{i=1}^{2n} i \tag{3.17}$$

Using the Hardy-Weinberg model for each of n genes or an n gene equilibrium model (as defined above for the two-gene equilibrium model) requires that much less information be carried forward. In both cases the number of values required increases arithmetically with n. For Hardy-Weinberg n + 1 values are needed, while the n gene equilibrium model requires 2n + 1 values. Unfortunately, as has been demonstrated for the two-gene case, these approximations produce inaccurate estimates in some circumstances.

An alternative modeling methodology exists, with relevance where a single trait is governed by a large number of genes, each of which has a small but additive effect (referred to as a quantitative trait). These models can be described as quantitative resistance models. An example of their use is found in Gardner et al.[10]

3.4.2.8 Spatial Models of Herbicide Resistance

The models that have been described take no account of spatial pattern in the populations of competing plants. This assumption is incorrect and may have important implications in some cases. It is possible to construct models that do account for spatial variability. This can be done by dividing an area into small sub-areas that can be assumed to be uniform, running a separate population model for each sub-area, and applying transfer functions to account for movement of seed and pollen.

3.5 SIMULATION MODELS AS TOOLS FOR HERBICIDE-RESISTANCE MANAGEMENT

Thus far, we have discussed in detail the intrinsic population dynamic (Section 3.2) and population genetic (Section 3.3) processes that regulate weed population densities and, more importantly, the genetic structure of these populations with respect to their herbicide-resistance status. In Section 3.4 approaches to modeling these phenomena were reviewed and compared. This final section uses a two-gene, explicit genotype model, described in Section 3.4.2.6, to explore predicted rates of resistance evolution under a range of integrated management and herbicide use scenarios.

The model structure and parameterization is based on the life cycle and population dynamics of *Lolium rigidum* (annual ryegrass), the ubiquitous and resistance-prone grass weed of southern Australian cropping systems. Annual ryegrass is a completely outcrossing species and completely random, assortative mating is assumed. Resistance is conferred by single, nuclear, dominant genes.

As discussed in Sections 3.4.1.1 and 3.5 the life cycle of an annual plant population can be represented from a conceptual and modeling perspective as a number of transitions from one life history stage to the next. The fraction of the population which survives each life history stage is dependent on intrinsic density-dependent and -independent population processes (seedling mortality, herbivory, dispersal) and on agronomic practices (herbicides, tillage, seed capture at harvest). For example, the transition from established seedlings to mature plants ($F_{ie \to p}$, Eq. 3.2) is dependent upon control by preseeding nonselective and postemergent selective herbicides and tillage control (Figure 3.1). As such, in a model which explores the effect of weed management strategies on weed densities and herbicide-resistance evolution, $F_{ie \to p}$ is the product of the number of seedlings that become established and the efficacy of these control strategies, expressed as proportional survival of seedlings. In addition, the proportion of established seedlings that is subjected to each of these control strategies is dependent on their time of emergence. While this model does not have an explicit cohort structure it does allow for a proportion of seedlings to emerge after crop seeding and hence avoid preseeding control strategies.

Default values for key life cycle and genetic parameters are given in Table 3.5 and results from simulation runs are presented in Figures 3.12A to D. All simulations are based on a 30-year cropping rotation with results presented for total ryegrass seed production and proportion of resistant (Aa and aa) seed produced. Deviations from default parameter values are given in the legend to Figure 3.12 for each of the scenarios.

TABLE 3.5
Default Values for Parameters Used in the Two-Gene Model for Simulations Described in Section 3.5

Parameter	Symbol	Default Value
Total initial ryegrass seed bank density (seeds m^{-2})	P_{stg-yj}	100
Crop density (plants m^{-2})	P_{cp}	100
The fraction of the gene pool of the initial seed that is allele a (resistant)	$F_{qastg-y0}$	10^{-6}
Fraction of seed from the seed pool that germinates	$F_{istg-\to g}$	0.8
Fraction of ungerminated seed that survives in the seed bank from germination to seed set	$F_{istg+\to stp}$	0.9
Fraction of seed bank that survives from seed set to germination	$F_{istp\to stg-}$	0.9
Fraction of germinated seed that would establish as seedlings in the absence of preseeding tillage and nonselective herbicide	$F_{ig\to e}$	1
Fraction of germinating seedlings which survive preseeding nonselective herbicide control	$F_{ig\to e\ knock}$	0.05
Fraction of germinating seedlings which survive preseeding tillage control	$F_{ig\to e\ till}$	0.7
Fraction of established seedlings which emerge after crop sowing	$F_{ig\to e\ post}$	0.4
Fraction of phenotypically susceptible seedlings (AA) that survive to maturity	$F_{AAe\to p}$	0.05
Fraction of phenotypically resistant seedlings (Aa and aa) that survive to maturity	$F_{Aae\to p}$ $F_{aae\to p}$	1
Fraction of mature seed not removed by the harvest operation	$F_{isp\to stp}$	0.1

3.5.1   SCENARIO 1: CONTINUOUS CROPPING WITH INTENSIVE USE OF SINGLE MODE OF ACTION HERBICIDE

Scenario 1 simulates a continuous cropping rotation with almost exclusive reliance on herbicides (preseeding nonselective and in-crop selective) for ryegrass control. The crop (based on continuous wheat) is seeded at standard southern Australian rates (100 plants m^{-2}) with a zero-tillage seeding system. Delayed seeding is not practiced. Weed seed capture at harvest is not simulated. Postemergent selective herbicidal control is based on continuous use of a single mode of action and 95% control of susceptible ryegrass is assumed. Results are presented in Figure 3.12A.

The model predicts the evolution of resistance within 4 to 5 years. As expected, this is accompanied by a rapid increase in total ryegrass seed production. This strategy with almost complete reliance on herbicide control and little integrated weed management represents the "worst case scenario" and is clearly unsustainable. In southern Australian cropping systems, continued use of the ACCase- and ALS-inhibitor herbicides for ryegrass control has often resulted in complete failure of these herbicides over this timescale.

3.5.2   SCENARIO 2: CONTINUOUS CROPPING WITH LESS INTENSIVE USE OF SINGLE MODE OF ACTION HERBICIDE

The Scenario 2 strategy is identical to that outlined for Scenario 1, the only difference being that 80% control of susceptible ryegrass is assumed ($F_{AAe \to p} = 0.8$). In practice, this can be achieved in a number of ways. These include using herbicides with less persistence such that weeds emerging after postemergent herbicide applications are not controlled. Alternatively, timing applications so that late emerging weeds are not controlled will similarly reduce the effective selection pressure for resistance traits. This strategy allows susceptible individuals to remain in the population and has been discussed by Maxwell and Roush et al. in terms of a "resistance avoidance threshold."[71,72] This has clear implications for long-term management as it will result in replenishment of the seed bank and increased potential for future weed infestations. Results from this simulation are shown in Figure 3.12B.

Clearly, reducing the selection pressure for resistance has, as we might expect, slowed the rate of resistance evolution compared to Scenario 1. This, however, has come at an unacceptable price with total seed production escalating within 2 to 3 years to unsustainable levels. Where weed densities are this high it is of little concern to a grower whether these populations are resistant or susceptible, as yield reduction will be the result regardless. For this strategy to have any chance of success additional weed control strategies must be practiced to reduce the return of weed seed to the seed bank.

3.5.3   SCENARIO 3: CONTINUOUS CROPPING WITH LESS INTENSIVE USE OF SINGLE MODE OF ACTION HERBICIDE AND INCREASED CULTURAL CONTROL (INTEGRATED WEED MANAGEMENT)

The output from the simulation run in Scenario 2 clearly demonstrates that where herbicides are the major means of weed control, reducing the efficacy of selective in-crop herbicides results in a rapid increase in weed densities. This negates any benefit from reduced rates of resistance evolution. For this strategy to have any benefit, seed-set and hence return to the seed bank by surviving plants must be reduced. This can be achieved by implementing a more integrated control strategy with increased reliance on cultural weed control.

Scenario 3 continues to rely on a single herbicide mode of action with 80% efficacy; however, a number of nonchemical strategies are included. Crop seeding rates are increased (150 plants m^{-2}) to ensure a more competitive crop. Delayed seeding is practiced to enable greater control from preseeding nonselective and a direct-drilling as opposed to a zero-tillage seeding regime is implemented. In addition, weed seed capture at harvest reduces the return of mature seed to the seed bank. These practices are reflected in changes to the default parameterization of the model (see Figure 3.12C legend).

While having little effect on the rates of resistance evolution (8 to 10 years), a more integrated management approach has increased the sustainability of the system beyond 10 years, at which point weed seed production begins to rapidly increase to unacceptable levels.

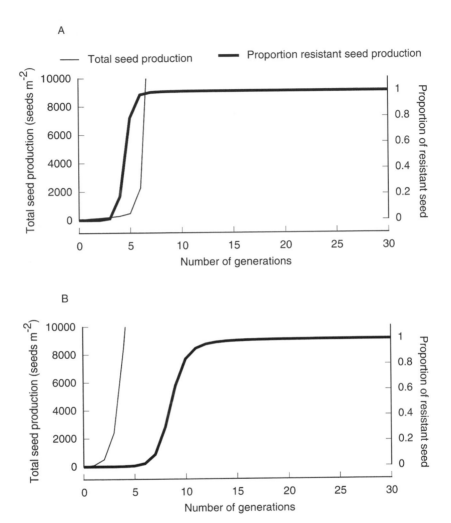

FIGURE 3.12 Predicted rates of herbicide resistance evolution and total ryegrass seed production from a two-gene simulation model.

A.   Scenario 1: Continuous cropping with intensive use of single mode of action herbicide.

B.   Scenario 2: Continuous cropping with less intensive use of single mode of action herbicide.

$$F_{AAe \to p} = 0.8$$

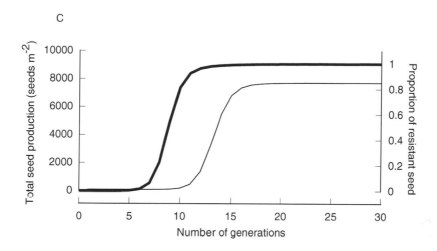

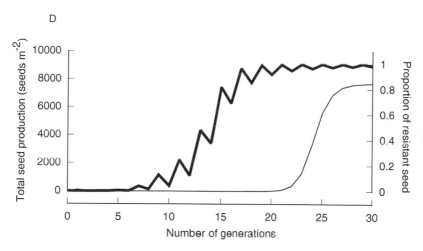

FIGURE 3.12 (CONTINUED)

C. Scenario 3: Continuous cropping with less intensive use of single mode of action herbicide and increased cultural control (integrated weed management).

$$F_{AAe \to p} = 0.2$$

Fraction of established seedlings which survive at seeding cultural control = 0.3
Fraction of established seedlings which emerge after crop sowing = 0.3
Fraction of mature seed removed by the harvest operation = 0.3

D. Scenario 4: Continuous cropping with rotation of selective herbicide modes of action and integrated weed management.

$$F_{AAe \to p} = 0.2$$
$$F_{BBe \to p} = 0.2$$

Fraction of established seedlings which survive at seeding cultural control = 0.3
Fraction of established seedlings which emerge after crop sowing = 0.3
Fraction of mature seed removed by the harvest operation = 0.3

3.5.4 SCENARIO 4: CONTINUOUS CROPPING WITH ROTATION OF SELECTIVE HERBICIDE MODES OF ACTION AND INTEGRATED WEED MANAGEMENT

To date simulated management scenarios have relied on a single herbicide mode of action for ryegrass control. Another central tenet of integrated weed management is the rotation of herbicide modes of action. This is easily simulated with the two-gene resistance model, which is able to track the evolution of multiple-resistance to two herbicide groups.

In Scenario 4, the integrated management strategies introduced above are maintained, but in-crop selective herbicides are rotated on an annual basis with both herbicides having 95% efficacy against susceptible (AA and BB) and 0% efficacy against resistant seedlings (Aa, aa and Bb, bb). Results are shown in Figure 3.12D. As resistance to both herbicides is conferred by single, dominant genes and initial allele frequencies and selection pressures are the same, rates of resistance evolution to both herbicides are identical and represented by the same line in Figure 3.12D.

Even at the high selection pressures imposed (95% efficacy) rotating herbicides delays the evolution of resistance (10 to 15 years for both herbicides) compared to previous scenarios. The benefits in terms of reduced ryegrass seed production are even greater as weed numbers do not begin to increase rapidly until resistance to both herbicides becomes fixed in the population. At this point, multiple-resistance ensures that ryegrass is not controlled by either herbicide. The success of this strategy may be compromised where ryegrass is able to develop nontarget-site cross-resistance to both herbicide groups. This was assumed not to be the case for the herbicides used in this simulation.

The simulations presented in Figures 3.12A to D have shown how a two-gene model of herbicide resistance can be used to make predictions for the rate of herbicide-resistance evolution under a range of management scenarios. The four scenarios presented are by no means exhaustive of the range of management options available. They have, however, clearly demonstrated the benefits of these models for increasing understanding of the population genetic and dynamic factors which drive resistance evolution and for exploring "on the ground" management strategies for preventing or delaying the onset of resistance in broadacre cropping rotations.

3.6 SUMMARY

This chapter has presented a review of the processes of population dynamics and population genetics that underpin the evolution of resistance in weed species (Sections 3.2 and 3.3). Section 3.4 discussed and compared approaches to modeling these phenomena, from simple cases to the development of more complex explicit multigene models. Finally, we used a two-gene model to explore management options for delaying or preventing the evolution of herbicide resistance.

The widespread utility of these models discussed in Section 3.1.3 has hopefully been demonstrated here. Simulation models are vital tools in ongoing attempts to understand the processes which drive resistance evolution and formulate effective strategies for its containment. In the future, multigene models should be developed further so that the complex patterns of cross- and multiple-resistance exhibited by species such as *Lolium rigidum* and *Alopecurus myosuroides* can be better understood. In addition, spatially explicit models of herbicide resistance should be developed to investigate processes of gene flow and dispersal and spread of resistance genes within and between populations and to study competitive interactions in patchily distributed weed populations.

3.7   APPENDIX

TABLE A3.1
Variable Types

Symbol	Description	Units
P_i	Number of plants of type i	Plants/m^2
K_i	Competition scaling factor for plant type i	
M_i	Maximum seed production for plant type i	Plants/m^2
F_i	A fraction of plants or gametes of type i	
G_n	Distinct genotypes for n genes	

TABLE A3.2
Subscripts Used in Variables

Symbol	Description
s	Ungerminated seed in the ground
g	Germinated seed
e	Established seedling
p	Mature plant
q	Gamete (pollen and or ovules)
d	Denotes addition from outside
w	Denotes a weed
c	Denotes a crop plant
i or j = 1...n	Type index where n is the number of instances of a class of entity such as species, genotypes, generations, or herbicides
sp	Seed on plants
sipj	Seed of type i on plants of type j
yi	A generation in a sequence of generations
tg-	Time of the life cycle immediately before germination
tg+	Time of the life cycle immediately after germination
tp	Time of the life cycle after seed-set and entry of new seed into the pool
tg-yi	The time immediately before germination in generation i
tg-yi-1	The time immediately before germination in the generation before generation i
tg0	The start time, immediately before the first germination event
A or B	The allele of gene A or B that confers susceptibility to 1 or more herbicides (generally more common at tg0)

TABLE A3.2 (CONTINUED)
Subscripts Used in Variables

Symbol	Description
a or b	The allele of gene A or B that confers resistance to one or more herbicides (generally less common at tg0)
AA or BB	Denotes a plant that is homozygous susceptible in gene A or B
aa or bb	Denotes a plant that is homozygous resistant in gene A or B
aA, Aa, bB or Bb	Denotes a plant that is heterozygous in gene A or B
aAbB	Denotes a plant that is heterozygous in gene A and B where both resistance genes come from the same parent
aABb	Denotes a plant that is heterozygous in gene A and B where one resistance gene comes from each parent
$u \rightarrow v$	A transition from stage u to stage v where u and v are stages in the life cycle of a plant type

REFERENCES

1. Warwick, S. I., Genetic variation in weeds — with particular reference to Canadian agricultural weeds, in *Biological Approaches and Evolutionary Trends in Plants*, Kawano, S., Ed., Academic Press, London, 1990, 3.
2. Maxwell, B. D. and Mortimer, A. M., Selection for herbicide resistance, in *Herbicide Resistance in Plants: Biology and Biochemistry*, Powles, S. B. and Holtum, J. A. M., Eds., Lewis Publishers, Boca Raton, FL, 1993, 1.
3. Jasieniuk, M., Brûlé-Babel, A. L., and Morrison, I. N., The evolution and genetics of herbicide resistance in weeds, *Weed Sci.*, 44, 176, 1996.
4. Warwick, S. I., Herbicide resistance in weedy plants: physiology and population biology, *Annu. Rev. Ecol. Syst.*, 22, 95, 1991.
5. Mortimer, A. M., Ulf-Hansen, P. F., and Putwain, P. D., Modelling herbicide resistance — a study of ecological fitness, in *Achievements and Developments in Combating Pesticide Resistance*, Denholm, I., Devonshire, A. L., and Hollomans, D. W., Eds., Elsevier, Amsterdam, 1993, 148.
6. Gressel, J. and Segel, L. A., The paucity of plants evolving genetic resistance to herbicides: possible reasons and implications, *J. Theor. Biol.*, 75, 349, 1978.
7. Putwain, P. D. and Mortimer, A. M., The resistance of weeds to herbicides: rational approaches for the containment of a growing problem, in *Proc. Brighton Crop Prot. Conf. — Weeds*, British Crop Protection Council, Farnham, 1989, 285.
8. Maxwell, B. D., Roush, M. L., and Radosevich, S. R., Predicting the evolution and dynamics of herbicide resistance in weed populations, *Weed Technol.*, 4, 2, 1990.
9. Gressel, J. and Segel, L. A., Modelling the effectiveness of herbicide rotations and mixtures as strategies to delay or preclude resistance, *Weed Technol.*, 4, 186, 1990.
10. Gardner, S. N., Gressel, J., and Mangel, M., A revolving dose strategy to delay the evolution of both quantitative vs major monogene resistances to pesticides and drugs, *Int. J. Pest Manage.*, 44, 161, 1998.
11. Hall, L. M., Holtum, J. A. M., and Powles, S. B., Mechanisms responsible for cross resistance and multiple resistance, in *Herbicide Resistance in Plants: Biology and Biochemistry*, Powles, S. B. and Holtum, J. A. M., Eds., Lewis Publishers, Boca Raton, FL, 1993, 243.

12. Richter, O. and Zwerger, P., Temporal and spatial dynamics of herbicide resistance: a new approach combining population dynamics, population genetics and cellular automata, in *Human and Environmental Exposure to Xenobiotics, Proceedings of the XI Symposium Pesticide Chemistry*, Cremona, Italy, 1999, 599.

13. Cousens, R. and Mortimer, A. M., *Dynamics of Weed Populations*, Cambridge University Press, Cambridge, U.K., 1995.

14. Merrell, D. J., *Ecological Genetics*, University of Minnesota Press, Minneapolis, 1981, 500.

15. Jasieniuk, M., Brûlé-Babel, A. L., and Morrison, I. M., The evolution and genetics of herbicide resistance in weeds, *Weed Sci.*, 44, 176, 1996.

16. Darmency, H., Genetics of herbicide resistance in weeds, in *Herbicide Resistance in Plants: Biology and Biochemistry*, Powles, S. B. and Holtum, J. A. M., Eds., Lewis Publishers, Boca Raton, FL, 1993, 263.

17. Saari, L. L., Cotterman, J. C., and Thill, D. C., Resistance to acetolactate synthase inhibiting herbicides, in *Herbicide Resistance in Plants: Biology and Biochemistry*, Powles, S. B. and Holtum, J. A. M., Eds., Lewis Publishers, Boca Raton, FL, 1993, 83.

18. Rainey, P. and Moxon, R., Unusual mutational mechanisms and evolution, *Science*, 260, 1958, 1993.

19. Lenski, R. E. and Mittler, J. E., The directed mutation controversy and Neo-Darwinism, *Science*, 259, 188, 1993.

20. Bridges, B. A., Hypermutation under stress, *Nature*, 387, 557, 1997.

21. Bettini, P., McNally, S., Savignac, M., Darmency, H., Gasquez, J., and Dron, M., Atrazine resistance in *Chenopodium album*: low and high levels of resistance to the herbicide are related to the same chloroplast psbA gene mutation, *Plant Physiol.*, 84, 1442, 1987.

22. Matthews, J. M. and Powles, S. B., Aspects of the population dynamics of selection for herbicide resistance in *Lolium rigidum* (Gaud), *Proc. 1st Int. Weed Cont. Congr.*, Melbourne, Australia, 1992, 318.

23. Darmency, H. and Gasquez, J., Appearance and spread of triazine resistance in common lambsquarter (*Chenopodium album*), *Weed Technol.*, 4, 173, 1990.

24. Christoffers, M. J., Genetic aspects of herbicide resistant weed management, *Weed Technol.*, 13, 647, 1999.

25. Powles, S. B. and Matthews, J. M., Multiple herbicide resistance in annual ryegrass (*Lolium rigidum*): a driving force for the adoption of integrated weed management, in *Resistance 91: Achievements and Developments in Combating Pesticide Resistance*, Denholm, I., Devonshire, A. L., and Holloman, D. W., Eds., Elsevier Applied Science, New York, 1992, 75.

26. Islam, A. K. M. R. and Powles, S. B., Inheritance of resistance to paraquat in Barley grass *Hordeum vulgare* Steud, *Weed Res.*, 28, 393, 1988.

27. Gould, F., Comparisons between resistance management strategies for insects and weeds, *Weed Technol.*, 9, 830, 1995.

28. Hirschberg, J. A. and McIntosh, L., Molecular basis for herbicide resistance in *Amaranthus hybridus*, *Science*, 22, 1346, 1983.

29. Darmency, H. and Gasquez, J., Inheritance of triazine resistance in *Poa annua*: consequences for population dynamics, *New Phytol.*, 89, 487, 1981.

30. Holliday, R. J. and Putwain, P. D., Evolution of herbicide resistance in *Senecio vulgaris*: variation in susceptibility to simazine between and within populations, *J. Appl. Ecol.*, 17, 799, 1980.

31. Via, S., Quantitative genetic models and the evolution of pesticide resistance, in *Pesticide Resistance: Strategies and Tactics for Management*, National Academy Press, Washington, D.C., 1986, 222.

32. Brent, K. J., Holloman, D. W., and Shaw, M. W., Predicting the evolution of fungicide resistance, in *Managing Resistance to Agrochemicals*, Green, M. B., LeBaron, H. M., and Moberg, W. K., Eds., ACS Symp. Ser. 421, American Chemical Society, Washington, D.C., 1990, 303.

33. Murray, J. M., An example of the effects of drug resistance on the optimal schedule for a single drug in cancer chemotherapy, *IMA J. Math. Appl. Med. Biol.*, 12, 55, 1995.

34. Gressel, J. and Segel, L. A., Modelling the effectiveness of herbicide rotations and mixtures as strategies to delay or preclude resistance, *Weed Technol.*, 4, 186, 1990.

35. Brown, A. H. D. and Burdon, J. J., Mating systems and colonising success in plants, in *Colonization, Succession and Stability*, Gray, A. J., Crawley, M. J., and Edwards, P. J., Eds., Blackwell Scientific, Oxford, 1987, 115.

36. Levin, D. A. and Kerster, H. W., Gene flow in plants, in *Evolutionary Biology*, Dobzhansky, T., Hecht, M. K., and Steere, W. D., Eds., Plenum Press, New York, 1974, 139.

37. Rieger, M. A., Preston, C., and Powles, S. B., Risks of gene flow from transgenic herbicide-resistant canola (*Brassica napus*) to weedy relatives in southern Australian cropping systems, *Aust. J. Agric. Res.*, 50, 115, 1999.

38. Zemetra, R. S., Hansen, J., and Mallory-Smith, C. A., Potential for gene transfer between wheat (*Triticum aestivum*) and jointed goatgrass (*Aegilops cylindrica*), *Weed Sci.*, 46, 313, 1998.

39. Martinez-Ghersa, M. A., Ghersa, C. M., Vila-Aiub, M. M., Satorre, E. H., and Radosevich, S. R., Evolution of resistance to diclofop-methyl in ryegrass (*Lolium multiflorum*): investigation of the role of introgression with related species, *Pestic. Sci.*, 51, 305, 1997.

40. Smeda, R. J., Currie, R. S., and Rippee, J. H., Fluazifop-P resistance as a dominant trait in Sorghum (*Sorghum bicolor*), *Weed Technol.*, 14, 397, 2000.

41. Scheffler, J. A., Parkinson, R., and Dale, P. J., Evaluating the effectiveness of isolation distance for field plots of oilseed rape (*Brassica napus*) using a herbicide resistance transgene as a selectable marker, *Plant Breed.*, 114, 317, 1995.

42. Timmons, A. M., O'Brien, E. T., Charters, Y. M., Dubbels, S. J., and Wilkinson, M. J., Assessing the risks of wind pollination from fields of genetically modified *Brassica napus* spp. Oleifera, *Euphytica*, 85, 417, 1995.

43. Levin, D. A., Dispersal versus gene flow in plants, *Ann. Mo. Botanical Garden*, 68, 233, 1981.

44. Levin, D. A. and Kerster, H. W., Gene flow in seed plants, *Evol. Biol.*, 7, 139, 1974.

45. Stallings, G. P., Thill, D. C., and Mallory-Smith, C. A., Pollen-mediated gene flow of sulfonylurea-resistant Kochia (*Kochia scoparia* (L.) Schrad.), *Weed Sci. Soc. Am. Abstr.*, 33, 60, 1993.

46. Maxwell, B. D., Predicting gene flow from herbicide resistant weeds in annual agricultural systems, *Bull. Ecol. Soc. Am. Abstr.*, 73, 264, 1992.

47. Ellstrand, N. C. and Marshall, D. L., Interpopulation gene flow by pollen in wild radish, *Raphanus sativus*, *Am. Nat.*, 126, 606, 1985.

48. McCanny, S. J. and Cavers, P. B., Spread of proso millet (*Panicum miliaceum*) in Ontario, Canada 2, Dispersal by combines, *Weed Res.*, 28, 67, 1988.

49. Thomas, A. G., Gill, A. M., Moore, P. W. M., and Forcella, F., Drought feeding and dispersal of weeds, *J. Aust. Inst. Agric. Sci.*, 50, 103, 1984.

50. Lande, R., The response to selection on minor and major mutations affecting a metrical trait, *Heredity*, 50, 47, 1983.

51. Thompson, C. R., Thill, D. C., and Shafii, B., Growth and competitiveness of sulfonyl-urea-resistant and -susceptible Kochia (*Kochia scoparia*), *Weed Sci.*, 42, 172, 1994.
52. Gill, G. S., Cousens, R. D., and Allan, M. R., Germination, growth and development of herbicide resistance and susceptible populations of rigid ryegrass (*Lolium rigidum*), *Weed Sci.*, 44, 252, 1996.
53. Wiederholt, R. J. and Stoltenberg, D. E., Similar fitness between large crabgrass (*Digitaria sanguinalis*) accessions resistant or susceptible to acetyl-coenzyme A carboxylase inhibitors, *Weed Technol.*, 10, 42, 1996.
54. Wiederholt, R. J. and Stoltenberg, D. E., Absence of differential fitness between giant foxtail (*Setaria faberi*) accessions resistant and susceptible to acetyl-coenzyme A carboxylase inhibitors, *Weed Sci.*, 44, 18, 1996.
55. Holt, J. S. and Thill, D. C., Growth and productivity of resistant plants, in *Herbicide Resistance in Plants: Biology and Biochemistry*, Powles, S. B. and Holtum, J. A. M., Eds., Lewis Publishers, Boca Raton, FL, 1993, 299.
56. Warwick, S. I. and Black, L. D., Relative fitness of herbicide resistant and susceptible biotypes of weeds, *Phytoprotection*, 75, 37, 1994.
57. Gressel, J. and Ben-Sinai, G., Low intraspecific competitive fitness in a triazine-resistant, nearly nuclear-isogenic line of *Brassica napus*, *Plant Sci.*, 1, 29, 1985.
58. McCloskey, W. B. and Holt, J. S., Effect of growth temperature on biomass production of nearly isonuclear triazine-resistant and -susceptible common groundsel (*Senecio vulgaris* L.), *Plant Cell Environ.*, 14, 699, 1991.
59. Holt, J. S. and Radosevich, S. R., Differential growth of two common groundsel (*Senecio vulgaris*) biotypes, *Weed Sci.*, 31, 112, 1983.
60. Firbank, L. G. and Watkinson, A. R., On the analysis of competition within two species mixtures of plants, *J. Appl. Ecol.*, 22, 503, 1985.
61. Thompson, C. R., Thill, D. C., and Shafii, B., Germination characteristics of sulfonyl-urea-resistant and -susceptible Kochia (*Kochia scoparia*), *Weed Sci.*, 42, 50, 1992.
62. Kremer, E. and Lotz, L. A. P., Germination and emergence characteristics of triazine susceptible and resistant biotypes of *Solanum nigrum*, *J. Appl. Ecol.*, 35, 302, 1998.
63. Kremer, E. and Lotz, L. A. P., Emergence depth of triazine susceptible and resistant *Solanum nigrum* seeds, *Ann. Appl. Biol.*, 132, 277, 1998.
64. Kremer, E., Fitness of Triazine Susceptible and Resistant *Solanum nigrum* L. in Maize, Ph.D. thesis, University of Wageningen, the Netherlands, 1998.
65. Cousens, R., A comparison of empirical models relating crop yield to weed and crop density, *J. Agric. Sci.*, 105, 513, 1985.
66. De Wit, C. T., On competition, *Versl. Landbouwkd. Onderz. Rijkslandproefstn.*, 66, 8, 1960.
67. Kroppf, M. J. and van Laar, H. H., Eds., *Modelling Crop–Weed Interactions*, CAB International, Wallingford, U.K., 1993.
68. Pannell, D. J., Stewart, V., Bennett, A., Monjardino, M., Schmidt, C., and Powles, S., A bioeconomic model for ryegrass (*Lolium rigidum*) management in Australian cropping, *Agric. Syst.*, in press.
69. Falconer, D. S., *Introduction to Quantitative Genetics*, 4th ed., Burnt Mill, Harlow, U.K., 1996.
70. Wright, S., Coefficients of inbreeding and relationship, *Am. Nat.*, 56, 330, 1972.
71. Maxwell, B., Weed thresholds: the space component and considerations for herbicide resistance, *Weed Technol.*, 6, 205, 1992.
72. Roush, M. L., Radosevich, S. R., and Maxwell, B., Future outlook for herbicide-resistance research, *Weed Technol.*, 4, 208, 1990.

4 World Maize/Soybean and Herbicide Resistance

Micheal D. K. Owen

CONTENTS

4.1   INTRODUCTION

The evolution of herbicide-resistant (HR) weeds in maize and soybean has been a major production issue during the last decade. While HR weed populations have been a problem predominantly in developed countries, the potential exists for resistance to evolve in developing nations as weed control tactics change. Generally, maize and soybean production systems have more flexibility, tools, and alternative strategies than other crop production systems. However, due to factors currently affecting agriculture — many of which are not related to actual production practices but rather to socioeconomic issues — producers have utilized few of the available weed management tools. The focus for weed management, particularly in developed countries, has been on herbicides, often to the exclusion of all other available tactics. While the use of herbicides has been historically effective and efficient, resultant changes in the agroecosystem, specifically in the weed community, have become an issue. Further, legitimate concerns about the impact of herbicide tactics on the environment have been expressed.[1] Thus, it is important to determine how herbicides impact maize and soybean production systems and the weed communities.

4.1.1 PLANT ADAPTATION TO AGRICULTURE

Each maize and soybean agroecosystem is unique and thus has unique genetic and ecological characteristics resulting in the evolution of an associated pest complex that is capable of considerable morphological adaptation.[2] The weed community that develops in response to agricultural practices evolves continuously from nonweedy ancestors and associated weedy plants.[3–5] Vegetative mimicry of the crop has been an important adaptive response in weeds, and is particularly effective where hand weeding is a management tactic.[6,7] More recently, seed mimicry has evolved and has been an important characteristic increasing the success of weed populations in crops.[7] However, there are few examples describing this type of morphological mimicry in weed populations in maize and soybean agroecosystems. On the other hand, a third type of mimicry, biochemical mimicry, has become prevalent in maize and soybean in developed countries.[8]

4.1.2 PREDICTION OF HERBICIDE RESISTANCE IN WEEDS

The development of HR weed biotypes was predicted before herbicides were widely used in maize and soybean.[9] Generally, shifts in weed communities were thought to be greater problems than herbicide resistance.[10] Recent reports stress that resistance of pests to chemical control is the most important threat to world food and fiber production.[11,12] The reported increase in the relative importance of HR in weed populations is due, in part, to single-strategy weed management.[13] The failure of agronomists to recognize that management of a weed population reflects the management of a genetically renewable resource has resulted in tactics that favor a specifically adapted genotype.[12] The costs of poor management result in increased production costs, pest control failures, and the possible loss of the herbicide as a management tool.[11] However, two important questions need to be asked in order to understand the practical implementation of resistant weed management in maize and soybean. Has any herbicide been "lost" due to a HR weed biotype? Is the cost of HR weed populations that great to the producer?

4.1.3 HERBICIDES AND WEED POPULATIONS

Changes in weed populations occur from the agricultural manipulation necessary to produce maize and soybean.[14] It is important to recognize that while HR in weeds has become more common, particularly in the last decade, a greater problem is the failure to maintain weed populations below the level that will not reduce crop yield. Changes in weed communities are more difficult to document but may be more important than HR weed biotypes.[15] Importantly, there has been no indication that any weeds have been eliminated from the agroecosystem despite more than 50 years of herbicide use.[15] Regardless, HR weed populations are an important agricultural problem in maize and soybean.[8] A survey conducted by Heap[16] indicates 233 HR weed biotypes located throughout the world. These biotypes encompass 150 plant species and are found in 47 countries. Developed countries, with intense agriculture, account for more than 90% of the HR weed biotypes. However, the actual number of HR weed biotypes may be greater, as not all instances of resistance have been reported.

4.1.4   Overview of Agriculture and Pest Management

Maize and soybean generally have a greater genetic potential for yield than what is actually realized.[17] Much of the yield loss is attributable to an unfavorable physio-chemical environment. Weeds contribute to this unfavorable environment and can reduce crop yields by more than 80%.[18] It is important to understand that weeds behave differently than other pests in that weeds do not directly attack the crop; instead, they compete for resources, thus reducing crop vigor and yield quantity and quality.[19] The lack of visible damage to the crop often results in poor timing of specific management tactics until after the yield loss has occurred. The annual losses attributable to weeds in the United States are estimated to be $18 to $20 billion.[20,21] An estimated 41% of herbicides applied in the United States are applied to maize.[22] The percent of soybean area treated with herbicides is similar to corn. Given the high usage of herbicides in maize and soybean, concerns for ground and surface water contamination also represent significant, indirect costs.[23] Indirect costs of pesticide use in the United States were estimated to be an additional $1 billion.[24] Pesticide use in the Far East, Europe, and the United States accounts for 86% of the world use.[25]

4.1.5   Overview of Herbicides and Weed Management

The development of new herbicides has provided growers with tools with which to manage weeds.[26] This has resulted in effective and efficient control of weeds to the extent that herbicides are now considered an essential component of high production agriculture.[27] Weed management has become the management of herbicide technology and the importance of understanding interactions between weeds and crops has been minimized.[28] As a result, the loss of some active ingredients and cancellation of registrations may limit weed management options.[29] Furthermore, changes in the agricultural chemical industry, such as mergers, grower programs, and herbicide management philosophy, may have a dramatic impact on the management of HR weeds.[27]

   Clearly, plant breeders have made good use of evolution to exploit the high yield potential in maize and soybean. However, it is also evident that weed scientists have been less than effective in managing weed evolution.[30] Weed populations have adapted physiologically to herbicides much more rapidly than biological wisdom might predict, particularly when considering the rapid increase in HR weed populations.[2] The rapid evolution of HR weed populations during the 1990s has been of particular concern for agricultural producers who recognize the need for efficient weed control as a means to reduce resource consumption and achieve sustainability.[31]

It can be argued that weed science has been extremely successful in managing the most detrimental and consistent pest complex that impacts world food production. It can also be argued that weed science has failed to effectively manage weeds, given the increasing costs of weed control, the continued loss of yield due to weeds, the potential environmental costs associated with herbicide use, and the widespread evolution of HR weed populations. Regardless, it is clear that major changes in the systems should be considered.[32]

The objectives of this chapter are to review the current world situation in maize and soybean production and provide an update on the development of HR weeds in these agroecosystems. The current HR weed management strategies will be discussed and evaluated. Alternative tactics will be proposed and a realistic assessment of these tactics provided. The reader should consider these questions while reading this chapter. Are HR weeds actually a serious threat in maize and soybean agroecosystems? Are current weed management strategies and specific tactics inappropriate, given the socioeconomic climate in developed countries? Should developing countries adopt new weed management strategies, given the problems that developed countries have experienced?

4.2 DESCRIPTIONS OF MAJOR MAIZE AND SOYBEAN AGROECOSYSTEMS

It is impossible to provide a clear and definitive description of the maize and soybean agroecosystems throughout the world. While there are general similarities to the agroecosystems in large geographical regions, the specific characteristics vary widely. Thus, maize and soybean agroecosystems will be characterized in terms encompassing large geographical land areas and countries. North and South America and China represent the primary soybean production areas.[33] North and South America and Europe produce most of the maize. Africa does not have a large production area of either soybean or maize, although South Africa and Zimbabwe have 3.2 million ha and 1.6 million ha of maize, respectively.[33]

Maize and soybean are major crops in the world and rank fourth and fifth, respectively, behind wheat, oilseeds, and rice in hectares grown.[33] World maize production represented an estimated 140 million ha in 1999 and a yield of 4.29 t/ha. World soybean production occurred on an estimated 71 million ha with a yield of 2.2 t/ha. Overall production in 1999, compared to 1998, was down 0.6 and 2.9% for maize and soybean, respectively. Importantly, in the absence of management practices, weeds caused greater yield reductions than any other pest complex.[19]

The average reduction in world crop yield loss attributable to weed interference was estimated to be 11.5%, with 25% losses occurring in developing countries and an average of 5% losses in developed countries.[34] Weed control requires up to 70% of the labor invested in crop production in some traditional cropping systems, yet the yield reductions are still extremely high.[35] A major consideration is the lack of technical weed management expertise in the developing world and the reliance on manual labor.[36,37] However, in developed countries where the crop production systems are technologically rich and the loss of yield relatively low, the use of herbicides may contribute to the crop yield loss by increasing the sensitivity of the crop to other pest complexes or by injuring rotational crops.[38]

World herbicide use represents 44% of the chemical pesticides purchased, but only 10% of the world herbicide use is in developing countries.[25] In developed countries, maize and soybean production are heavily dependent on herbicides due, in part, to the unavailability and the lack of cost effectiveness of hand labor.[39] Further, a majority of maize and soybean hectares are treated with herbicides that represent very few mechanisms of action.[40] An emphasis on herbicidal control of weeds has allowed the adoption of reduced tillage systems that lessen soil erosion.[36,41,42] The combination of herbicides and reduced tillage has made producers less likely to use alternative weed management tactics, thus increasing selection pressure on the agroecosystems for weed communities that are well adapted to the system.[43]

The weed communities that have developed in maize and soybean agroecosystems in response to management strategies favor annual species.[44] While the agroecosystems in different regions may have different weed communities, generally the major weeds demonstrate adaptability to a wide range of biotic and abiotic factors.[3] In fact, the dominant weed species may be characterized as "pre-adapted" to human activity.[18] The weed community will change in response to the selection pressures applied by weed management practices, either gradually, as is the case for cropping systems in developing countries,[7] or relatively rapidly, as is the case in developed countries.[45]

Historically much of the change in weed communities in maize and soybean agroecosystems was the result of the accidental introduction of exotic species.[6,10] Generally, weed communities have become more abundant and diverse in response to modern agriculture.[46] Once introduced, the selective forces of crop production modify both the genetic and species diversity within the weed community.[45,47] There is also a concern for the introgression of traits, particularly herbicide resistance, from crops to the weed community, although this concern is minimal for maize and soybean.[48] This results in weed communities in maize and soybean agroecosystems that are well adapted to the production systems and thus difficult to manage.[45] Importantly, the agroecosystems have become less coupled to natural feedback mechanisms that exist in natural ecosystems, and more responsive to social and economic considerations.[45]

4.2.1 North and South America

North and South America account for approximately 39% of the world maize production and 57% of the soybean production.[33] Brazil and Argentina are the primary producers of maize and soybean in South America.[33] Generally, these agroecosystems are based on high inputs, and weed control is with herbicides. The use of post-planting mechanical strategies is minimal.[49] Triazine herbicides have been widely used in maize production systems.[50] Dinitroaniline (DNA) and acetyl coenzyme A carboxylase (ACCase)-inhibitor herbicides have been widely used in soybean.[51] Rotation systems are simple, particularly in North America where maize is often grown continuously or in rotation with soybean. In South America, soybean is often grown continuously. Central America produces relatively little maize and soybean and that which is grown typically is on small production systems.

4.2.1.1 United States, Mexico, and Canada

In 1999, approximately 29 million ha of maize was grown in the United States. Mexico produced over 8 million ha and Canada grew slightly more than 1 million ha.[33] The United States grew almost 30 million ha of soybean in 1999, Canada produced 1 million ha, and Mexico grew 0.1 million ha.[33] Iowa and Illinois account for approximately one third of maize and soybean production in the United States and the province of Ontario is the primary producer of these grains in Canada. Interestingly, less than 2% of the U.S. population lives on farms.[27,52] Thus, most of the U.S. voting public does not understand the technical and economic issues facing maize and soybean production.

While crop rotations are typically very simple in North America (e.g., maize–soybean) and likely have limited utility in regard to weed management, crop rotation was the most important integrated pest management (IPM) practice reported by growers.[53] The lack of crop rotation diversity is attributable, in part, to the increased mechanization of agriculture.[54] Further, the availability and use of herbicides and nitrogen fertilizers lessen the need for diverse rotations. However, growers will alter crop rotations when specific pest problems dictate the need for alternative tactics.[55]

The specific weed communities in the United States have changed due to production practices.[7] However, there are a number of weed species that are seemingly ubiquitous over broad geographical areas. For example, foxtails (*Setaria* spp.) are found throughout the maize and soybean production areas of the United States and Canada. Amaranthaceae, Asteraceae, Chenopodiaceae, and Malvaceae are also important "weedy" families for North America. A large percentage of important weeds in North America, compared to Europe, were imported from Europe and Asia.[56] However, it is often very difficult to determine the location from which the weed originated.[3] Generally, all maize and soybean production systems in North America have infestations of annual grasses and broadleaf weeds that require management despite historic and continuous efforts by producers to control these weeds.

Tillage is practiced on most of the maize and soybean hectares. For example, in Iowa, two tillage trips are made on every field.[57] However, tillage does not typically include a moldboard plow and at least 25% crop residue remains on the soil surface on most hectares. Some form of reduced tillage was used on approximately 20% of Ontario farms in 1991.[58] It is likely that this percentage has increased significantly in the last decade. Importantly, the adoption of conservation tillage systems increased herbicide use, particularly under no-tillage systems.[59] If ridge-tillage systems are used, less herbicide is applied. However, these systems required mechanical tactics to replace herbicidal weed control. No-tillage systems that used the most herbicides were the most energy-efficient weed management systems.[58]

Herbicide use has changed dramatically in the last decade from preplant incorporated herbicides to postemergence or soil-applied/postemergence sequential herbicide applications.[60] More recently, total postemergence strategies have been used on a majority of the soybean hectares and it is likely that herbicide use in maize will follow a similar trend.[61] The shift toward postemergence applications as the sole tactic for weed management has been influenced by the introduction of HR cultivars. During the 1990s, acetolactate synthase (ALS)-inhibitor herbicides were applied to a majority of the soybean hectares and a significant percentage of the maize hectares.[62] However, there has been a major change in herbicide usage in soybean due to the availability of glyphosate-resistant soybean varieties. An estimated 12 to 14 million ha of glyphosate-resistant soybean was planted in the United States in 1999,[63] and glyphosate was used on approximately 50% of the soybean hectares in the United States as the sole means of weed control.[61] The adoption of HR maize has been less despite the availability of glyphosate-, glufosinate-, and imidazolinone-resistant hybrids.[61] Questions have been posed as to the yield of HR maize and soybean cultivars when compared to traditional cultivars.[48]

Availability of HR maize and soybean may reduce the utilization of alternative weed management tactics. For example, only 73% of the maize hectares in Iowa was cultivated and 18% rotary hoed in 1998.[57] Producers assessed these mechanical tactics as very effective only 24 to 31% of the time. Thus, maize and soybean agroecosystems in North America will continue to depend on herbicides for weed management.[39] It does not appear likely that integrated weed management (IWM) systems will be adopted unless governmental policy changes.[64] Past governmental policies supported low diversity crop production systems that were dependent on herbicides for weed control. Producers feel that integrated systems represent a greater economic risk compared to the present systems, and unless the policy makers develop a strategic plan that allows environmentally safe and economically sound management of weeds, it is not likely that maize and soybean production systems will move toward IWM.[31,32]

4.2.1.2 Brazil

Brazil is a major exporter of soybean and had almost 13 million ha of soybean in 1999.[33] The soybean production area is in southern Brazil. The Brazilian Cerrados had 6.5 million ha of soybean and 50% was grown under no-tillage production systems.[65] Glyphosate-resistant soybean accounted for approximately 30% of the 1999 crop.[66] Brazil also had over 12 million ha of maize in 1999.[33] Approximately 80% of the maize and soybean hectares is treated with herbicides.[65]

Brazilian weed communities have changed significantly within the last 25 years and reflect changes in crop production strategies.[67] There are approximately 500 weedy species in Brazil, of which 80% are indigenous species. Currently, *Euphorbia heterophylla* is reported to be the most important weed species in Brazilian soybean production.[67]

4.2.1.3 Argentina

Argentina had almost 8 million ha of soybean and over 3 million ha of maize in 1999, and was a major exporter of both crops.[33] The trend toward conservation tillage systems in Argentina is similar to Brazil. Glyphosate-resistant soybean is planted on 90% of the hectares and glyphosate has been widely used as the primary herbicide for weed control.[68] The weed communities in Argentina have also changed dramatically in response to agricultural practices.[69] Thirty exotic species were reported to be important weeds.

4.2.2 WESTERN AND EASTERN EUROPE

Eastern and Western Europe were major importers of maize while producing almost 11 million ha in 1999.[33] Eastern Europe had 6.8 million ha of maize and 1.5 million ha of maize grown for silage was reported for the Ukraine.[33,70] Soybean is not widely grown in Europe.[33] Crop rotation and herbicides are primary tactics for weed management.[71] The triazine herbicides have been widely used in European maize production.[77] Another important trend is the governmental mandates to use lower quantities of herbicide active ingredients.[72] However, the use of HR cultivars has not been widely adopted in Europe due to governmental regulations and concerns about potential environmental impact.[73] Conservation tillage is widely practiced in Europe.

Similar to other regions, weed communities in Europe have changed in response to production practices.[74] In Hungary, *Ambrosia artemisiifolia, Datura strumonium,* and *Xanthium strumarium* have become important weeds in maize production. Perennials are increasing in importance due, in part, to conservation tillage practices. European weed communities also contain exotic species of non-European origin.[56] Exotic species consist of an estimated 580 species, or approximately 37% of the European flora. Most weeds introduced into Europe originate from the Western Hemisphere and Asia. Countries differed in the number of alien species. For example, France had 479 exotic species while Turkey had only 42.[56]

4.2.3 CHINA

Very little information is available about China maize and soybean production practices. Maize was produced on almost 26 million ha in 1999, but the yield was approximately 60% of that reported for the United States.[33] Soybean was produced on approximately 8 million ha, with yields approximately 20% of the United States. Total Chinese crop production accounts for approximately 11% of the land area.[75] Most of the weed control is done by hand, although some soybean production is highly mechanized.[51] State farms use a wheat–wheat–soybean or wheat–maize–soybean rotational strategy.[75] On the state farms, herbicide use is becoming increasingly common. For example, triazine herbicides have been used continuously in Eastern Jilin Province maize production systems.[75]

4.3   OVERVIEW OF HERBICIDE-RESISTANCE DEVELOPMENT IN MAJOR MAIZE AND SOYBEAN AGROECOSYSTEMS

This section will not be a comprehensive review of resistance throughout the world, but rather an update of HR weeds in maize and soybean agroecosystems with an emphasis on the factors that affected the development of the HR populations. An international survey conducted in 1997 listed 216 HR weed biotypes in 45 countries.[76] These biotypes encompassed 145 weed species and were located primarily in developed countries where agriculture is input-intensive. Maize had 50 species reported with resistant biotypes and soybean had 22 resistant species reported.[77] While very few of the HR biotypes have proven unmanageable in maize and soybean agroecosystems, the problems highlight the concerns with the current approaches to weed control.[78–80] A general lack of understanding of the complex biological processes that affect the development of HR populations could ultimately increase the cost of weed management and result in the loss of cost-effective herbicides.[78,81]

The spread of HR weed populations has been relatively rapid in maize and soybean agroecosystems due, in part, to management factors associated with those crops.[82] It is interesting to note that HR weed populations typically consist of one species despite the similar exposure of all species within the weed community to selection by the herbicide.[83] This raises the possibility that certain weed species are more likely to evolve HR populations.[84] Regardless of whether some weeds are predisposed to the evolution of resistance, weed communities are genetically diverse and demonstrate considerable adaptability to agronomic practices. Thus, weeds will find a favorable niche in any agroecosystem irrespective of strategies used to manage the crop.[39]

4.3.1   NORTH AND SOUTH AMERICA

Resistance in weed populations is generally not a problem in most countries in the Western Hemisphere. While HR weed populations exist in some of the developing countries of Central and South America, typically HR weed populations do not occur in maize agroecosystems. For example, despite high triazine use in Mexico and Brazil, no triazine-resistant weed populations have been reported.[85] While ALS-inhibitor-resistant weeds are very prevalent in maize and soybean agroecosystems in the temperate regions of North and South America, the risk of resistance in the tropics is much lower.[86] However, ALS-inhibitor resistance has been reported in Costa Rican rice production.[87] Resistance to ACCase inhibitors has also been reported in *Eriochloa punctata*, a weed common in Bolivian maize and soybean agroecosystems.[88] Cross-resistance to many ACCase inhibitors was observed in one resistant biotype. Most HR weed populations have been reported in Brazil, Canada, and the United States. While there are no reports of HR weed populations in Argentina, given the relatively high use of herbicides in maize and soybean agroecosystems, it is likely that HR weed populations exist, particularly to ALS-inhibitor herbicides.[77]

4.3.1.1 United States and Canada

A survey conducted in Illinois indicated that growers were concerned primarily about the management of *Abutilon theophrasti*, *Setaria* spp., and *Chenopodium album*.[89] Also included in the top ten weeds were *Amaranthus* spp., *Xanthium strumarium*, *Ambrosia trifida*, and *Sorghum bicolor*. Each of these species has HR populations to at least one herbicide mechanism of action. The United States currently has 80 reported HR biotypes and Canada has 32 reported HR biotypes.[77] Through the 1980s, triazine resistance was the greatest problem.[90] However, when ALS-inhibitor herbicides were introduced and adopted in maize and soybean production, HR weed populations rapidly evolved. The reported number and occurrences of herbicide-resistant weed populations in the United States and Canada are shown in Table 4.1. It should be noted that the list is probably not complete because HR weed population-resistant biotypes are not included on the Web site unless reported by researchers.[77]

4.3.1.1.1 Acetolactate synthase (ALS)-inhibitor herbicide resistance

A number of weeds have developed resistance to ALS-inhibitor herbicides (Table 4.2). A South Dakota population of *Helianthus annuus* demonstrated cross-resistance to imidazolinone and sulfonylurea herbicides.[91] The resistance was target-site-based and the biotype had I_{50} values of 39× and 9× for imazethapyr and chlorimuron, respectively, when compared to the sensitive biotype. Acetolactate synthase-resistant *H. annuus* populations were also reported in Kansas.[92] Importantly, resistant populations were found in roadside areas as well as fields.

A population of *Xanthium strumarium* was identified as resistant after multiple applications of imazaquin over 4 years.[93] This population was cross-resistant to flumetsulam and chlorimuron. Resistant populations were also identified in Oklahoma, but the problem was not as severe as originally suspected.[94] A population of *X. strumarium* was found to be cross-resistant to imidazolinones, sulfonylureas, triazolopyrimidines, and pyrimidyl oxybenzoates.[95] Inheritance of cross-resistance via pollen was dominant to semidominant.[96] Iowa and Ohio populations of *X. strumarium* were found to be similarly cross-resistant to imazethapyr and imazaquin, but not to chlorimuron.[97] The ALS resistance was at the enzyme level and was due to a single dominant nuclear gene.

A number of *Amaranthus* species have ALS-inhibitor herbicide resistance. A population of *A. rudis* from southern Iowa was cross-resistant to imidazolinone and sulfonylurea herbicides.[98] The population had been selected after 4 years of ALS-inhibitor herbicide use. However, cross-resistance to ALS-inhibitor herbicides has not been consistent among *Amaranthus* spp.[99] Comparison of imidazolinone-resistant and -susceptible *A. hybridus* indicated that the susceptible population had greater biomass production when grown under noncompetitive conditions in the greenhouse.[100] However, under field conditions, no fitness differences were noted.

Resistance to ALS-inhibitor herbicides has been reported in *Setaria faberi*.[101] Resistant populations were identified in Minnesota and Wisconsin fields with a

TABLE 4.1
Reported Populations of Herbicide-Resistant Weeds in Canada and the United States

Weed	Herbicide MOA[a]	Year Reported	Estimated Number of Sites	Country
Abutilon theophrasti	PS II (Group 5)	1984	12	U.S.A.
A. hybridus	ALS (Group 2)	1992	>150	U.S.A.
	PS II (Group 5)	1972	>5600	U.S.A.
A. lividus	ALS (Group 2)	1993	1	U.S.A.
Amaranthus palmeri	ALS (Group 2)	1991	>1200	U.S.A.
	PS II (Group 5)	1993	>10	U.S.A.
	Microtubule (Group 3)	1989	>1000	U.S.A.
A. powellii	ALS (Group 2)	1996	>5	U.S.A.
		1998	>5	Canada
	PS II (Group 5)	1977	1	Canada
		1992	Unknown	U.S.A.
A. retroflexus	ALS (Group 2)	1995	Unknown	U.S.A.
		1998	1	Canada
	PS II (Group 5)	1980	>4000	U.S.A.
		1980	1000	Canada
A. rudis	ALS (Group 2)	1993	>4000	U.S.A.
	PS II (Group 5)	1994	>20	U.S.A.
Ambrosia artemisiifolia	PS II (Group 5)	1976	100	Canada
		1993	>500	U.S.A.
	ALS (Group 2)	1998	>500	U.S.A.
A. trifida	ALS (Group 2)	1998	>500	U.S.A.
Ammannia auriculata	ALS (Group 2)	1997	>50	U.S.A.
Anthemis cotula	ALS (Group 2)	1997	100	U.S.A.
Avena fatua	Unknown (Group 8)	1989	>1000	Canada
	Lipid (Group 8)	1989	>1000	Canada
	ACCase (Group 1)	1990	>1000	Canada
			>900	U.S.A.
	Lipid (Group 8)		>1100	U.S.A.
	Cell division (Group 15)		Unknown	U.S.A.
	Unknown (Group 8)	1993	>150	U.S.A.
	ALS (Group 2)	1994	>1000	Canada
	Unknown (Group 25)	1994	>1000	Canada
	ALS (Group 2)	1996	>10	U.S.A.
Brassica campestris	PS II (Group 5)	1977	5	Canada
Bromus tectorum	ALS (Group 2)	1997	>5	U.S.A.
Centaurea solstitialis	Auxin (Group 4)	1988	1	U.S.A.
Chenopodium album	PS II (Group 5)	1973	1000	Canada
		1977	>10000	U.S.A.
C. strictum	PS II (Group 5)	1976	100	Canada
Chloris inflata	PS II (Group 5)	1987	500	U.S.A.
	PS II (Group 7)			U.S.A.
Commelina diffusa	Auxin (Group 4)	1957	Unknown	U.S.A.

TABLE 4.1 (CONTINUED)
Reported Populations of Herbicide-Resistant Weeds in Canada
and the United States

Weed	Herbicide MOA[a]	Year Reported	Estimated Number of Sites	Country
Convolvulus arvensis	Auxin (Group 4)	1964	Unknown	U.S.A.
Conyza canadensis	PS I (Group 22)	1993	5	Canada
		1994	>50	U.S.A.
Cyperus difformis	ALS (Group 2)	1993	1000	U.S.A.
Datura stramonium	PS II (Group 5)	1992	>10	U.S.A.
Daucus carota	Auxin (Group 4)	1957	Unknown	Canada
		1993	50	U.S.A.
Digitaria ischaemum	ACCase (Group 1)	1996	1	U.S.A.
D. sanguinalis	ACCase (Group 1)	1992	>1	U.S.A.
Echinochloa crus-galli	PS II (Group 5)	1978	5	U.S.A.
		1981	500	Canada
	PS II (Group 7)	1990	>25	U.S.A.
	Auxin (Group 4)	1998	>5	U.S.A.
E. phyllopogon	ACCase (Group 1)	1998	>50	U.S.A.
	Lipid (Group 8)	1998	>50	U.S.A.
Eleusine indica	Microtubule (Group 3)	1973	>2500	U.S.A.
Galeopsis tetrahit	ALS (Group 2)	1995	1	Canada
	Auxin (Group 4)	1998	1	Canada
Galium spurium	ALS (Group 2)	1997	Unknown	Canada
	Auxin (Group 4)	1997	Unknown	Canada
Helianthus annuus	ALS (Group 2)	1996	>160	U.S.A.
Kochia scoparia	PS II (Group 5)	1976	>2500	U.S.A.
	ALS (Group 2)	1987	>8500	U.S.A.
		1988	>50	Canada
	Auxin (Group 4)	1995	>500	U.S.A.
Lactuca serriola	ALS (Group 2)	1987	>600	U.S.A.
Lepidium virginicum	PS I (Group 22)	1993	5	Canada
Lolium multiflorum	ACCase (Group 1)	1987	>2500	U.S.A.
	ALS (Group 2)	1995	>500	U.S.A.
L. perenne	ALS (Group 2)	1989	>5	U.S.A.
L. persicum	ACCase (Group 1)	1993	1	U.S.A.
L. rigidum	EPSPS (Group 9)	1998	5	U.S.A.
Neslia paniculata	ALS (Group 2)	1998	1	Canada
Panicum capillare	PS II (Group 5)	1981	500	Canada
Poa annua	PS II (Group 5)	1994	>60	U.S.A.
	PS II (Group 7)	1994	>50	U.S.A.
	Lipid (Group 16)	1994	>50	U.S.A.
	Microtubule (Group 3)	1997	5	U.S.A.
Polygonum pennsylvanicum	PS II (Group 5)	1990	10	U.S.A.
Portulaca oleracea	PS II (Group 7)	1991	5	U.S.A.
Sagittaria montevidensis	ALS (Group 2)	1993	1000	U.S.A.

TABLE 4.1 (CONTINUED)
Reported Populations of Herbicide-Resistant Weeds in Canada and the United States

Weed	Herbicide MOA[a]	Year Reported	Estimated Number of Sites	Country
Salsola iberica	ALS (Group 2)	1987	>3000	U.S.A.
Scirpus mucronatus	ALS (Group 2)	1997	>50	U.S.A.
Senecio vulgaris	PS II (Group 5)	1970	>500	U.S.A.
	PS II (Group 6)	1995	>1	U.S.A.
Setaria faberi	PS II (Group 5)	1984	>5	U.S.A.
	ACCase (Group 1)	1991	>2	U.S.A.
	ALS (Group 2)	1996	>2	U.S.A.
S. glauca	PS II (Group 5)	1981	1000	Canada
		1984	>1	U.S.A.
S. lutescens	ALS (Group 2)	1997	>1	U.S.A.
S. viridis	Microtubule (Group 3)	1988	1000	Canada
		1989	>1000	U.S.A.
	ACCase (Group 1)	1992	>1000	Canada
	ALS (Group 2)	1999	>1	U.S.A.
S. viridis var. *robusta-alba* Schreiber	ALS (Group 2)	1996	>1	U.S.A.
Sida spinosa	ALS (Group 2)	1995	Unknown	U.S.A.
Sinapis arvensis	Auxin (Group 4)	1991	10	Canada
	ALS (Group 2)	1992	5	Canada
	PS II (Group 5)	1994	1	Canada
	ALS (Group 2)	1999	>5	U.S.A.
Solanum americanum	PS I (Group 22)	1985	100	U.S.A.
S. ptycanthum	ALS (Group 2)	1999	>5	U.S.A.
Sonchus asper	ALS (Group 2)	1992	5	Canada
Sorghum bicolor	ALS (Group 2)	1994	>5	U.S.A.
S. halpense	ACCase (Group 1)	1991	>20	U.S.A.
	Microtubule (Group 3)	1992	Unknown	U.S.A.
Stellaria media	ALS (Group 2)	1992	>500	Canada
Xanthium strumarium	Unknown (Group 17)	1985	>1100	U.S.A.
	ALS (Group 2)	1989	>1200	U.S.A.

[a] Herbicide MOA = mechanism of action. Herbicide groups are described according to classifications by the Weed Science Society of America (*Weed Technol.*, 11, 384, 1997).

Source: Adapted from Heap, I. M., International survey of herbicide-resistant weeds, http://www.weedscience.com, 2000.

history of ALS-inhibitor herbicide use in maize and soybean. Acetolactate synthase-inhibitor herbicide resistance was also reported for *Sorghum bicolor* populations in Nebraska.[102] Iowa populations also have been discovered, as well as ALS-inhibitor-resistant *Eriochloa villosa* (M. Owen, unpublished data, 1999). Finally, a population of *Kochia scoparia* from Illinois was found to be multiple-resistant to ALS inhibitors and triazine herbicides.[103] This population demonstrated site-of-action mediated resistance to triazine (atrazine), imidazolinone (imazethapyr), and sulfonylurea (thifensulfuron and chlorsulfuron) herbicides. Generally, ALS-inhibitor-resistant weed populations were discovered within 5 years of the initial use of ALS-inhibitor herbicides.[104]

4.3.1.1.2 Triazine herbicide resistance

While resistance to triazine herbicides was first reported several decades ago and has been widely documented historically,[82] the occurrence of new triazine-resistant populations has not been as rapid as with other herbicide mechanisms of action during the last decade.[77] However, triazine resistance is still widespread in maize agroecosystems in Canada and the United States.[82] In Nebraska, 64% of all fields surveyed had triazine-resistant populations of *Amaranthus rudis* and 92% of the fields had resistant populations when growers initially suspected resistance.[105] Maize production practices accounted for the consistency of resistant populations from field to field and the distribution was likely attributable to equipment. Triazine-resistant *A. powellii* populations are also reported.[106]

Most of the triazine-resistant weed populations have a target-site mutation that confers resistance.[82] The target-site mutation results in a fitness penalty for the resistant population when compared to the susceptible population.[107] However, metabolism-based triazine resistance is reported in *Abutilon theophrasti* populations in Wisconsin.[108,109] Triazine-resistant populations of *Chenopodium album* were also discovered in Iowa despite the use of chloroacetamide herbicides in combination with atrazine.[110,111] Triazine-resistant *C. album* populations were suggested to be less likely when chloroacetamide herbicides are used.[112]

Overall, 27 triazine-resistant weed biotypes have been reported in the United States and Canada.[77] Triazine resistance has occurred in at least 33 states of the United States and four provinces of Canada.[113] However, given the changes in triazine herbicide use in maize and soybean agroecosystems, it is unlikely that triazine resistance will escalate as a problem.

4.3.1.1.3 Dinitroaniline (DNA) herbicide resistance

The number of DNA herbicide-resistant weed biotypes has never been great despite the widespread use of these herbicides. Dinitroaniline herbicides are registered in maize and soybean and have been important components of soybean agroecosystems. DNA herbicide-resistant *Eleusine indica* was reported in South Carolina, although the primary crop was cotton.[114] Populations of *Setaria viridis* in Manitoba were documented resistant to DNA herbicides.[115] *Amaranthus palmerii* populations resistant to DNA herbicides were also reported in the southeastern United States.[116] The *S. viridis* and *A. palmerii* populations were resistant to a number of DNA herbicides but were not resistant to herbicides with other mechanisms of action.[116,117] While DNA herbicide-resistant populations are found in other crop agroecosystems, the weeds are common to maize and soybean. Thus, it is likely that resistant populations could exist in maize and soybean agroecosystems.

TABLE 4.2
Resistance to ALS^a-Inhibitor Herbicides in Canada and the United States

Weed Species[b]	Location	Selective Herbicide	Reported in Maize or Soybean	Cross-Resistance	Multiple-Resistance
Amaranthus hybridus	KY, VA	Various	Both	Yes	NR[c]
A. lividus	NJ	Imazethapyr	No	NR	NR
A. palmeri	KS, AR, NC, SC	Imazethapyr	Both	Yes	NR
A. powellii	OH	Imazaquin	Soybean	NR	NR
	Canada	Imazethapyr	No	NR	NR
A. retroflexus	AR, MD	Imazaquin	Soybean	NR	NR
	Canada	Imazethapyr	No	NR	NR
A. rudis	IA, IL, KS, MO, OH	Imazethapyr	Both	Yes	NR
Ambrosia artemisiifolia	IL, IN, MI, MN, OH	Cloransulam-methyl	Both	Yes	NR
A. trifida	IA, IL, IN, OH	Cloransulam-methyl	Soybean	Yes	NR
Ammannia auriculata	CA	Bensulfuron-methyl	No	NR	NR
Anthemis cotula	ID	Chlorsulfuron	No	NR	NR
Avena fatua	MT, ND	Various	No	Yes	Yes
Bromus tectorum	OR	Primisulfuron-methyl and sulfosulfuron	No	Yes	NR
Cyperus difformis	CA	Bensulfuron-methyl	No	NR	NR
Galeopsis tetrahit	Canada	Metsulfuron-methyl	No	NR	NR
Galium spurium	Canada	Metsulfuron-methyl	No	NR	NR
Helianthus annuus	IA, KS, MO, SD	Imazethapyr	Both	Yes	NR
Kochia scoparia	Canada, CO, ID, IL, IN, KS, MN, MT, ND, NM, OK, OR, SD, UT, WA, WI, WY	Chlorsulfuron, metsulfuron-methyl	Both	Yes	Yes
Lactuca serriola	ID, OR, WA	Chlorsulfuron	No	Yes	NR
Lolium multiflorum	MS	Sulfometuron-methyl	No	NR	NR

L. perenne	CA, TX	Sulfometuron-methyl	No	NR	NR
Neslia paniculata	Canada	Metsulfuron-methyl	No	NR	NR
Sagittaria montevidensis	CA	Besulfuron-methyl	No	NR	NR
Salsola iberica	CA, ID, MT, OR, WA	Chlorsulfuron	No	Yes	NR
Scirpus mucronatus	CA	Bensulfuron-methyl	No	NR	NR
Setaria faberi	MN, WI	Nicosulfuron	Both	Yes	NR
S. lutescens	MN	Imazethapyr	Soybean	NR	NR
S. viridis	WI	Imazamox	Both	NR	NR
S. viridis var. *robusta-alba* Schreiber	MN	Nicosulfuron	Both	Yes	NR
Sida spinosa	NR	Imazaquin	No	NR	NR
Sinapis arvensis	Canada, ND	Cloransulam-methyl, imazethapyr, thifensulfuron, methyl, chlorsulfuron, ethametsulfuron-methyl	Soybean	Yes	NR
Solanum ptycanthum	ND, WI	Imazamoz, imazethapyr	Soybean	NR	NR
Sonchus asper	Canada	Metsulfuron-methyl	No	NR	NR
Sorghum bicolor	IA, KS, NB	Primisulfuron-methyl	Both	Yes	NR
Stellaria media	Canada	Chlorsulfuron	No	NR	NR
Xanthium strumarium	AR, IA, KS, MD, MN, MO, MS, OH, OK, TN	Imazaquin, imazethapyr	Both	Yes	NR

[a] ALS = acetolactate synthase or acetohydroxyacid synthase (AHAS).

[b] Adapted from Heap, I. M. International survey of herbicide-resistant weeds, http://www.weedscience.com, 2000.

[c] NR = not reported.

4.3.1.1.4   Acetyl coenzyme A carboxylase (ACCase)-inhibitor herbicide resistance

While these herbicides are not commonly used in maize agroecosystems, they are applied in soybean. Populations of *Setaria faberi*, *S. viridis*, and *Digitaria sanguinalis* have been reported resistant to ACCase inhibitors in Canada and Wisconsin.[118–120] Cross-resistance was reported in these weed species to cyclohex-anedione and aryloxyphenoxypropionate herbicides. Resistant populations of *S. faberi* did not demonstrate any fitness penalty when compared to susceptible *S. faberi* populations.[121] The mechanism for resistance was an alteration of the target enzyme, ACCase, in Iowa and Manitoba populations of *S. faberi* and *S. viridis*.[122] While there were no reported effects on growth characteristics for the resistant *Setaria* spp., *Avena fatua*, or *D. sanguinalis* populations, populations of *Lolium multiflorum* that were resistant to diclofop-methyl demonstrated a greater degree of dormancy compared to susceptible populations.[123]

4.3.1.1.5   Glyphosate herbicide resistance

There have been a number of indications that resistance to glyphosate could evolve despite the historic perspective that this was an unlikely consequence of glyphosate use.[124] For example, two varieties of *Festuca*, *F. rubra* and *F. longifolia*, evolved resistance to glyphosate through a regime of recurrent selections.[125] Examples have been reported demonstrating target-site resistance for glyphosate.[126] Glyphosate resistance in *Eleusine indica* was attributed to an altered EPSPS target site. Glyphosate resistance was also reported in Australian populations of *Lolium rigidum* after 15 years of use.[127] No specific mechanism to explain the glyphosate resistance in this species has been published.

None of the aforementioned examples of glyphosate-resistant weeds are likely to become important problems in maize and soybean agroecosystems. However, recent reports from the midwestern United States suggest that resistance to glyphosate can develop in maize and soybean agroecosystems. *Amaranthus rudis* has become a serious weed problem throughout the Midwest within the last decade.[128] Importantly, *A. rudis* is dioecious and thus has considerable potential for genetic diversity. Iowa growers reported difficulties controlling *A. rudis* in glyphosate-resistant soybean systems, even with multiple applications of glyphosate (M. Owen, unpublished data, 2000). Samples were collected from the field and evaluated for shikimate accumulation.[129] Putative resistant phenotypes accumulated at least five times less shikimate than putative susceptible phenotypes, suggesting that differences in EPSPS may account for the different phenotypes.[130] Seedling assays of different *A. rudis* populations revealed considerable variation within populations for glyphosate response. The variation for GR_{50} ranged from 0.3 to 8.01 mM glyphosate. Recurrent selection has resulted in genetically stable glyphosate-resistant *A. rudis* phenotypes. Given the widespread use of glyphosate, the ability for *A. rudis* to exhibit differential response to glyphosate, and the widespread distribution of this weed, glyphosate resistance may become an issue in maize and soybean agro-ecosystems in the near future.

4.3.1.2 Brazil

A limited number of references cite the existence of HR weed populations in Brazil. However, given the high use of herbicides in Brazilian soybean agroecosystems, HR populations should evolve. Recent reports of resistance to ALS inhibitors and ACCase inhibitors support this supposition. In Paraná State, located in southern Brazil, *Brachiaria plantaginea* populations evolved resistance to ACCase inhibitors.[131] These populations were found to be cross-resistant to aryloxyphenoxypropionate and cyclohexanedione herbicides including butroxydim, clethodim, diclofop-methyl, fenoxaprop-*p*-ethyl, fluazifop-*p*-butyl, haloxyfop-methyl-R, and sethoxydim.[132] Different HR populations demonstrated variable GR_{50} values for different herbicides.

Resistance to ALS inhibitors has been reported in several weed species. Populations of *Bidens pilosa* from São Paulo State received seven annual applications of imazaquin and were found to be resistant.[133] This population was cross-resistant to other imidazolinone and sulfonylurea herbicides at levels ranging from 370-fold higher to 12-fold higher, depending on the herbicide, when compared to susceptible populations. Other *Bidens* spp. and *Euphrobia heterophylla* from Brazil were found to be resistant to ALS inhibitors.[132] These populations were cross-resistant to chlorimuron-ethyl, flumetsulam, imazapyr, imazaquin, imazethapyr, metsulfuron, and nicosulfuron.

Given the increase in glyphosate use in South America and the occurrence of *Amaranthus* spp. in soybean agroecosystems, it is anticipated that management problems will develop as has been experienced in the midwestern U.S. maize belt. Further, with continued use of ALS-inhibitor herbicides, more resistant weed species will be identified and the frequency of HR populations will increase.

4.3.2 WESTERN AND EASTERN EUROPE

In Western and Eastern Europe, 168 HR weed biotypes have been reported in 19 countries.[77] Resistance to triazine herbicides accounted for approximately 66% of the HR weed biotypes (Table 4.3). Resistance has also been recorded for ACCase inhibitors, ALS inhibitors, DNA herbicides, synthetic auxin herbicides, urea/amide herbicides, and bypiridilium herbicides.[77] There are at least 35 broadleaf weed species and 17 grass weed species with HR populations.[134] The countries where HR populations occur most frequently include France with 30 HR populations, Spain with 24 HR populations, the United Kingdom with 19 HR populations, Belgium and Germany with 15 HR populations, respectively, and Switzerland with 14 HR weed populations.[77] Generally, there are fewer HR weed populations in the former Eastern Bloc countries due to the relatively low past usage of herbicides in the maize agroecosystems.[78] Most of the HR weed populations in maize agroecosystems are resistant to triazine herbicides[78] (Table 4.3). The other herbicide mechanisms of action had HR weed populations evolve in other crop agroecosystems.

The plant families that are most commonly represented in the HR weed populations are *Poaceae* (17 biotypes), *Amaranthaceae* and *Asteraceae* (8 biotypes each), *Polygonaceae* (5 biotypes), and *Caryophyllaceae* and *Chenopodiaceae*

TABLE 4.3
Reported Populations of Herbicide-Resistant Weeds in Western and Eastern Europe

Weed	Herbicide MOA[a]	Country and (Herbicide Group)	Reported in Maize or Soybean
Agrostis stolonifera	Pigment synthesis (Group 11)	Belgium	No
Alopecurus myosuroides	ACCase (Group 1), ALS (Group 2), PSII inhibitors (Group 7), Microtuble (Group 3)	Belgium (1, 7), France (1, 2), Germany (7), Netherlands (7), Spain (7), Switzerland (7), United Kingdom (1, 2, 3, 7)	No
Alisma plantago-aquatica	ALS (Group 2)	Italy, Portugal	No
Amaranthus albus	PS II (Group 5)	Spain	No
A. blitoides	PS II (Group 5)	Hungary, Spain	Corn
A. bouchonii	PS II (Group 5)	Hungary	Corn
A. chlorostachs	PS II (Group 5)	Hungary	Corn
A. cruentus	PS II (Group 5)	Spain	No
A. hybridus	PS II (Group 5)	France, Italy, Spain, Switzerland	Corn
A. lividus	PS II (Group 5)	France, Switzerland	Corn
A. powellii	PS II (Group 5)	Czech Republic, France, Switzerland	Corn
A. retroflexus	PS II (Group 5), PS II (Group 7)	Austria (5), Bulgaria (7), Czech Republic (5), France (5), Germany (5, 7), Hungary (5), Poland (5), Spain (5), Switzerland (5)	Corn
Ambrosia artemisifolia	PS II (Group 5)	Hungary	Corn
Apera spica-venti	PS II (Group 7)	Switzerland	No
Arenaria serpyllifolia	PS II (Group 7)	France	Corn
Atriplex patula	PS II (Group 5)	Germany	Corn
Avena fatua	ACCase (Group 1)	Belgium, France, United Kingdom	No
A. sterilis	ACCase (Group 1)	United Kingdom	No
A. sterilis ludoviciana	ACCase (Group 1)	France, Italy	No
Bidens tripartite	PS II (Group 5)	Austria	Corn
Bromus tectorum	PS II (Group 5), PS II (Group 7)	France (5), Spain (5, 7)	Corn
Capsella bursa-pastoris	PS II (Group 5)	Poland	No

Species	Resistance	Countries	Corn
Chamomilla suaveolens	PS II (Group 5)	United Kingdom	No
Chenopodium album	PS II (Group 5), PS II (Group 7)	Belgium (5), Bulgaria (5), Czech Republic (5), France (5), Hungary (5), Germany (5), Italy (5), Netherlands (5), Norway (5, 7), Poland (5), Slovenia (5), Spain (5), Switzerland (5), United Kingdom (5)	Corn
C. ficifolium	PS II (Group 5)	Germany, Switzerland	Corn
C. polyspermum	PS II (Group 5)	France, Germany, Hungary, Switzerland	Corn
C. strictum	PS II (Group 5)	Czech Republic	Corn
Cirsium arvense	Auxin (Group 4)	Hungary, Sweden	No
Conyza bonariensis	PS II (Group 5)	Spain	No
C. canadensis	PS II (Group 5), PS II (Group 7), PS I (Group 22)	Belgium (5, 22), Czech Republic (5), France (5, 7), Hungary (5), Poland (5), Spain, (5), Switzerland (5), United Kingdom (5)	Corn
Digitaria sanguinalis	PS II (Group 5)	France, Poland	Corn
Echinochloa crus-galli	Microtuble (Group 3), PS II (Group 5), PS II (Group 7)	Bulgaria (3), France (5), Greece (7), Poland (5), Spain (5)	Corn
Epilobium adenocaulon	PS II (Group 5), PS I (Group 22)	Belgium (5, 22), Poland (5), United Kingdom (5, 22)	No
E. tetragonum	PS II (Group 5)	France, Germany	Corn
Fallopian convolvulus	PS II (Group 5)	Austria, Germany	Corn
Galinsoga ciliata	PS II (Group 5)	Germany, Switzerland	Corn
Lolium multiflorum	ACCase (Group 1), ALS (Group 2)	France (1), Italy (1, 2), United Kingdom (1)	No
L. rigidum	ACCase (Group 1), PS II (Group 5), PS II (Group 7)	France (1), Greece (1), Spain (1, 5, 7)	No
Matricaria matricarioides	PS II (Group 5)	United Kingdom	No
M. perforata	Auxin (Group 4)	France, United Kingdom	No
Panicum dichotomiflorum	PS II (Group 5)	Spain	Corn
Papaver rhoeas	ALS (Group 2), Auxin (Group 4)	Greece (2), Italy (2), Spain (2, 4)	No
Poa annua	PS II (Group 5), Pigment (Group 11), PS I (Group 22)	Belgium (5, 11, 22), Czech Republic (5), France (5), Germany (5), Netherlands (5), Norway (5), United Kingdom (5, 22)	No

TABLE 4.3 (CONTINUED)
Reported Populations of Herbicide-Resistant Weeds in Western and Eastern Europe

Weed	Herbicide MOA[a]	Country and (Herbicide Group)	Reported in Maize or Soybean
Polygonum aviculare	PS II (Group 5), Pigment (Group 11)	Belgium (11), Netherlands (5)	Corn
P. hydropiper	PS II (Group 5)	France	Corn
P. lapathifolium	PS II (Group 5)	Czech Republic, France, Germany, Spain,	Corn
P. persicaria	PS II (Group 5)	Czech Republic, France	Corn
Scirpus mucronatus	ALS (Group 2)	Italy	No
Senecio vulgaris	PS II (Group 5), PS II (Group 7)	Belgium (5), Czech Republic (5), France (5), Germany (5), Netherlands (5), Norway (5), Switzerland (5, 7), United Kingdom (5)	Corn
Setaria faberi	PS II (Group 5)	Spain	Corn
S. glauca	PS II (Group 5)	France, Spain	Corn
S. verticillata	PS II (Group 5)	Spain	Corn
S. viridis	PS II (Group 5)	France, Spain	Corn
S. viridis var. major	PS II (Group 5)	France	Corn
Solanum nigrum	PS II (Group 5)	Belgium, France, Germany, Italy, Netherlands, Poland, Spain, Switzerland, United Kingdom	Corn
Sonchus asper	PS II (Group 5)	France	Corn
Stellaria media	ALS (Group 2), Auxin (Group 4), PS II (Group 5)	Denmark (2), Germany (5), Sweden (2), United Kingdom (4)	Corn

[a] Herbicide MOA = mechanism of action. Herbicide groups are described according to classifications by the Weed Science Society of America. (*Weed Technol.*, 11, 384, 1997).

Source: Adapted from Heap, I. M., International survey of herbicide-resistant weeds, http://www.weedscience.com, 2000; Gressel, J., Ammon, H. U., Fogelfors, H., Gasquex, J., Kay, Q. O. N., and Kees, H., Discovery and distribution of herbicide-resistant weeds outside North America, in *Herbicide Resistance in Plants*, LeBaron, H. M. and Gressel, J., Eds., John Wiley & Sons, New York, 1982, 31; Hartmann, F., Lanszki, I., Szentey, L., and Toth, A., Resistant weed biotypes in Hungary, in *Proc. 3rd Int. Weed Science Congr.*, 3, Legere, A., Ed., International Weed Science Society, Corvallis, OR, 2000, 138.

(4 biotypes each).[134] A significant number of weed species have immigrated into Europe. For example, 274 plant species from North and South America have immigrated and adapted to conditions in France.[135] The "weedy" status in the country of origin was the best predictor of weediness potential in the introduced country. Given that many of the genera and species introduced from the Americas have HR populations, it is probable that more and varied HR weed populations will eventually develop in Europe.[135]

Triazine herbicide resistance has been estimated to cover more than 2 million ha in Western Europe and 1 million ha of maize production in Eastern Europe, primarily in Hungary.[136] A number of triazine-resistant populations of weeds have evolved, including several populations of *Amaranthus retroflexus*[137–139] and *Erigeron canadensis*.[138,139] Other triazine-resistant weeds in Hungary are *A. blitoides*, *A. chlorostachs*, *A. bouchonii*, *Chenopodium album*, *C. polyspermum*, and *Ambrosia artemisifolia*.[139] Triazine-resistant weed populations are also reported in the former Czechoslovakia.[140]

Two levels of atrazine resistance have been reported in *C. album* populations in France, depending on the lethal dose response of the biotypes.[141] The mutants demonstrating intermediate and high levels of resistance, based on atrazine dose response and fluorescence assay, had the same serine to glycine mutation at amino acid position 264 in the D1 protein that is the target site for triazines. The source of the variable response to triazine was unclear; however, the resistant populations evolved after only two generations of selection.[141] This rapid occurrence of triazine resistance may account for the widespread distribution of HR weed populations. Triazine resistance was also reported in *Solanum nigrum* populations in the Netherlands.[142] *Senecio vulgaris* populations resistant to atrazine were reported in the Czech Republic.[143]

Other resistant weed populations include 2,4-D and MCPA resistance in *Cirsium arvense*, and urea and paraquat resistance in *Conyza canadensis*.[139] Glyphosate-resistant weed populations have not evolved even after 18 years of use in annual crops in Spain.[71] It should be noted that the use of glyphosate in the maize agro-ecosystems was a single preplant application. Further, given the concerns about transgenic crops in Europe, it is unlikely that glyphosate resistance will evolve in the foreseeable future. However, variable response to glyphosate was observed in *Sorghum halepense* populations found in Greece.[144] While variable responses were observed in selected callus cultures *in vitro*, similar differences were observed *in vivo* on donor plants.

4.3.3 China

Herbicide-resistant weed populations have evolved widely in China, but only anecdotal reports in maize or soybean agroecosystems are noted.[75] There is widespread resistance to butachlor and thiobencarb in *Echinochola crus-galli* in rice agroecosystems.[145] However, *E. crus-galli* is a common weed in maize and soybean in China[75] and has evolved resistance to atrazine elsewhere.[78] Use patterns of atrazine, simazine, and trifluralin suggest that some HR weed populations are likely in Heilongjiang and Jilin Provinces.[75] Interestingly, there are reports that the use rate of atrazine has roughly doubled in the last 30 years, suggesting that resistance evolution may be under way (W. Ahrens, personal communication, 2000). The apparent lack of resistant populations in maize and soybean may result from the prevalent use of hand weeding in these crops.

4.3.4 AFRICA

Three HR populations have been documented in South Africa: two weed populations resistant to ACCase inhibitors and one population resistant to triazine herbicides.[77] This report was current as of July 2000. A 1982 report indicated that no HR weed populations had been reported in South Africa.[78] Generally speaking, herbicide use in Africa is not widespread enough to facilitate the evolution of HR weed populations. Hand weeding is still the prevalent tactic for weed management. However, recent development of HR maize cultivars and the use of imidazolinone herbicides as seed treatments suggest that ALS-resistant *Striga* spp. populations could evolve in maize.[146]

Using the imidazolinone herbicides as seed coatings in imidazolinone-resistant maize has dramatically increased the harvest index for *Striga*-infested maize.[147,148] However, unless alternative management strategies are employed, HR populations are predicted to evolve within 6 years.[146] As in other developing countries, hand weeding is the principal tactic used for weed management; thus, it is unlikely that HR weed populations will develop in Africa in the near future even with increased use of herbicides. *Striga* spp., however, represents a unique situation. Minimal herbicide technology is required for the seed-coating tactic. Further, this important parasitic weed cannot be controlled effectively by hand weeding. Unless alternative tactics are developed, ALS-inhibitor herbicide resistance in *Striga* spp. is likely.

4.4 MANAGEMENT TACTICS FOR HERBICIDE RESISTANCE IN MAIZE AND SOYBEAN AGROECOSYSTEMS

There have been major changes in herbicide use in the last three decades. Amide, carbamate, and triazine herbicides captured 75% of the herbicide market in 1982.[27] Newer chemistry in the 1990s required 80 times less active ingredient than herbicides used in the 1970s. Herbicide classes that appeared promising two decades ago (e.g., nitrophenols) have been banned by the U.S. Environmental Protection Agency.[27] Importantly, although not universally accepted, there is a movement to lessen or eliminate herbicide use.[149] Weed science research, particularly as related to HR weed populations, must accommodate a complex arena of technical, social, economic, and political considerations.[150] Management tactics for HR weed populations must be implemented despite a general lack of genetic and ecological knowledge about the target organisms.[151] Herbicide-resistant weed populations are sufficiently problematic to demand concerted effort in research and education to preserve herbicides as effective tools for maize and soybean production.[8] Unfortunately, remedial methods that have been used are largely ineffective when HR populations have become widely dispersed.[152]

Strategies to control HR weeds and the integration of HR crops into maize and soybean agroecosystems were identified by agricultural chemical dealers as top research priorities.[89] Concerns about HR weed populations resulted in formation of an industry-led group, Herbicide Resistance Action Committee (HRAC), to promote tactics to manage HR weed populations.[153,154] However, some scientists suggest that due to the inevitability of HR weed populations and other environmental and economic factors associated with herbicide use, herbicides should be discarded in favor of nonchemical control techniques.[12] Less radical approaches are likely the norm, but it is critical to lessen the dependency on herbicides for weed management in maize and soybean to maintain their effectiveness.[29] It should be recognized that weeds are also capable of adaptation to alternative strategies.[155] Thus, effective management tactics for HR weed populations must be based on an understanding of plant physiology, genetics, and ecology.[44,80,156] Methods must be considered in regard to the overall agroecosystem. Current philosophy suggests that only through a diverse management program will HR weed populations be minimized. Integrated weed management (IWM) has become the primary message from weed scientists to agricultural practitioners.[62,157–159] Single tactics, intended to control every weed all of the time, consistently fail.[160] The agricultural chemical industry and national and local extension have not provided the information needed to delay the onset of HR weed populations.[84] Weed science must learn from the experiences of entomologists where resistance management is critically important and insecticide resources are diminishing.[161]

Perhaps IWM does not go far enough. All factors should be considered when developing weed management tactics including, for example, the impact of herbicide labeling and negative cross-resistance.[162,163] Further, methods to manage/control other pest complexes must be considered.[164] Possibly, plans should focus not on the actual control of HR weed populations, but rather on the general ecological principles that have resulted in the change of the weed community.[10] Understanding the dispersal of HR weeds within the entire agroecosystem is more important from a management perspective than developing single control tactics.[165] For alternative strategies to be successful and actually implemented by practitioners, the risks (economic, ecological, and environmental) associated with the tactics must be considered. Practitioners will not preemptively adopt complicated HR weed management programs, particularly if they are costly.[152]

Historically, the risk associated with HR weed populations has been ignored due to the availability of alternative herbicides that would control the weed problem.[166] Also, there has not been a mechanism to assess the risk of HR weed populations.[167] For example, the risk associated with IWM for HR weed populations must account for effects on yield, economic returns, time, and labor management.[43] If IWM tactics represent a longer-term approach to HR weed population management, practitioners will be hesitant to adopt the tactics unless there is an immediate economic return.[168] Further, agricultural practitioners question who should be responsible for the implementation of tactics other than herbicides.[31] A benefit/risk assessment of alternative tactics for HR weed populations should account for the potential of the tactic(s) to increase crop yield, the ecological implications of the HR weed population on the agroecosystem, the impact of the tactic(s) on the agroecosystem, and the availability of strategies less disruptive to the agroecosystem.[169] This section reviews the tactics used to manage HR weed populations in maize and soybean agroecosystems, briefly discusses alternative strategies, and assesses why alternative strategies are not likely to be adopted by practitioners.

4.4.1   MANAGEMENT TACTICS USED HISTORICALLY FOR HERBICIDE-RESISTANT WEED POPULATIONS

The goal of weed management tactics is to limit or eradicate weed populations locally from specific cropping fields.[18] The primary strategy for the last three decades has focused on using herbicides to the functional exclusion of all other tactics.[60] Maize and soybean producers either change herbicides or change crop cultivars and herbicides when an HR weed population develops. The use of new herbicides, while development has been slowed due to company mergers and increased development costs, is still viewed as a primary strategy for the management of HR weed populations.[156] Given that every approach to control weeds will open an ecological niche for other weed populations, it is no great surprise that HR weed populations have developed widely in maize and soybean agroecosystems.[170]

Efforts by private and public sector scientists to curtail the development of HR weed populations have been largely unsuccessful.[60] In part, the lack of success in North America is due to the fact that the problem of HR weeds is perceived to be relatively small and manageable.[104] Herbicides provide consistent weed control when compared to alternative tactics.[171] While pervasive evidence exists that alternative tactics provide more stability in the weed community dynamics, herbicides will continue to be the primary tactic for weed management in maize and soybean agroecosystems.[172] Growers who adopt alternative strategies for weed control incur a number of risks, including economic issues, time management, and relative efficacy, compared to the "traditional" use of herbicides.[169]

Thus, the response to the question about the tactics used historically to manage HR weed populations is that there have not been any proactive adjustments made to weed management programs. Herbicides have been, and are likely to remain, the primary tactic for weed control and the management of HR weed populations in the foreseeable future. In fact, one must question whether in maize and soybean agroecosystems, when compared to other crop agroecosystems, HR weed populations actually represent a serious threat to production.[94] Generally, control of HR weed populations is not a major issue in maize and soybean production except in specific fields. However, individual producers with HR weed populations will be economically affected by the problem. Nonetheless, producers do not adjust management tactics until HR weed populations develop, based on discussions with growers and evidence from field experiences with widely dispersed triazine- and ALS-resistant weed populations.

An informal survey conducted by the author with extension personnel across the United States maize belt reinforces the premise that HR weeds are not seen as serious issues for growers, despite the inevitability of their development. For example, *Ambrosia trifida* populations that have evolved resistance to ALS-inhibitor herbicides are so widely dispersed in Ohio that the utility of these herbicides is seriously threatened (M. Loux, personal communication, 2000). However, Ohio producers have done little to preserve these important herbicides. Obviously, there are other issues of greater economic importance. Further, the agricultural chemical industry continues to aggressively promote its proprietary products in the same manner despite the threat of HR weed populations and the ultimate loss of market due to HR weed populations.

The availability of HR maize and soybean cultivars has been proposed as an important tactic for the management of HR weed populations.[173] For example, the adoption of glyphosate-resistant soybean has been widespread, despite possible negative consequences of its overuse selecting for resistant weed populations.[174] This tactic is functionally no different than changing herbicides, which has proven to be only a short-term resolution to the development of HR weed populations.

Surveys demonstrate that growers are aware of the risks associated with the evolution of HR weed populations.[94,175-177] While triazine-resistant weed populations were first identified in Ontario, Canada in 1974, producers did not make major changes in weed management tactics and, as a result, more than 75% of maize fields have at least two triazine-resistant weed species.[178] Extension programs that emphasize the rotation of herbicides resulted in high levels of grower awareness of HR weed populations, but only limited "trial" adoption of remedial tactics.[179]

Management plans and weed identification publications are widely distributed in the U.S. maize belt (e.g.,[128,180]). However, HR weed populations continue to develop. Agricultural scientists must recognize that historic approaches to HR weed population management ignore the ecological implications of the problem and thus perpetuate the selection of HR weed populations.[80] At best, present methods only maintain the frequency of the resistant gene(s) in the weed population but do nothing to reverse the evolution of resistance.[30] Major changes in maize and soybean production practices are necessary to effectively manage HR weed populations. However, the question is whether these changes will be implemented. Perhaps HR weed populations in maize and soybean agroecosystems are not economically or ecologically serious enough to induce major changes in weed management tactics.

4.4.2 ALTERNATIVE MANAGEMENT TACTICS FOR HERBICIDE-RESISTANT WEED POPULATIONS

There is general agreement that single, "simple" approaches to weed management, including HR weed populations, have been largely unsuccessful from an ecological perspective.[136,160] When diversity is reduced and agricultural practices are dependent on herbicides to manage weeds, weed populations shift to more tolerant species and HR weed populations evolve. Contrarily, it can be argued that herbicides represent the most successful tactic to economically manage a pest complex.[28,39] The issue of weed management becomes a question of "value." Is the "value" of alternative strategies greater than the "value" of traditional tactics economically, environmentally, and ecologically? Management of HR weed populations requires more than herbicides.[84] Alternative IWM tactics must be functionally integrated such that each tactic complements other tactics, thus increasing the negative impact on HR weed populations.[181] While the focus of this discussion is on HR weed populations, it should be noted that IWM concepts are also applicable to prevent weed population shifts, in general.[182]

Using alternative weed management strategies will help minimize input costs for weed control and may identify unnecessary costs.[183] IWM will promote on-farm biological diversity and thus create an agroecosystem less prone to catastrophic changes in pest populations such as the development of HR weeds.[184] Regardless, the key to HR weed population management is to use multiple strategies that enhance the competitiveness of the crop against the weed population.[13] Cultural and mechanical practices will be most effective when combined with the judicious use of herbicides in an integrated system.[185] However, to be adopted, IWM tactics must be economically sustainable.[159,186] Diversity is the key to successful IWM. Tactics may include, but are not limited to, HR crops,[187] weed seed removal or exclusion from the field,[48,186,188] site-specific management,[189] the use of green manures,[149] or more traditional cultural techniques.[190] Possible alternative strategies will be described in the following sections. Together, these tactics can be used to develop an IWM system to manage HR weed populations in maize and soybean agroecosystems.

4.4.2.1   Herbicide Tactics to Manage Herbicide-Resistant Weed Populations

Herbicides continue to be the first, and perhaps most widely utilized, means to manage HR weed populations.[171] Even where mandates exist that require the reduction of pesticides, switching from high-dose to low-dose herbicides or the removal of inactive isomers will ensure the continued importance of herbicides for weed management.[191] This is particularly true for maize and soybean where there are many more herbicides available. Understanding the herbicide mechanism of action and rotating to a different, albeit efficacious, herbicide with a different mechanism of action is critically important to manage an HR weed population.[186,192–194] Alternative tactics based on herbicides to manage HR weeds may include the use of innovative herbicide mixtures,[29] herbicide rotations,[195] alternative application techniques,[196] different herbicide application timings,[197] and the use of herbicides that alter weed seed viability.[198] The temporary abandonment of specific herbicides has been suggested as a strategy but has not proven to be worthwhile due to weed refugia, seedbanks, and seed dormancy.[152] Importantly, while HR weed populations have evolved in response to herbicide use, no herbicide has actually been lost to maize and soybean agroecosystems due to resistance.[153] It is also important to recognize that while the management options described in this section are based on innovative herbicide use, the greatest impact on HR weed populations will be accrued when the herbicide(s) is combined with other alternative strategies.[195]

4.4.2.1.1   Utilization of herbicide tank mixtures
The basis for using herbicide tank mixtures to manage HR weed populations focuses on the need to include herbicides with different mechanisms of action that similarly control the target weed and have similar persistence in the agroecosystem.[152,199] However, this option theoretically requires that growers purchase two or more herbicides that essentially control the same weed. Typically monocotyledonous and

dicotyledonous weeds will infest a field at economically important levels. Consequently, herbicide tank mixtures are designed to improve the spectrum of weed control rather than diminish herbicide selection pressure. For example, glyphosate was a necessary addition in tank mixtures with bentazon or thifensulfuron to improve control of *Abutilon theophrasti* and *Chenopodium album*.[200] However, neither bentazon nor thifensulfuron has activity on *Amaranthus rudis* or annual grass weeds. Thus, if resistance to glyphosate were a concern in these species, the herbicide tank mixture of either bentazon or thifensulfuron with glyphosate would not meet the criteria for HR weed population management.[199]

4.4.2.1.2 Alteration of herbicide application timing

Herbicide application timing can be adjusted to improve the management of HR weed populations from several perspectives. Fundamentally, herbicides should be applied when the greatest exposure to the target is likely. *Amaranthus rudis* tends to germinate and emerge considerably later than most annual broadleaf weeds in the midwestern United States.[201] Resistant populations of this weed have evolved and to effectively reduce the population, herbicide application timing should be later in the growing season than is typical.[98] Another consideration is that weeds are more sensitive to herbicides during the early stages of development.[13] An effective tactic would be to target herbicide applications during the most sensitive stage of HR weed development. For example, control of *Xanthium strumarium* was better when herbicides were applied postemergence compared to soil applications.[202]

4.4.2.1.3 Alteration of herbicide application rates

Using reduced rates of efficacious herbicides along with other management tactics to effectively manage HR weed populations may be a viable management option. However, the description of a reduced herbicide rate is somewhat misleading. Herbicide rates should be described relative to the minimum lethal dose rather than what is printed on the product label.[160] Further, reduced rates may provide economic control of the target weed population, but the consistency of activity may be somewhat variable. Thus, when reduced rates of herbicides are used, typically other strategies such as cultivation and rotary hoeing[203–205] or narrow rows[206] should be included. In each of these examples cited, control of weeds was similar to that provided by the full rate of herbicide. The importance for the management of HR weed populations is likely more a factor of multiple tactics rather than the reduction of herbicide rate. Also, the consistency of reduced herbicide rates is influenced more by environmental conditions than higher herbicide rates.[13] Alternatively, increased herbicide rates may be used to manage difficult weed populations but may increase the selection for individuals within the weed population that express resistance.[85]

4.4.2.1.4 Alternative herbicide use strategies

Another way to reduce the selection pressure that herbicides place on weed populations is to minimize the area of the field that is treated. This may provide weed "refugia" within the field, similar to the primary tactic used to manage insecticide resistance, and thus maintain susceptibility within the field with minimal impact on yield. Perhaps the simplest alternative use pattern for herbicides is banding over the crop row. Banding herbicides reduces the amount of active ingredient per hectare without reducing the effective herbicide application rate.[207] Banded herbicides generally provide acceptable weed control while reducing cost.[208] Because only a band over the crop row is treated, banding must be combined with mechanical cultivation for broad-spectrum weed management. Banding may be an excellent option for the management of HR weed populations.

The application of herbicides as a seed dressing also has weed management utility. While only a small region immediately surrounding the germinating crop seed is weed free, interference by weeds is sufficiently delayed to allow the application of other tactics such as cultivation or postemergence herbicide applications. Because only a small area receives the initial herbicide, selection for HR weed populations is low and the evolution of HR weed populations should be delayed. This application strategy has great potential for the management of parasitic weeds in developing countries.[148] The technique has limitations, as the herbicide concentration near the crop seed is quite high, and unless HR crops are used, phytotoxicity is a concern (M. Owen, unpublished data, 1999).

Site-specific application of herbicides is an excellent alternative use tactic for the management of HR weed populations. As weeds occur in aggregated spatial patterns, treating only those patches that have weeds will significantly reduce the amount of herbicide applied to a field.[209] This could reduce the spread of HR weed populations while also reducing overall selection pressure from herbicides. By limiting the treated area in the field, there is a better chance that herbicide-sensitive weed genotypes will continue to exist in the field, thus serving to dilute the selection process for HR weed populations. The long-term benefits of patch spraying will depend on the initial distribution of the target weed population.[210] Field-scale demonstrations of patch herbicide applications have been successful and the economic savings can be considerable.[211]

Other alternative herbicide application strategies for the management of HR weed populations include the use of synergists either to enhance herbicide efficacy or to lessen herbicide metabolism in the target weed population.[212] Negative cross-resistance has been identified in some HR weed populations and may represent an opportunity to improve the management of HR weed populations. The inclusion of alternative herbicides such as pyridate or bentazon in a weed management program can be an effective tactic to control triazine-resistant weed populations.[142] Other possibilities for alternative herbicides that may be beneficial in managing HR weed populations are plant-derived natural products or phytotoxic chemicals from microbes.[213,214] While these alternative herbicides exist, there are no current examples of compounds that effectively control HR weed populations.

4.4.2.2   Herbicide-Resistant Traits in Maize and Soybean as a Tactic to Manage Herbicide-Resistant Weed Populations

The development and adoption of HR crop cultivars represents a major scientific achievement but also has become one of the most contentious issues facing world agriculture. Despite considerable investment of private and governmental funds, significant questions concerning genetically modified crops exist, which science has not answered to the satisfaction of the general world community.[215] While the industry believes that sufficient oversight and regulation exist within governments to safely implement plant biotechnological products, questions about the risks associated with transgenic plants continue to be voiced.[216,217] The general concept of HR crops suggests that adoption of this technology will dramatically improve weed management and provided flexibility to crop production systems.[218] Specifically, HR crops will provide options for weed management in minor crops, allow the use of environmentally benign herbicides, improve the economics of crop production, and serve as an effective tactic for control of HR weed populations.[219] Importantly, given the difficulty and expense of developing new herbicides, the ability to alter the crop genome to allow the application of existing effective herbicides without fear of phytotoxicity promises to increase profitability of agriculture and improve weed management options for growers.[220]

In maize and soybean, resistance to imidazolinones, sulfonylureas, glufosinate, and glyphosate are most important.[61,187] Transgenic maize systems using glufosinate were successful, but only when multiple applications, residual herbicides, or mechanical control were included.[221] While numerous successful examples for transgenic crop systems can be cited, the HR crop systems are not infallible. Another concern of the seed industry is the time required for inclusion of the HR trait into the crop genome and the limited life of maize and soybean cultivars.[222]

A primary issue for HR crops is the potential movement of the HR trait to weedy plants.[218] There is little likelihood of this occurring with HR maize and soybean, given that there are no closely related weeds for these crops.[173] Another concern with HR crops is the discovery of herbicide residues in the HR crop grain.[223] Glyphosate is not metabolized in HR maize and soybean. Topical applications during soybean flowering resulted in 0.1 to 3.3 ppm glyphosate in soybean grain. Glyphosate may also affect phytoalexin synthesis.[224] Phytoalexins (e.g., glyceollin) are important secondary metabolite products of the shikimate pathway that serve to protect the plant from diseases.

Finally, there are examples of volunteer crops having a different phenotype compared to the original crop. These volunteers could become a serious weed problem,[228] but this is highly unlikely for HR maize and soybean. In fact, the use of glyphosate-resistant soybean and glyphosate may actually enhance the microenvironment in the field. Greater diversity of beneficial insects, Carabidae (ground beetles) and Gryllidae (field crickets), were found in HR soybean treated with glyphosate compared to conventional weed management practices.[226] Collembola populations and diversity also were greater in HR soybean. The delayed application of glyphosate allowed weeds to grow longer and act as food sources and refugia for these insect populations. Generally, HR maize and soybean represent a positive tactic for the management of HR weed populations; however, there are negative factors that must be considered.

4.4.2.3 Tillage and Mechanical Management Tactics
to Manage Herbicide-Resistant Weed Populations

Tillage, whether primary techniques used for soil inversion or secondary techniques such as cultivation, is mainly a weed management tactic.[227] With regard to HR weed population management, mechanical tactics are the most important considerations for the reduction of herbicides when compared to other options included in IWM systems.[228] Mechanical cultivation can lessen herbicide selection pressure by replacing herbicide applications, and thus deter the evolution of HR weed populations.[62] However, tillage systems that result in high levels of plant residue on the surface may lessen the effectiveness of secondary mechanical strategies.[41] It is important to recognize that weed populations can also adapt (become "resistant") to tillage.[229]

The primary tillage system has a major impact on weed population dynamics. *Setaria viridis* populations were highest in chisel plow and no-tillage systems compared to conventional tillage systems.[230] Similarly, *Chenopodium album* and *Amaranthus retroflexus* populations were also higher in reduced tillage systems. These species have numerous HR populations, so choice of tillage system could predispose a field to the evolution of resistance to herbicides. Wind-dispersed weed species are also associated with reduced tillage systems.[231] Seeds that are dispersed by wind can move considerable distances and rapidly infest fields. As *Conyza canadensis* has evolved HR populations, resistant populations may spread rapidly in new fields.

Tillage affects the distribution of weed seeds in the soil, and thus the control by herbicides.[232] Reduced tillage systems result in 85% of weed seeds in the upper 5 cm of soil, while more aggressive tillage systems evenly distribute the seeds throughout the depth of tillage. Tillage also impacts the accessibility of weed seeds to soil-applied herbicides. More weed seeds were found in the unaggregated fraction of soil in reduced tillage systems and thus were more likely to come into contact with herbicides.[232] Increased herbicide efficacy on weed species increases selection for HR populations.

The primary tillage system affects the weed management tactics available to growers as certain herbicides or application techniques may be precluded in some tillage systems.[233] The less tillage that is included in the cropping system, the greater the importance of herbicides for weed management and the greater the potential evolution of HR weed populations. The combination of herbicides and mechanical strategies provided more effective and consistent weed management than either tactic alone.[234–236] Furthermore, selection pressure on the weed population for the evolution of HR populations is reduced. Better still is the inclusion of multiple mechanical techniques such as cultivation and rotary hoeing.[43,237] However, reliance on mechanical tactics alone will result in variable weed management causing relatively high economic risk.[237]

Integration of mechanical strategies with reduced herbicide rates improves weed management and lessens the selection pressure for HR weed populations. The key to the optimization of mechanical tactics relates to the timeliness of the operation.[238,239] Timeliness relates not only to the size of weed and crop, but also to soil conditions.[185] Effective management of HR weed populations with tillage and mechanical techniques requires better management skills, particularly where reduced tillage systems are employed. Regardless, the inclusion of mechanical strategies will improve HR weed population management.

4.4.2.4 Cultural Tactics to Manage Herbicide-Resistant Weed Populations

The key to better weed management is greater diversification of the agroecosystem and more innovative strategies.[240] While some cultural tactics are integrated in all maize and soybean production systems, many producers are not consciously aware that many of the agronomic practices are actually weed management options.[49] Understanding the cropping system and optimizing cultural tactics will greatly enhance the management of HR weed populations.[171,227] Further, adopting cultural practices that negatively affect man-caused weed seed movement will reduce the likelihood of the introduction of HR weeds into a field.[241] Perhaps there are cultural strategies that can inhibit the dispersion of HR traits via pollen movement. Cultural weed management can reduce the importance of exotic weeds in an agroecosystem.[242] Examples of cultural strategies include manipulating crop density, crop diversity, crop cultivar, planting density, planting configuration, and fertilization strategy.[54] However, for the greatest and most consistent impact on HR weed populations, multiple cultural tactics must be employed.

4.4.2.4.1 Crop rotations

The concept of using crop rotations to manage HR weed populations is extremely simple; diversify the crop sequence and spread the risk of economic loss in any one crop.[26,171] Diverse crop rotation creates an unfavorable environment for specific weeds and thus delays or deters the adaptation of that population to the agroecosystem.[64] Crop rotation is a preventative tactic that reduces the weed seedbank size and diversity.[181] However, while research has clearly demonstrated a positive "rotation effect" on crop production, and has offered possible reasons for the effect such as suppression of pest, the observations have never been satisfactorily explained.[243,244]

In regard to the management of HR weed populations, crop rotation is generally considered more from the opportunity to use alternative herbicides than the ecological perspective.[54,62] However, the longer-term benefits of crop rotation as a tactic to manage HR weed populations should focus on the ecological principles.[171,245] Rotation systems in the Netherlands include different crops to provide opportunities to improve soil fertility, crop health, and weed management without increasing the needs for external inputs.[160]

Senna obtusifolia populations were reduced due to the integration of crop rotation and herbicides.[246] A diverse rotation tactic that included *Arachis hypogae*, *Mucuna aterrima*, and maize consistently reduced the weed seedbank compared to simple rotation systems.[247] The weed community diversity increased with a maize/soybean rotation, compared to a continuous maize rotation, and weed management opportunities improved.[248]

The key issues in crop rotation as a management tactic for HR weed populations are economics and weed population response time. Typically, weed population response requires several cycles of the rotation scheme, while growers need immediate control, from an economic perspective. However, the benefits of crop rotation for the management of HR weed populations are significant, particularly when herbicide use is integrated into the tactic.

4.4.2.4.2 Adjustment of seeding time

Using seeding time as a cultural tactic for the management of HR weeds has considerable benefits in some agroecosystems.[186] Adjusting crop seeding time has been widely utilized for the management of many pest populations.[22] However, the effect of seeding time on reducing HR weed populations requires the integration of other cultural and herbicide tactics. Including spring grains in a crop rotation strategy can seriously disrupt weed growth and development.[54]

Sustainable growers in Iowa delayed planting 1 week or more in maize and soybean agroecosystems.[249] While there was some crop yield penalty attributable, in part, to the delayed seeding, input costs for weed control were lower. Thus, net profit per hectare was similar for the sustainable grower compared to the conventional farmer.

The potential yield penalty due to delayed planting depends on the crop, the variety, and the environmental conditions.[250–252] Soybean does not demonstrate a significant loss of yield potential when planting is delayed over a long time span.[250] Relative yields were similar for soybean planted in Iowa during late April through May. Yields declined significantly when soybean was planted in early June. However, maize planted after early May in Illinois tended to lose yield rapidly.[252]

Weed management was improved by integrating tillage into the system after weed germination occurred. Tillage effectively controlled more weeds in the sustainable system than was experienced in the conventional system that was seeded, and thus tilled, earlier and before weed germination had occurred. Altered seeding time, particularly delayed seeding, was an effective tactic in soybean; however, in maize, there was a potential for reduced yields. Thus, the best manner in which to use seeding time as a cultural tactic for HR weed management is to integrate a more diverse crop rotation system that includes crops with different seeding times compared to maize and soybean.

4.4.2.4.3 Adjustment of seeding rate

Managing HR weeds by adjusting the crop seeding rate is a cultural tactic that increases the competitive ability of the crop at the expense of the weed population.[186] Increasing crop competitive ability may reduce the selection pressure imposed by the herbicide on the agroecosystem and may also reduce weed seed production.[62] Increasing soybean seeding rate to approximately 1 million seeds ha^{-1} reduced weed interference and resulted in higher soybean yields.[253] When maize seeding rate was doubled, weed control for herbicide treatments applied at 25% reduced rates was similar for standard rates.[254] Greater maize populations reduced *Setaria* spp. growth and seed production.[255] The effect of adjusting seeding rate on HR weed populations is attributable to shading.[62] *Abutilon theophrasti* seed production and dormancy were decreased when shading was increased from 30 to 76%.[256] Thus adjusting crop seeding rate may diminish HR weed populations and also negatively affect weed seed quality.

However, the increased cost of seed should be evaluated with regard to benefits of weed management. For example, full soybean yields were demonstrated at plant populations from approximately 173,000 to 346,000 plants ha^{-1}.[250] This is an increase in seed cost of approximately \$40 ha^{-1} for a glyphosate-resistant soybean variety (K. Whigham, personal communication, 2000). The estimated cost for an application of glyphosate at 2.34 l ha^{-1}, assuming a cost of \$10.60 l^{-1} for the herbicide and \$12.35 ha^{-1}, is approximately \$37 ha^{-1}. Increasing maize seeding rate from 74,100 seeds ha^{-1} to 86,450 seeds ha^{-1} will add approximately \$22 ha^{-1} in seed costs (D. Farnham, personal communication, 2000). However, it is difficult to economically quantify the resultant improvement in weed control attributable to the increased seeding rate.

4.4.2.4.4   Alternative planting configuration

Changing crop planting configuration by narrowing interrow spacing improves the competitive ability of the crop.[62] The effect on HR weed population is similar to that caused by increasing crop seeding rates. The crop reduces light availability to weeds, thus improving control.[228,254,257] Further, improved competitive ability for crops allows the reduction of herbicide rate, and thus selection pressure on the weed population.[206,257] In Argentina, soybean planted in narrow rows reduced *Avoda cristata* biomass and seed production when reduced rates of glyphosate were applied.[258]

Seed production by *Solanum ptycanthum* was dramatically reduced when soybean was planted in 18-cm rows compared to 76-cm rows.[259] Narrow-row (19-cm) soybean reduced weed biomass 65 and 78% when compared to 57- and 95-cm row spacing.[260] The soybean canopy closed much earlier in the 19-cm rows compared to the 95-cm rows where canopy closure occurred 12 weeks after planting. Importantly, 19-cm rows demonstrated high levels of weed control even when herbicide inputs were reduced.

However, sustainable growers in Iowa typically planted maize and soybean in 96-cm rows compared to 76-cm rows for growers who used herbicides as the primary tactic for weed management.[249] Wider row spacing was necessary for the mechanical tactics used for weed management. Also, predation of *Amaranthus* spp. and *Chenopodium album* was greater in wider soybean rows.[261]

4.4.2.4.5   Crop cultivars

Conceptually, the choice of crop cultivar may have an important role in the management of HR weed populations.[186] Improved competitive ability for crops could also reduce dependence on herbicides for weed management further aiding in HR weed management.[262] Differences in crop genetics resulting in improved competitive ability against weeds could be one of the criteria used for crop breeding programs. Using a highly competitive crop cultivar, when integrated with other cultural tactics, would result in a robust HR weed management system. Yield potential, rather than competitive ability, typically under ideal growth conditions, is a major criterion for cultivar selection. Improved crop cultivars with high yield potential may be less competitive with weeds than older cultivars.[35] Maize yield losses attributable to *Abutilon theophrasti* competition for light were lower for older hybrids compared with newer hybrids.[262] However, *A. theophrasti* seed production was similar for all hybrids. Overall maize yield potential was higher for newer hybrids. Thus, the economic considerations of cultivar selection relative to weed management must be considered carefully. The cultivar traits of leaf area index and overall height correlated with high crop yield and weed suppression and could be useful indicators for the selection of crop cultivar with regard to the management of HR weed populations.

4.4.2.4.6   Cover crops, mulches, and intercrop systems

The inclusion of plants other than the crop represents a potentially important tactic for the management of HR weed populations. This tactic includes cover crops, mulches, living mulches, smother crops, and intercropping systems.[54] The goal of this tactic is to replace an unmanageable weed population with a manageable plant population.[263] Further, the inherent weed suppression that results from this tactic allows the reduction of herbicide use. However, the elimination of herbicide use is not realistic solely from the use of these alternative plantings. Typically, the suppression of weed populations is not high or long enough and thus this tactic is not widely used.[264] Residues from cover crops, mulches, living mulches, and smother crops may also inhibit crop emergence.[265]

The residues from cover crops, mulches, and living mulches physically suppress weed emergence.[263] The most important consideration concerning the suppression was the area that the plant residue covered.[266] Weed seed size was inversely correlated with the suppressive ability of the cover crop residues. Cover crops and living mulches reduced the weed seedbank 44 and 50%, respectively, in Nigerian agroecosystems.[267] Another aspect of cover crops is the potential for allelochemicals (i.e., chemicals released from cover crop and mulch residues that inhibit weed germination and development).[245] Legumes (e.g., *Trifolium incarnatum*) and nonleguminous species (e.g., *Secale cereale* and *Brassica napus*) are reported to have allelochemicals that improve weed management.[268,269]

Intercropping represents a spatially diverse agroecosystem where multiple crops are planted together as a way to increase crop production per unit land area.[54] Intercropping systems are widely utilized in Latin America, Asia, and Africa. As the component crops for intercrop systems are selected more from the perspective of profit rather than weed suppressive ability, their effectiveness for managing HR weeds will vary, depending on the specific intercrop system.

4.4.2.4.7   Management of weed seedbanks

The management of HR weed populations ultimately focuses on the management of the weed seedbank. Several general tactics can be considered for weed seedbank management. Obviously, the best strategy is to keep HR weed populations from ever developing, thus eliminating the seedbank. Removing the weed seeds during harvest, thus keeping them from returning to the field, is an innovative technique used by growers in specific agroecosystems.[186] This concept may have applicability in maize and soybean agroecosystems.

Unfortunately, HR weed populations are typically discovered after the seedbank contains high levels of resistant seed. Once established, given seed dormancy and the relatively long life of weed seedbanks, it is important to keep the HR seedbank small through manipulation of the agroecosystem. If the weed seedbank maintains the susceptible genotype at a relatively high level, the evolution of the HR weed population will be significantly delayed.[155] However, if even a small number of HR weeds reproduce, the HR seedbank can increase rapidly.[270] When an economic threshold level of *Abutilon theophrasti* was allowed to reproduce, the seedbank increased 24-fold the next year.

Management of the weed seedbank integrates many techniques. Tillage is a primary agronomic technique that significantly influences the weed seedbank, both on the physical placement of weed seeds and on the edaphic conditions that affect weed seed germination.[233] Tillage changes the soil–seed microsite characteristics and thus affects germination, which may account for the different responses of weed species to tillage practices.[271] Primary tillage by moldboard plowing caused a rapid decline in the weed seedbank.[272] Weed seed placement is relatively uniform under an aggressive tillage regime; however, weed seeds are concentrated near the soil surface under reduced tillage.[273] *Setaria* spp. was more affected by tillage than *Chenopodium album* and *Amaranthus retroflexus*. Large-seeded annual broadleaf weeds such as *Xanthium strumarium* and *Helianthus annuus* are not well adapted to reduced tillage systems and populations diminish under this type of tillage.

Tillage also impacts the likelihood that seeds will be infected with fungi, thus affecting the active weed seedbank. *S. faberi* and *S. viridis* demonstrated greater fungal colonization under reduced tillage regimes and seeds that were located nearer the soil surface were more likely to be infected.[274] Thus, as a greater number of weed seed will be located near the soil surface in reduced tillage, and a greater percentage of seeds near the soil surface will be attacked or killed by fungi, the weed seedbank may decline rather quickly. Herbicides and tillage combined with autumn fallow were effective in reducing *S. viridis* and *Digitaria sanguinalis* seedbanks in maize agroecosystems.[275] An important question is whether the HR weed seedbank responds differently to tactics other than herbicide when compared to the susceptible seedbank. While different species may respond differently, evidence suggests that HR does not impact the behavior of the weed seedbank.[276]

4.4.2.4.8 Adjustment of nutrient use

Subtle changes in nutrient use may impact the relationship between crops and weeds. Increased fertility tends to favor the weed community and often losses of crop yield from pest infestations increase as nutrients are added.[22] Soybean yield loss attributable to weed interference was greater when nitrogen fertilizer was used in the previous maize crop.[277] However, other examples show that nitrogen fertilization increased the competitiveness of crops and caused a decline in weed population density.[278]

Data indicate that manipulation of fertility affects the temporal aspect of weed interference. It may be possible to adjust the timing of nutrient availability in order to shift the competitive advantage for nutrient capture to crops.[54] Nitrogen is the primary nutrient that can be adjusted. The critical period of weed management in maize was affected by nitrogen fertilization.[279] Lower levels of nitrogen resulted in the weed community interfering with maize sooner and lengthened the period of weed control necessary to protect the yield. Nitrogen form may also impact the timing of nitrogen availability. Nitrogen from legume residues will become available more slowly than from synthetic sources. Also, delayed release formulations of nitrogen could also be used. The benefits of nutrient manipulation with regard to HR weed population management may be greater than the risks of lowering crop yields. However, this trade-off must still be investigated.

4.4.2.4.9   Biological control strategies

Perhaps no other weed management tactic has been touted as much as the potential for using biological strategies to control weeds. HR weed populations could be key targets for the development of biological weed management tactics. There are many examples of using seed predators and pathogens to destroy the weed seedbank.[280,281] Other examples of introducing exotic pathogens or predators have resulted in the dramatically successful control of *Chondrilla juncea, Cardus nutans,* and *Senecio jacobaea.*[282] More than 100 microorganisms have been identified as potential bio-control agents.[283]

Weed seed predation is a promising tactic for biological weed management. Seeds of *Ambrosia artemisiifolia, Amaranthus retroflexus, Cassia obtusifolia,* and *Datura stramonium* were predated by Carabidae (ground beetles), Gryllidae (field crickets), and Formicidae (ants) 2.3 times greater in no-tillage soybean agroecosystems, when compared to conventional tillage agroecosystems.[284] Weeds are allowed to grow longer in some transgenic soybean varieties, and predator populations were found to be greater in transgenic soybean systems than in conventional herbicide treatments.[226] The difference in predator populations was likely attributable to changes in the microhabitat facilitated by the delayed application of the herbicide.

The use of biotechnology to alter the genomes of potential mycoherbicides may increase the successes of biological control in the future.[283] A number of agricultural chemical companies indicate that the commercial and ecological correctness of mycoherbicides make them suitable targets of developmental efforts for use in agriculture.[285]

There are several important possibilities for the use of plant pathogens as mycoherbicides. Herbicides can predispose target weeds to secondary infection and ultimate control by a pathogen.[286] Imazaquin reduced the dry weight of *C. obtusifolia* but did not kill the plants.[287] *Alternaria cassiae* provided only 58% control while the combination of herbicide and mycoherbicide controlled 96% of the weeds. Other possibilities for biological weed control are the use of fungal combinations or arthropods plus pathogenic fungi.[285] Obviously, given the limited spectrum of biological control, the management of HR weed populations will require the optimization of control agents.

Another possible opportunity for the biological control of HR weed populations exists if the basis or existence of weed-suppressive soils can be identified and understood. Changes in crop management can alter the soil microbial community.[288] Thus, it is conceptually possible to manipulate the agroecosystems such that a diverse microbial population develops that can suppress the weed community.[284] Weed-suppressive soils will likely require the integration of herbicide tactics to meet the objectives of weed management in maize and soybean agroecosystems, particularly for HR weed populations. Fungi have been identified as synergists for glyphosate activity.[289] If the interactions between herbicides and soil microorganisms can be identified, reduced herbicide rates may be more consistently effective, thus reducing the selection pressure on HR weeds.

4.4.2.5 The Use of Models to Manage Herbicide-Resistant Weed Populations

Predictive models for weed management may provide economic and environmentally acceptable recommendations for appropriate weed control.[19,183] Models have been developed to predict the effects of tillage on weed emergence,[290] economic weed thresholds,[291] the evolution of weed resistance,[81,156] and the impact of herbicide rotations and tank mixtures on HR weed populations.[112,195] Models could be a key component of HR weed management.[80] However, models are typically based on an economic threshhold (i.e., level of weed infestation that is of economic importance).[292] Factors that affect the economic threshold include diversity of weed community, weed population density, environmental conditions, cost of control, and value of crop.

However, one of the key considerations is the impact of economic thresholds on future weed populations. Thus, an economic optimum threshold is a better indicator than a strict economic threshold because it theoretically accounts for future weed population changes and management practices to ensure long-term profitability.[293] When economic optimum thresholds are compared to economic thresholds, typically the acceptable population of weeds declines significantly.[294] The economic optimum thresholds for *Abutilon theophrasti* and *Helianthus annuus* were predicted to be 7.5- and 3.6-fold lower, respectively, than the economic threshold. Nevertheless, variability attributable to location and environment suggests that the concept of economic optimum threshold is of questionable value.[270] Thus a working definition of thresholds is extremely difficult.[12]

4.4.3 WHY HISTORIC TACTICS DO NOT WORK AND ALTERNATIVE STRATEGIES ARE NOT ACCEPTABLE FOR THE MANAGEMENT OF HERBICIDE-RESISTANT WEED POPULATIONS

Producers make weed management decisions based on an approach of risk avoidance.[295] When decisions incur some level of risk (i.e., the development of HR weed populations), producers will accept the risk only when increases in economic return are reasonably assured in a relatively short time period. Herbicides have provided agriculture with an effective, economic, and consistent technique to control the most important and widely dispersed pest complex.[39] Importantly, the use of herbicides has increased the level of crop production efficiency, allowing fewer farmers to produce maize and soybean on more hectares. The need for manual labor in maize and soybean production has generally been minimized because of the widespread use of herbicides.[39] Given the general success of herbicide use, the dependence on these important tools will continue until nonchemical weed tactics are developed such that their use does not incur increased risk.[39] An assessment of the benefits and risks of HR weed management tactics is listed in Table 4.4.

TABLE 4.4
Assessment of Benefits and Risks of Alternative Tactics for Herbicide-Resistant Weed Management

Tactic	Benefits	Risks	Likelihood of Adoption and Impact on HR Weed Management
Herbicide MOA rotation	Reduction of selection pressure, control of HR weed populations	Unavailability of different MOA, phytotoxicity, increased cost, inappropriate weed spectrum	Excellent
Herbicide tank mixtures	Reduction of selection pressure, improved weed spectrum	Poor efficacy on HR species, increased cost, phytotoxicity	Excellent
Herbicide application timing	Improved efficacy on HR species, efficient use of herbicide	Lack of herbicide residual activity, post-application may result in yield loss if not timely, increased number of applications	Good to excellent
Herbicide application rates	Better control of target species	Increased selection pressure (higher rates), selection for non-target site resistance (lower rates), inconsistent efficacy (lower rates)	Poor to fair
Herbicide banding	Reduced cost, consistent efficacy, reduced selection pressure	Need for mechanical control, specialized equipment for application, increased application time	Poor
Herbicide seed coating	Decreased selection pressure, reduced costs	Crop phytotoxicity, minimal area of influence, inconsistent efficacy, requirement for supplemental tactics, lack of supportive research	Poor
Site-specific herbicide application	Decreased cost of herbicide, reduced selection pressure	Increased cost of application, unavailability of consistent weed population maps, poor understanding of weed seedbank dynamics, increased variability of efficacy	Poor
Herbicide synergists and alternative products	Improved efficacy, reduced herbicide amount, new MOA	No research base, inconsistent efficacy, lack of available products	Poor

Strategy	Advantages	Disadvantages	
Herbicide-resistant crops	No crop phytotoxicity, different MOA, reduced herbicide amount, application timing less of a factor	Increased cost of seed, need for more applications per season, increased selection pressure from the MOA used, poor agronomic characteristics of the HR crop, movement of HR trait into weeds, volunteer HR crops as weeds	Excellent
Primary tillage	Decreased selection pressure, excellent and consistent efficacy, depletion of weed seedbank	Increased time requirement, increased soil erosion, increased cost, requires additional tactics	Excellent
Mechanical strategies	Decreases selection pressure, consistent efficacy, inexpensive	Increased time requirement, high level of management skill needed, requires additional tactics, potential for crop injury	Fair to poor
Crop rotation	Changes agroecosystem, allows different herbicide tactics (MOA etc.) may facilitate other alternative strategies	Economic risk of alternative rotation crop, lack of adapted rotation crop, rotation crop not dissimilar and thus minimal impact on weed community, requirement for herbicides	Good to fair
Adjustment of seeding time	Improved efficacy on target weeds, reduction of selection pressure	Requires alternative strategies (primary tillage or herbicide application), potential yield loss, need for increased rotation diversity	Fair to poor
Adjustment of seeding rate	Reduced selection pressure, improved competitive ability for the crop	Increased seed cost, potentially increased pest problems, increased intraspecific competition/reduced yields	Fair
Alternative planting configuration	Improved competitive ability for the crop, reduced selection pressure	Unavailability of mechanical strategies, emphasis on herbicides	Good
Crop cultivars	Improved competitive ability for the crop, reduced selection pressure	Lack of research base, inconsistent impact on HR weed population	Good to fair
Cover crops, mulches, and intercrop systems	Improved competitive ability, reduced selection pressure, improved system diversity, allelopathy	Inconsistent effect on HR weed populations, lack of understanding about the systems, limited research base, potential crop yield loss, need for herbicide to manage the cover crop, lack of good cover crop species	Poor

TABLE 4.4 (CONTINUED)
Assessment of Benefits and Risks of Alternative Tactics for Herbicide-Resistant Weed Management

Tactic	Benefits	Risks	Likelihood of Adoption and Impact on HR Weed Management
Management of seedbanks	Reduced HR weed pressure, reduced selection pressure	Lack of understanding about weed seedbank dynamics, requires aggressive tillage, emphasis on late herbicide applications, requires high level of management skills	Good to fair
Adjustment of nutrient use	Improved competitive ability for the crop, efficient use of nutrients, reduced selection pressure	Lack of research base, inconsistent results, potential crop yield loss	Poor
Biological control	Reduced selection pressure, reduced cost of HR weed management, long-term impact on HR weed population	Unavailability of biocontrol agents, crop yield loss due to slow efficacy, inconsistent results, typically targets one weed species	Poor
Weed management models	More efficient use of herbicides, reduced cost of HR weed control	Inconsistent results, lack of weed seedbank data, loss of crop yield	Fair to poor

However, the development of HR weed populations is attributable to the prevalence of herbicides as the primary strategy for weed control. For example, triazine-resistant weed populations are found in most of the world maize and soybean agroecosystems. Furthermore, the inevitability of triazine resistance without alternative tactics was predictable.[81,112,156,195] Regardless, triazine herbicide use in maize agroecosystems has increased in the U.S. maize belt.[49] Producers are more concerned with the immediacy of weed control and economic viability than the risk of HR weed populations. If producers proactively eliminated triazine use in maize so as to eliminate the risk of triazine resistance, maize and soybean yields were predicted to decline 3 and 20%, respectively, and net farm income by 38%.[296] Generally, producers rely entirely on herbicides for weed management, and deal with resultant problems by applying another herbicide.[49]

Alternative herbicide use tactics, while supported by promising research, have only served to delay the development of HR weed populations (Table 4.4). Tank mixtures of herbicides with different mechanisms of action were strongly supported by industry as an important strategy to curb HR weed populations. However, the redundancy of essentially using two herbicides to control the same weed does not meet the immediate weed management needs of maize and soybean producers.[199] Further, using two highly effective herbicides, in a tank mixture, could select for nontarget-site mechanisms of resistance.[171] Adjusting herbicide rate has not proven an effective deterrent for the development of HR weed populations. Lower herbicide doses may increase weed survival and allow individuals in the population that have weaker resistance mechanisms to increase the resistance gene pool.[171] The use of high herbicide rates has resulted in the rapid selection of ALS resistance in *Amaranthus rudis*.[98] High rates of glyphosate resulted in the discovery of differential response within *A. rudis* populations.[130] Generally, adjustment of herbicide application rates should not be considered an effective management tactic for HR weed populations (Table 4.4).

The use of HR crops for the management of HR weeds is not uniformly recommended by weed scientists. While many scientists believe that HR crops will improve the management of HR weed populations, others fear that HR weed populations will increase as HR crops are adopted.[39,187] Further, the adoption of HR crop systems may result in producers using fewer alternative strategies to manage weed populations.[187] Producers are likely to focus solely on the herbicide for which the crop has resistance. While the herbicide may initially be effective, the selection pressure brought to bear on the agroecosystem will likely result in the development of an HR weed population or a shift in the weed population to more tolerant species.[174]

The use of mechanical strategies for weed control is effective when practiced correctly. However, many maize and soybean producers are not able to cultivate in a timely manner, thus reducing the effectiveness of these tactics.[237] Importantly, untimely mechanical practices can cause crop stand loss.[238] Labor availability and the field working days also influence the risk of using mechanical weed control strategies.[43] Generally, weed management programs with herbicides minimize the time in the field for weed control and thus are deterrents to the adoption of mechanical practices (Table 4.4).

The issues of risk, labor availability, and time management impact the adoption of cultural tactics for weed management.[238] For example, while crop rotation represents an important cultural strategy for weed management, there are no definitive data describing the actual effect of crop rotation on weed populations.[2] Another concern is the limitation of adapted crops in certain agroecosystems, thus hindering the implementation of an alternate crop rotation system in order to manage an HR weed population.[227] Importantly, any alternative rotation crop must be economically viable before it will be accepted by producers, regardless of the positive impact on HR weed populations (Table 4.4).

Altering planting time may be beneficial for HR weed management, but it represents a significant risk to maize growers due to loss of potential yield, thus hampering the adoption of this strategy.[238,252] Changing planting configuration by narrowing row spacing has been widely adopted for soybean but it results in the loss of mechanical tactics for weed management.[55] Biological control opportunities are also diminished in narrow row culture.[261] Increasing seeding rates may improve HR weed control, but actual effect on weed populations is difficult to quantify. Further, the increased seed cost may be close to the cost of a herbicide application. Using competitive crop cultivars to reduce HR weed populations is not an attractive alternative due to the lack of information about the competitive ability of maize and soybean cultivars.[62] Cover crops, mulches, and smother crops have considerable risks associated with their adoption as a means to manage HR weed populations.[264,265]

While research suggests that adjustment of nutrients may provide some opportunity to manage HR weeds, the results are not consistent and growers are likely to perceive this strategy as extremely risky due to possible negative effects on crop yield.[43] Given the similarity of nutrient requirements for crops and weeds, adjustment of nutrients may not be a particularly effective strategy to manage HR weed populations (Table 4.4). Furthermore, the weed community is typically well adapted to the agroecosystems, so differences in yields attributable to changes in fertility and subsequent increased weed interference are unlikely.[35]

Considerable efforts have been expended in biological control of weeds (Table 4.4). However, the use of biological control of weeds in row crop agroecosystems has only limited success.[32] The discovery of soil microorganisms that are effective biocontrol agents is promising but, to date, there are no successes in maize and soybean agroecosystems.[283] While the diversity of soil microbes is quite high, there is limited information about them. Further, natural control encompasses a balance of predator–prey and parasite–host organisms and thus may not be appropriate to eliminate HR weed populations.[22] Biological tactics must be addressed in the context of the agroecosystems and will likely require considerable skill to manipulate the system to increase communities of pathogen and predators that will attack weeds.[181] In developing countries, biological control should not be a major priority of research efforts, given the lack of personnel and resources.[282] However, considerable efforts have been expended toward biological control tactics in Latin America, primarily the management of insect pests.

Attempts to manage weeds based on predictive models have been largely unsuccessful despite considerable research in this area (Table 4.4). When model-based weed control tactics were utilized, results were variable when compared to standard weed control recommendations.[248] More important, the standard recommendations resulted in a smaller weed seedbank when compared to the model-based results.[297] Furthermore, there are economic considerations about weed thresholds that are not biologically defendable.[49] Producers are concerned about harvest problems attributable to weed thresholds, weed seed production, and the general appearance of fields when threshold models are used as management tools.[298] Given the lack of understanding when considering weed seedbank dynamics, and the importance of eliminating HR weed populations, models may not be useful tools for maize and soybean agroecosystems.

The management of HR weed populations must focus on limiting gene flow into fields.[30] Given the current considerations of time and scale in maize and soybean agroecosystems, producers may not have the ability to effectively manage weed seed movement. Fewer producers are managing larger agricultural enterprises. Moving tillage and harvesting equipment between fields has resulted in the relatively rapid spread of weeds (e.g., *Eriochloa villosa*) in the midwestern United States. Considering time limitations, it is far easier and faster to control weeds with repeated applications of herbicides rather than proactively manage the weed problem. Concerns for "aesthetically acceptable" weed control also precipitate the applications of additional herbicides despite the selection pressure for the evolution of HR weed populations these applications represent.[21]

The adoption of alternative tactics, while effectively diminishing the potential for HR weed populations, requires better management skills and labor.[150] A serious problem hampering the possible adoption of alternative IWM techniques is a lack of research that documents the assumed benefits compared with traditional practices.[183] Much of the research that supports alternative tactics as appropriate strategies to manage HR weed populations has been conducted in crops other than maize and soybean. Often, there is little information about the economic benefits of these strategies compared to standard practices, or any indication of the risks that producers must accept. For example, while greater diversity in crop rotation is potentially a powerful tactic to manage HR weed populations, little if any research has investigated the ecological or economic fit of the alternative rotation crops.

Researchers must recognize that while it is important to identify specific strategies to manage HR weeds, the adoption of these strategies requires that the benefits on HR weeds are not offset by risks of poor profitability. Research on alternative strategies should include an economic assessment, as well as the biological impact of the tactic on HR weeds. While mechanical weed control is an effective alternative technique to manage HR weeds, growers may perceive the risks of time utilization to be too risky to adopt the strategy.[239] Essentially, research on alternative strategies must include an assessment of the effect of tactic on the overall maize and soybean agroecosystems.

Another consideration is that alternative IWM systems may have greater weed community diversity.[184] Producers may presume that greater weed community diversity represents a serious management problem. Increasing cropping system diversity may eliminate the likelihood of HR weed populations but may not be economically acceptable. While the adoption of alternative IWM tactics to manage HR weed populations may be warranted by "science," socioeconomic factors are more important

(Table 4.4). There is little question that maize and soybean agroecosystems would improve if management decisions were based on science and objectivity.[27] However, pest management problems are the responsibility of producers who have little impetus to make environmentally better, long-term strategic plans.[31] Given the socioeconomic climate that currently surrounds agriculture, unless government regulations are created that mandate that producers manage HR weed populations with alternative strategies, weed management that relies solely on herbicides will continue. The short-term economics and risks are more important to maize and soybean producers than the development of HR weed populations.

4.5   SUMMARY AND CONCLUSIONS

The evolution of HR weed populations is greatest in maize and soybean agroecosystems compared to other agroecosystems because there is more reliance on herbicides as the sole tactic for weed management resulting in extremely high selection pressure.[76] Furthermore, the cost of repeat herbicide applications is cheaper and typically more effective than alternative tactics. Factors such as aesthetics, resulting in biologically questionable herbicide applications, may increase the speed at which HR weed populations evolve. Industry also imparts considerable pressure on producers to use herbicides rather than alternative tactics to manage weeds. Despite these considerations, the importance of HR weed populations in maize and soybean agroecosystems is questionable. Given the availability of tactics for weed management in these agroecosystems, the evolution of HR populations will likely have a minor economic impact. However, it should be assumed that the active development of HR weed management strategies is only beginning and the ramifications of future HR weeds may move producers toward proactive HR weed management.[299]

Current approaches to weed management have shifted toward the use of HR crops, particularly those with resistance to glyphosate. Multiple applications of glyphosate increase the probability that either weed populations will shift or glyphosate-resistant weed populations will evolve.[300] Maize and soybean producers believe that HR crops provide a more cost-effective and environmentally acceptable tactic for weed management. They will not willingly adopt alternative tactics proactively despite evidence that problems will evolve in the agroecosystem.[20] However, while HR crops serve to reduce the rate of HR weed evolution, this tactic will not ultimately eliminate the risk of HR weeds.[30] Major changes in the agroecosystems focusing on diversification of crops and management tactics must occur to preclude future problems with HR weeds. However, this diversification will not be employed unless the alternative agroecosystems are economically viable.

Unfortunately, considerations other than sound science often are more important factors influencing decisions implemented by governments.[301] Some suggest that the environmental and sociological costs of pesticide use issues be transferred to the industry.[38] Regardless, the complexity of issues surrounding HR weed evolution and management must be addressed.[44] A key component to proactive HR weed management is an assessment of the actual costs to society for HR weed populations.

REFERENCES

1. Perkins, J. H. and Pimentel, D., Society and pest control, in *Symp. Environmental, Socioeconomic, and Political Aspects of Pest Management Systems,* Am. Assoc. Adv. Sci. 43, Pimentel, D., Ed., Westview Press, Boulder, CO, 1980, 1.
2. Gould, F., The evolutionary potential of crop pests, *Am. Sci.,* 79, 496, 1991.
3. Baker, H. G., The continuing evolution of weeds, *Econ. Bot.,* 45(4), 445, 1991.
4. De Wet, J. M. J. and Harlan, J. R., Weeds and domesticates: evolution in the man-made habitat, *Econ. Bot.,* 29, 99, 1975.
5. Hiko-Ichi, O. and Wen-Tsai, C., The impact of cultivation on populations of wild rice, *Oryza sativa* f. *spontanea,* OΨTON, 13(2), 105, 1959.
6. Baker, H. G., The evolution of weeds, in *Annual Review of Ecology and Systematics,* 5, Johnston, R. F., Frank, P. W., and Michener, C. D., Eds., Annual Reviews, Palo Alto, CA, 1974, 1.
7. Barrett, S. C. H., Crop mimicry in weeds, *Econ. Bot.,* 37(3), 255, 1983.
8. Holt, J. S. and LeBaron, H. M., Significance and distribution of herbicide resistance, *Weed Technol.,* 4(1), 141, 1990.
9. Harper, J. L., The evolution of weeds in relation to resistance to herbicides, in *Proc. 3rd Brit. Weed Cont. Conf.,* 3(1), 179, 1956.
10. Parker, C., Prediction of new weed problems especially in the developing world, in *Symp. Brit. Ecol. Soc. Origins of Pest, Parasite, Disease, and Weed Problems,* 17, Cherrett, J. M. and Sagar, G. R., Eds., Blackwell Scientific, Oxford, 1976, 249.
11. Moberg, W. K., Understanding and combating agrochemical resistance: a chemist's perspective on an interdisciplinary challenge, in *Managing Resistance to Agrochemicals: From Fundamental Research to Practical Strategies,* Green, M. B., LeBaron, H. M., and Moberg, W. K., Eds., American Chemical Society, Washington, D.C., 1990, 1.
12. Hueth, D. and Regev, U., Optimal agricultural pest management with increasing pest resistance, *Am. J. Agric. Econ.,* 56, 543, 1974.
13. Dieleman, J. A. and Mortensen, D. A., Influence of weed biology and ecology on development of reduced dose strategies for integrated weed management systems, in *Integrated Weed and Soil Management,* Hatfield, J. L., Buhler, D. D., and Stewart, B. A., Eds., Ann Arbor Press, Chelsea, MI, 1998, 333.
14. Haas, H. and Striebig, J. C., Changing patters of weed distribution as a result of herbicide use and other agronomic factors, in *Herbicide Resistance in Plants,* LeBaron, H. M. and Gressel, J., Eds., John Wiley & Sons, New York, 1982, 57.
15. Holt, J. S., Impact of weed control on weeds: new problems and research needs, *Weed Technol.,* 8, 400, 1994.
16. Heap, I., Relationship between agronomic practices and the development of herbicide resistance, in *Proc. 3rd Int. Weed Science Congr.,* 3, Legere, A., Ed., International Weed Science Society, Corvallis, OR, 2000, 137.
17. Boyer, J. S., Plant productivity and environment, *Science,* 218, 443, 1982.
18. Cousens, R., *Dynamics of Weed Populations,* Cambridge University Press, Cambridge, 1995, 1.
19. Hurle, K., Present and future developments in weed control — a view from weed science, *Pflanzenschutz Nachr. Bayer,* 51, 109, 1998.
20. Tsaftaris, A., The development of herbicide-tolerant transgenic crops, *Field Crops Res.,* 45, 115, 1996.
21. Bridges, D. C., Impact of weeds on human endeavors, *Weed Technol.,* 8, 398, 1994.

22. Pimentel, D., The ecological basis of insect pest, pathogen and weed problems, in *Symp. Brit. Ecol. Soc. Origins of Pest, Parasite, Disease, and Weed Problems*, 18, Cherrett, J. M. and Sagar, G. R., Eds., Blackwell Scientific, Oxford, 1977, 3.

23. Liebman, M. and Dyck, E., Weed management — a need to develop ecological approaches, *Ecol. Appl.*, 3(1), 39, 1993.

24. Pimentel, D., Andow, D., Gallahan, D., Schreiner, I., Thompson, T. E., Dyson-Hudson, R., Jacobson, S. N., Irish, M. A., Kroop, S. F., Moss, A. M., Shepard, M. D., and Vinzant, B. G., Pesticides: environmental and social costs, in *Symp. Environmental, Socioeconomic, and Political Aspects of Pest Management Systems,* Am. Assoc. Adv. Sci. 43, Pimentel, D., Ed., Westview Press, Boulder, CO, 1980, 99.

25. Stephenson, G. R., Herbicide use and world food production: risks and benefits, in *Proc. 3rd Int. Weed Science Congr.*, 3, Legere, A., Ed., International Weed Science Society, Corvallis, 2000, 240.

26. Norton, G. A. and Conway, G. R., The economic and social context of pest, disease and weed problems, in *Symp. Brit. Ecol. Soc. Origins of Pest, Parasite, Disease, and Weed Problems*, 18, Cherrett, J. M. and Sagar, G. R., Eds., Blackwell Scientific, Oxford, 1977, 205.

27. Ellis, J. F., Herbicide development and marketing of weed control in the United States of America, in *Proc. 1st Int. Weed Science Congress*, 1, Combellack, J. H., Levick, K. J., Parsons, J., and Richardson, R. G., Eds., International Weed Science Society, Corvallis, OR, 1992, 74.

28. Zimdahl, R. L., Weed science in sustainable agriculture, *Am. J. Alt. Agric.*, 10, 138, 1995.

29. Gressel, J., Addressing real weed science needs with innovations, *Weed Technol.*, 6, 509, 1992.

30. Christoffers, M. J., Genetic aspects of herbicide-resistant weed management, *Weed Technol.* 13, 647, 1999.

31. McPhee, G., Resistant pests: a producer's perspective, *Phytoprotection*, 75 (Suppl.), 91, 1994.

32. Lewis, W. J., van Lenteren, J. C., Phatak, S. C., and Tumlinson, J. H., III, A total system approach to sustainable pest management, *Proc. Natl. Acad. Sci.*, 94, 12243, 1997.

33. Anon., World crop production estimates, www.fas.usda, 1999.

34. Parker, C. and Fryer, J. D., Weed control problems causing major reductions in world food supplies, *FAO Plant Prot. Bull.*, 23, 83, 1975.

35. Koch, W., Beshir, M. E., and Unterladstatter, R., Crop loss due to weeds, *FAO Plant Prod. Prot.*, 31, 103, 1983.

36. Hartmans, E. H., International agricultural research centers in weed management in the developing countries, *FAO Plant Prod. Prot.*, 31, 97, 1983.

37. Holm, L., The role of weeds in world food production, in *Proc. North Cent. Weed Cont. Conf.*, 30, Ross, M. and Lembi, C., Eds., North Central Weed Control Conference, Champaign, IL, 1975, 18.

38. Goldberg, R., Rissler, J., Shand, H., and Hassebrook, C., *Biotechnology's Bitter Harvest*, Biotechnology Working Group, 1990, 73 pp.

39. Burnside, O. C., An agriculturalist's viewpoint of risks and benefits of herbicide-resistant cultivars, in *Herbicide-Resistant Crops — Agricultural, Environmental, Economic, Regulatory, and Technical Aspects*, Duke, S. O., Ed., Lewis Publishers, Boca Raton, FL, 1996, 391.

40. Owen, M. D. K., Producer attitudes and weed management, in *Integrated Weed and Soil Management*, Hatfield, J. L., Buhler, D. D., and Stewart, B. A., Eds., Ann Arbor Press, Chelsea, MI, 1996, 43.

41. Buhler, D. D., Influence of tillage systems on weed population dynamics and management in corn and soybean in the Central USA, *Crop Sci.*, 35, 1247, 1995.
42. Vidal, R. A., Fleck, N. G., and Merotto, A., Jr., Past and current challenges of soybean weed management, in *Proc. 3rd Int. Weed Science Congr.*, 3, Legere, A., Ed., International Weed Science Society, Corvallis, OR, 2000, 249.
43. Gunsolus, J. L. and Buhler, D. D., A risk management perspective on integrated weed management, in *Expanding the Context of Weed Management*, Buhler, D. D., Ed., Haworth Press, Binghamton, NY, 1999, 167.
44. Buhler, D. D., Development of alternative weed management strategies, *J. Prod. Agric.*, 9(4), 501, 1996.
45. Ghersa, C. M., Roush, M. L., Radosevish, S. R., and Cordray, S. M., Coevolution of agroecosystems and weed management, *Bioscience*, 44(2), 85, 1994.
46. Forcella, F. and Harvey, S. J., Relative abundance in an alien weed flora, *Oecologia*, 59, 292, 1983.
47. Naylor, J. M. and Jana, S., Genetic adaptation for seed dormancy in *Avena fatua*, *Can. J. Bot.*, 54, 306, 1976.
48. Thill, D. C., The spread of herbicide resistance, in *Herbicide-Resistant Crops — Agricultural, Environmental, Economic, Regulatory, and Technical Aspects*, Duke, S. O., Ed., Lewis Publishers, Boca Raton, FL, 1996, 331.
49. Buhler, D. D., Doll, J. D., Proost, R. T., and Visocky, M. R., Integrating mechanical weeding with reduced herbicide use in conservation tillage corn production systems, *Agron. J.*, 87, 507, 1995.
50. Ampong-Nyarko, K., Weed management in tropical cereals: maize, sorghum and millet, in *Weed Management for Developing Countries*, Labrada, R., Caseley, J. C., and Parker, C., Eds., FAO Press, Rome, 1994, 264.
51. de la Cruz, R., Ampong-Nyarko, K., Labrada, R., and Merayo, A., Weed management in legume crops: bean, soybean, and cowpea, in *Weed Management for Developing Countries*, Labrada, R., Caseley, J. C., and Parker, C., Eds., FAO Press, Rome, 1994, 273.
52. Anon., 1987 Census of Agriculture AC87-A25, 1, United States Department of Commerce Bureau of the Census, U.S. Government Printing Office, Washington, D.C., 1989.
53. Anon., Crop rotation used more than any other IMP practice, *Cons. Technol. Inf. Cen.*, 17(4), 8, 1999.
54. Liebman, M. and Dyck, E., Crop rotation and intercropping strategies for weed management, *Ecol. Appl.*, 3(1), 92, 1993.
55. Griffin, J. L. and Harger, T. R., Red rice (*Oyrza sativa*) and junglerice (*Echinochloa colonum*) control in solid-seeded soybeans (*Glycine max*), *Weed Sci.*, 34, 582, 1986.
56. Weber, E. F., The alien flora of Europe: a taxonomic and biogeographic review, *J. Veg. Sci.*, 8, 565, 1997.
57. Owen, M. D. K., Hartzler, R. G., Buhler, D. D., and Duffy, M. D., What influences weed management decisions? in *Proc. Weed Sci. Soc. Am.*, 38, Murry, D. S., Ed., Allen Press, Lawrence, KS, 1998, 31.
58. Swanton, C. J., Murphy, S. D., Hume, D. J., and Clements, D. R., Recent improvements in the energy efficiency of agriculture: case studies from Ontario, Canada, *Agric. Syst.*, 52, 399, 1996.
59. Day, J. C., Hallahan, C. B., Sandretto, C. L., and Lindamood, W. A., Pesticide use in U.S. corn production: does conservation tillage make a difference? *J. Soil Water Conserv.*, 54(2), 477, 1999.

60. Owen, M. D. K., Risks and benefits of weed management technologies, in *Weed and Crop Resistance to Herbicides*, De Prado, R., Jorrín, J., and García-Torres, L., Eds., Kluwer Academic Press, London, 1996, 291.

61. Owen, M. D. K., Current use of transgenic herbicide-resistant soybean and corn in the USA, *Crop Prot.*, in press.

62. Boerboom, C. M., Nonchemical options for delaying weed resistance to herbicides in Midwest cropping systems, *Weed Technol.*, 13, 636, 1999.

63. Lawton, K., Roundup of a market, *Farm Ind. News*, 2, 4, 1999.

64. Ikerd, J. E., Government policy options: implications for weed management, *J. Prod. Agric.*, 9(4), 491, 1996.

65. Pereira, R. C. and Carmona, R., Soybean weed management in the Cerrados (Brazilian savannas), in *Proc. 3rd Int. Weed Science Congr.*, 3, Legere, A., Ed., International Weed Science Society, Corvallis, OR, 2000, 251.

66. Anderson, B. B., Factor Europe into soybean management decisions, *Iowa Soybean Rev.*, 10, 12, 1999.

67. Lorenz, H., The Brazilian weed flora: taxonomy and general aspects, in *Proc. 3rd Int. Weed Science Congr.*, 3, Legere, A., Ed., International Weed Science Society, Corvallis, OR, 2000, 46.

68. Vitta, J., Tuesca, D., Puricelli, E., Nisensohn, L., Faccini, D., and Leguizamon, E., Glyphosate-tolerant soybean and weed management in Argentina: present and prospects, in *Proc. 3rd Int. Weed Science Congr.*, 3, Legere, A., Ed., International Weed Science Society, Corvallis, OR, 2000, 163.

69. Suarez, S. A., Ghersa, C. M., De Le Fuente, E. B., and Leon, R. J. C., Shifts of species groups in crop-weed communities of the Pampas during 1926 to 1999, in *Proc. 3rd Int. Weed Science Congr.*, 3, Legere, A., Ed., International Weed Science Society, Corvallis, OR, 2000, 41.

70. Borona, V. and Zadorozhny, V., Damage from the weeds and optimal terms for their control in silage corn, in *Proc. 3rd Int. Weed Science Congr.*, 3, Legere, A., Ed., International Weed Science Society, Corvallis, OR, 2000, 73.

71. Costa, J., From research to practise: staying ahead of the problem, in *Weed and Crop Resistance to Herbicides*, De Prado, R., Jorrín, J., and García-Torres, L., Eds., Kluwer Academic Press, London, 1996, 315.

72. Darmency, H. and Gasquez, J., Fate of herbicide resistance genes in weeds, in *Managing Resistance to Agrochemicals: From Fundamental Research to Practical Strategies*, Green, M. B., LeBaron, H. M., and Moberg, W. K., Eds., American Chemical Society, Washington, D.C., 1990, 353.

73. Waters, S., The regulation of herbicide-resistant crops in Europe, in *Herbicide-Resistant Crops — Agricultural, Environmental, Economic, Regulatory, and Technical Aspects*, Duke, S. O., Ed., Lewis Publishers, Boca Raton, FL, 1996, 347.

74. Toth, A., Hartmann, F., Lanszki, I., and Szentey, L., Changes of weed populations in Hungarian fields based on four country-wide weed surveys, in *Proc. 3rd Int. Weed Science Congr.*, 3, Legere, A., Ed., International Weed Science Society, Corvallis, OR, 2000, 44.

75. Su, S. Q. and Ahrens, W. H., Weed management in Northeast China, *Weed Technol.*, 11, 817, 1997.

76. Heap, I. M., The occurrence of herbicide-resistant weeds worldwide, *Pestic. Sci.*, 51, 235, 1997.

77. Heap, I. M., International survey of herbicide-resistant weeds, http://www.weed-science.com, 2000.

78. Gressel, J., Ammon, H. U., Fogelfors, H., Gasquex, J., Kay, Q. O. N., and Kees, H., Discovery and distribution of herbicide-resistant weeds outside North America, in *Herbicide Resistance in Plants*, LeBaron, H. M. and Gressel, J., Eds., John Wiley & Sons, New York, 1982, 31.

79. Duke, S. O., Christy, A. L., Hess, F. D., and Holt, J. S., *Herbicide-Resistant Crops*, CAST, Ames, IA, 1991, 1.

80. Roush, M. L., Radosevich, S. R., and Maxwell, B. D., Future outlook for herbicide-resistance research, *Weed Technol.*, 4, 208, 1990.

81. Maxwell, B. D., Roush, M. L., and Radosevich, S. R., Predicting the evolution and dynamics of herbicide resistance in weed populations, *Weed Technol.*, 4, 2, 1990.

82. Bandeen, J. D., Stephenson, G. R., and Cowett, E. R., Discovery and distribution of herbicide-resistant weeds in North America, in *Herbicide Resistance in Plants*, LeBaron, H. M. and Gressel, J., Eds., John Wiley & Sons, New York, 1982, 9.

83. Moss, S. R. and Rubin, B., Herbicide-resistant weeds: a worldwide perspective, *J. Agric. Sci.*, 120, 141, 1993.

84. Gressel, J., Burgeoning resistance requires new strategies, in *Weed and Crop Resistance to Herbicides*, De Prado, R., Jorrín, J., and García-Torres, L., Eds., Kluwer Academic Publishers, London, 1996, 3.

85. Rubin, B., Herbicide resistance outside North America and Europe: causes and significance, in *Weed and Crop Resistance to Herbicides*, De Prado, R., Jorrín, J., and García-Torres, L., Eds., Kluwer Academic Publishers, London, 1996, 39.

86. Reed, W. T. and Chang, S. H., Resistance in weeds to sulfonylurea herbicides and its relevance to tropical weed control, in *Proc. 3rd Tropical Weed Science Conference*, 3, Lee, S. A. and Kon, K. F., Eds., Malaysian Plant Protection Society, Kuala Lumpur, 1996, 479.

87. Valverde, B. E., Chowes, L., Gonzalez, J., and Garita, I., Field-evolved imazapyr resistance in *Ixophorus unisetum* and *Eleusine indica* in Costa Rica, *Brit. Crop Prot. Conf.*, 3, 1189, 1997.

88. Franco, P., Caseley, J. C., Kim, D. S., Riches, C. R., and Miyasato, Y., Evaluation of response of fluazifop-p resistant carib grass (*Eriochloa punctata*) to other ACCase inhibitors and herbicides with other modes of action, in *Proc. Weed Sci. Soc. Am.*, 39, Wilcutt, J. W., Ed., Allen Press, Lawrence, KS, 1999, 52.

89. Stoller, E. W., Wax, L. M., and Alm, D. M., Survey results of environmental issues and weed science research priorities within the Corn Belt, *Weed Technol.*, 7, 763, 1993.

90. Heap, I. M., Herbicide resistant weeds in the USA, in *Proc. Weed Sci. Soc. Am.*, 39, Wilcutt, J. W., Ed., Allen Press, Lawrence, KS, 1999, 54.

91. White, A. D., Owen, M. D. K., Hartzler, R. G., and Cardina, J., Evaluation of common sunflower (*Helianthus annuus* L.) resistance to acetolactate synthase inhibiting herbicides, in *Proc. North Cent. Weed Sci. Soc.*, Ross, M. and Lembi, C., Eds., 53, 96, 1998.

92. Baumgartner, J. R., Al-Khatib, K., and Currie, R. S., Survey of common sunflower (*Helianthus annuus*) resistance to imazethapyr and chlorimuron in northeast Kansas, in *Proc. Weed Sci. Soc. Am.*, 39, Wilcutt, J. W., Ed., Allen Press, Lawrence, KS, 1999, 33.

93. Schmitzer, P. R., Eilers, R. J., and Cseke, C., Lack of cross resistance of imazaquin-resistant *Xanthium strumarium* acetolactate synthase to flumetsulam and chlorimuron, *Plant Physiol.*, 103, 281, 1993.

94. Heap, K. T. and Peeper, T. F., Survey of herbicide resistant weed management programs in Oklahoma, in *Proc. South. Weed Sci. Soc.*, 51, Dusky, J., Ed., Southern Weed Science Society, Champaign, IL, 1998, 38.

95. Kendig, J. A. and Barrentine, W. L., Identification of ALS-cross-resistant common cocklebur (*Xanthium strumarium*) in the mid-south, in *Proc. South. Weed Sci. Soc.*, 48, Street, J., Ed., Southern Weed Science Society, Champaign, IL, 1995, 173.

96. Ohmes, G. A. and Kendig, J. A., Inheritance of an ALS-cross-resistant common cocklebur (*Xanthium strumarium*) biotype, *Weed Technol.*, 13, 100, 1999.

97. Lee, J. M. and Owen, M. D. K., Comparison of acetolactate synthase enzyme inhibition among resistant and susceptible *Xanthium strumarium* biotypes, *Weed Sci.*, 48, 286, 2000.

98. Hinz, J. R. R. and Owen, M. D. K., Acetolactate synthase resistance in a common waterhemp (*Amaranthus rudis*) population, *Weed Technol.*, 11, 13, 1997.

99. Saari, L. L., Cotterman, J. C., and Thill, D. C., Resistance to acetolactate synthase inhibiting herbicides, in *Herbicide Resistance in Plants: Biology and Biochemistry*, Powles, S. B. and Holtum, J. A. M., Eds., CRC Press, Boca Raton, FL, 1994, 83.

100. Poston, D. H., Wilson, H. P., Hines, T. E., and Trader, B. W., Growth analyses in one imidazolinone-susceptible and four resistant smooth pigweed (*Amaranthus hybridus*) populations, in *Proc. Weed Sci. Soc. Am.*, 40, Wilcutt, J. W., Ed., Allen Press, Lawrence, KS, 2000, 62.

101. Volenberg, D. S., Stoltenberg, D. E., and Boerboom, C. M., Cross-resistance of giant foxtail (*Setaria faberi*) to ALS inhibitors, in *Proc. Weed Sci. Soc. Am.*, 40, Wilcutt, J. W., Ed., Allen Press, Lawrence, KS, 2000, 61.

102. Anderson, D. D., Lee, D. J., Martin, A. R., and Roth, F. W., DNA sequence comparison of the acetolactate synthase gene from ALS-resistant and susceptible shattercane, in *Proc. North Cent. Weed Cont. Conf.*, 52, Ross, M. and Lembi, C., Eds., 1997, 42.

103. Foes, M. J., Vigue, G., Stoller, E. W., Wax, L. M., and Tranel, P. J., A kochia (*Kochia scoparia*) biotype resistant to triazine and ALS-inhibiting herbicides, *Weed Sci.*, 47, 20, 1999.

104. Shaner, D. L., Herbicide resistance in North America: history, circumstances of development and current situation, in *Weed and Crop Resistance to Herbicides*, De Prado, R., Jorrín, J., and García-Torres, L., Eds., Kluwer Academic Publishers, London, 1996, 29.

105. Anderson, D. D., Roth, F. W., and Martin, A. R., Occurrence and control of triazine-resistant common waterhemp (*Amaranthus rudis*) in field corn (*Zea mays*), *Weed Technol.*, 10, 570, 1996.

106. Eberlein, C. V., Al-Khatib, K., Guttieri, M. J., and Fuerst, E. P., Distribution and characteristics of triazine-resistant Powell amaranth (*Amaranthus powellii*) in Idaho, *Weed Sci.*, 40, 507, 1992.

107. Ahrens, W. H. and Stoller, E. W., Competition, growth rate, and CO_2 fixation in triazine-susceptible and -resistant smooth pigweed (*Amaranthus hybridus*), *Weed Sci.*, 31, 438, 1983.

108. Gray, J. A., Balke, N. E., and Stoltenberg, D. E., Increased glutathione conjugation of atrazine confers resistance in a Wisconsin velvetleaf (*Abutilon theophrasti*) biotype, *Pestic. Biochem. Physiol.*, 55, 157, 1996.

109. Gray, J. A., Stoltenberg, D. E., and Balke, N. E., Productivity and intraspecific competitive ability of a velvetleaf (*Abutilon theophrasti*) biotype resistant to atrazine, *Weed Sci.*, 43, 619, 1995.

110. Owen, M. D. K., Lux, J. F., and Pecinovsky, K.T., Evaluation of triazine resistant common lambsquarters control in corn, Hopkinton, Iowa, 1992, in *Iowa State University Weed Science Research Report*, Owen, M. D. K., Ed., Iowa State University, Ames, 1992, 276.

111. Owen, M. D. K., Lux, J. F., and Pecinovsky, K.T., Evaluation of herbicides for triazine-resistant common lambsquarters management in corn, Hopkinton, Iowa, 1993, in *Iowa State University Weed Science Research Report*, Owen, M. D. K., Ed., Iowa State University, Ames, 1993, 358.

112. Gressel, J. and Segel, L. A., Modeling the effectiveness of herbicide rotations and mixtures as strategies to delay or preclude resistance, *Weed Technol.*, 4, 186, 1990.

113. LeBaron, H. M., Distribution and seriousness of herbicide-resistant weed infestations worldwide, in *Herbicide Resistance in Weeds and Crops*, Caseley, J. C., Cussons, G. W., and Atkin, R. K., Eds., Butterworth-Heinemann, Oxford, 1991, 27.

114. Midge, L. C., Gossett, B. J., and Murphy, T. R., Resistance of goosegrass (*Eleusine indica*) to dinitroaniline herbicides, *Weed Sci.*, 32, 591, 1984.

115. Morrison, I. N., Todd, B. G., and Hall, J. C., Confirmation of trifluralin-resistant green foxtail (*Setaria viridis*) in Manitoba, *Weed Technol.*, 3, 544, 1989.

116. Gossett, B. J., Murdock, E. C., and Toler, J. E., Resistance of Palmer amaranth (*Amaranthus palmeri*) to the dinitroaniline herbicides, *Weed Technol.*, 6, 587, 1992.

117. Beckie, H. J. and Morrison, I. N., Effects of ethafluralin and other herbicides on trifluralin-resistant green foxtail (*Setaria viridis*), *Weed Technol.*, 7, 6, 1993.

118. Stoltenberg, D. E. and Wiederholt, R. J., Giant foxtail (*Setaria faberi*) resistance to aryloxyphenoxypropionate and cyclohexanedione herbicides, *Weed Sci.*, 43, 527, 1995.

119. Heap, I. M. and Morrison, I. N., Resistance to aryloxyphenoxypropionate and cyclohexanedione herbicides in green foxtail (*Setaria viridis*), *Weed Sci.*, 44, 25, 1996.

120. Wiederholt, R. J. and Stoltenberg, D. E., Cross-resistance of a large crabgrass (*Digitaria sanguinalis*) accession to aryloxyphenoxypropionate and cyclohexanedione herbicides, *Weed Technol.*, 9, 518, 1995.

121. Wiederholt, R. J. and Stoltenberg, D. E., Absence of differential fitness between giant foxtail (*Setaria faberi*) accessions resistant and susceptible to acetyl-coA carboxylase inhibitors, *Weed Sci.*, 44, 18, 1996.

122. Shukla, A., Leach, G. E., and Devine, M. D., High-level resistance to sethoxydim conferred by an alteration in the target enzyme, acetyl-coA carboxylase in *Setaria faberi* and *Setaria viridis*, *Plant Physiol. Biochem.*, 35(10), 803, 1997.

123. Ghersa, C. M., Marinez-Ghersa, M. A., Brewer, T. G., and Roush, M. L., Selection pressures for diclofop-methyl resistance and germination time of Italian ryegrass, *Agron. J.*, 86, 823, 1994.

124. Bradshaw, L. D., Padgette, S. R., Kimball, S. L., and Wells, B. H., Perspectives on glyphosate resistance, *Weed Technol.*, 11, 189, 1997.

125. Johnston, D. T., Van Wijk, A. J. P., and Kilpatrick, D., Selection for tolerance to glyphosate in fine-leaved *Festuca* species, *Int. Turfgrass Res. Conf.*, 6, 103, 1989.

126. Dill, G., Baerson, S., Casagrande, L., Feng, Y., Brinker, R., Reynolds, T., Taylor, N., Rodriguez, D., and Teng, Y., Characterization of glyphosate resistant *Eleusine indica* biotypes from Malaysia, in *Proc. 3rd Int. Weed Science Congr.*, 3, Legere, A., Ed., International Weed Science Society, Corvallis, OR, 2000, 150.

127. Powles, S. B., Lorraine-Colwill, D. F., Dellow, J. J., and Preston, C., Evolved resistance to glyphosate in rigid ryegrass (*Lolium rigidum*) in Australia, *Weed Sci.*, 46, 604, 1998.

128. Pratt, D. B., Owen, M. D. K., Clark, L. G., and Gardner, A., *Identification of the Weedy Pigweeds and Waterhemps of Iowa*, PM-1786, Iowa State University, Ames, 1999, 19 pp.

129. Singh, B. K. and Shaner, D. L., Rapid determination of glyphosate injury to plants and identification of glyphosate-resistant plants, *Weed Technol.*, 12, 527, 1998.

130. Zelaya, I. A. and Owen, M. D. K., Differential response of common waterhemp (*Amaranthus rudis*) to glyphosate in Iowa, in *Proc. Weed Sci. Soc. Am.*, 40, Wilcutt, J. W., Ed., Allen Press, Lawrence, KS, 2000, 62.

131. Christoffoleti, P. J., Cortez, M. G., and Filho, R. V., Resistance of alexanderweed (*Brachiaria plantaginea*) to ACCase inhibitor herbicides in soybean from Paraná State — Brazil, in *Proc. Weed Sci. Soc. Am.*, 38, Murray, D. S., Ed., Allen Press, Lawrence, KS, 1998, 65.

132. Vidal, R. A. and Fleck, N. G., Three weed species with confirmed resistance to herbicides in Brazil, in *Proc. Weed Sci. Soc. Am.*, 37, Murray, D. S., Ed., Weed Science Society of America, Champaign, IL, 1997, 100.

133. Christoffoleti, P. J., Ponchio, J. A. R., Berg, C. V. D., and Filho, R. V., Imidazolinone resistant *Bidens pilosa* biotypes in the Brazilian soybean areas, in *Proc. Weed Sci. Soc. Am.*, 36, Murray, D. S., Ed., Weed Science Society of America, Champaign, IL, 1996, 10.

134. De Prado, R., Lopez-Martinez, N., and Gimenez-Espinosa, R., Herbicide-resistant weeds in Europe: agricultural physiological and biochemical aspects, in *Weed and Crop Resistance to Herbicides*, De Prado, R., Jorrín, J., and García-Torres, L., Eds., Kluwer Academic Publishers, London, 1996, 17.

135. Maillet, J. and Lopez-Garcia, C., What criteria are relevant for predicting the invasive capacity of a new agricultural weed? The case of invasive American species in France, *Weed Res.*, 40, 11, 2000.

136. Gressel, J., The potential role for herbicide-resistant crops in world agriculture, in *Herbicide-Resistant Crops — Agricultural, Environmental, Economic, Regulatory, and Technical Aspects,* Duke, S. O., Ed., Lewis Publishers, Boca Raton, FL, 1996, 231.

137. Solymosi, P., Kostyal, Z., and Lehoczki, E., Characterization of intermediate biotypes in atrazine susceptible populations of *Chenopodium polyspermum* and *Amaranthus retroflexus* in Hungary, *Plant Sci.*, 50, 173, 1989.

138. Lehoczki, E., Solymosi, P., Laskay, G., and Polos, E., Non-plastid resistance to diruon in triazine resistant weed biotypes, in *Herbicide Resistance in Weeds and Crops*, Caseley, J. C., Cussans, G. W., and Atkin, R. K., Eds., Butterworth-Heinemann, Oxford, 1991, 447.

139. Hartmann, F., Lanszki, I., Szentey, L., and Toth, A., Resistant weed biotypes in Hungary, in *Proc. 3rd Int. Weed Sci. Congr.*, 3, Legere, A., Ed., International Weed Science Society, Corvallis, OR, 2000, 138.

140. Chodova, D., Researching for atrazine resistant weed populations in Czechoslovakia, *Acta Herbol. Jugosl.*, 17, 37, 1988.

141. Bettini, P., McNally, S., Sevignac, M., Darmency, H., Gasquex, J., and Dron, M., Atrazine resistance in *Chenopodium album* — low and high levels of resistance to the herbicide are related to the same chloroplast PSBA gene for mutation, *Plant Physiol.*, 84, 1442, 1987.

142. Rotteveel, A. J. W., Naber, H., and Kremer, E., Changes in the triazine resistance ratio in a population of *Solanum nigrum* L. during a long term field experiment, in *Proc. Int. Symp. Weed and Crop Resistance to Herbicides*, De Prado, R., Jorrín, J., García-Torres, L., and Marshall, G., Eds., Kluwer Academic Publishers, London, 1996, 159.

143. Chodova, D., Mikulka, J., and Kocova, M., Comparing some differences of atrazine susceptible and resistant common groundsel (*Senecio vulgaris* L.), in *Proc. Int. Symp. Weed and Crop Resistance to Herbicides*, De Prado, R., Jorrín, J., García-Torres, L., and Marshall, G., Eds., Kluwer Academic Publishers, London, 1996, 144.

144. Kintzios, S., Markikis, M., Passadeos, K., and Economou, G., *In vitro* expression of variation of glyphosate tolerance in *Sorghum halepense, Weed Res.*, 39, 49, 1999.

145. Huang, B. and Gressel, J., Barnyardgrass (*Echinochloa crus-galli*) resistance to both butachlor and thiobencarb in China, *Res. Pest Manage.*, 9(1), 5, 1997.

146. Gressel, J., Segel, L., and Ranson, J. K., Managing the delay of evolution of herbicide resistance in parasitic weeds, *Int. J. Pest Manage.*, 42(2), 113, 1996.

147. Abayo, G. O., English, T., Eplee, R. E., Kanampiu, F. K., Ranson, J. K., and Gressel, J., Control of parasitic witchweeds (*Striga* spp.) on corn (*Zea mays*) resistant to acetolactate synthase inhibitors, *Weed Sci.*, 46, 459, 1998.

148. Ranson, J. K., Odhiambo, G. D., and Gressel, J., Seed dressing maize with imidazolinone herbicides to control *Striga hermonthica* (Del.) Benth., *Proc. Weed Sci. Soc. Am.*, 35, Kupatt, C., Ed., Weed Science Society of America, Champaign, IL, 1995, 5.

149. Strange, M. and Miller, C., *A Better Row to Hoe — The Economic, Environmental, and Social Impact of Sustainable Agriculture*, Northwest Area Foundation, St. Paul, MN, 1994, 39 pp.

150. Fryer, J. D., Research in weed management in the developing countries, *FAO Plant Prod. Prot.*, 31, 88, 1983.

151. Roush, R. T., Designing resistance management programs: how can you choose? *Pestic. Sci.*, 26, 423, 1989.

152. Gressel, J., Gardner, S. N., and Mangel, M., Prevention versus remediation in resistance management, in *Molecular Genetics and Evolution of Pesticide Resistance*, Brown, T. M., Ed., American Chemical Society, Washington, D.C., 1996, 169.

153. Graham, J. C., Role of the herbicide resistance action committee in weed resistance management, in *Weed and Crop Resistance to Herbicides*, De Prado, R., Jorrín, J., and García-Torres, L., Eds., Kluwer Academic Publishers, London, 1996, 17.

154. Anon., Guideline to the management of herbicide resistance, http://plantprotection.org/HRAC/Guideline.html, 2000.

155. Jordan, N. R. and Jannink, J. L., Assessing the practical importance of weed evolution: a research agenda, *Weed Res.*, 37, 237, 1997.

156. Radosevich, S. R., Maxwell, B. D., and Roush, M. L., Managing herbicide resistance through fitness and gene flow, in *Herbicide Resistance in Weeds and Crops*, Caseley, J. C., Cussons, G. W., and Atkin, R. K., Eds., Butterworth-Heinemann, Oxford, 1991, 129.

157. Jordan, N., Predicted evolutionary response to selection for tolerance of soybean (*Glycine max*) and intraspecific competition in a nonweed population of poorjoe (*Diodia teres*), *Weed Sci.*, 37, 451, 1989.

158. Thill, D. C., Lish, J. M., Callihan, R. H., and Bechinski, E. J., Integrated weed management — a component of integrated pest management: a critical review, *Weed Technol.*, 5, 648, 1991.

159. El Titi, A., Integrated farming: an ecological farming approach in European agriculture, *Outlook Agric.*, 21, 33, 1992.

160. Mortensen, D. A., Bastiaans, L., and Sattin, M., The role of ecology in the development of weed management systems: an outlook, *Weed Res.*, 40, 49, 2000.

161. Georghiou, G. P., Principles of insecticide resistance management, *Phytoprotection*, 75(Suppl.), 51, 1994.

162. Orson, J. H., The effect of labeling herbicides with their mode of action: a European perspective, *Weed Technol.*, 13, 653, 1999.

163. Gadamski, G., Ciarda, D., Gressel, J., and Gawronski, S. W., Negative cross-resistance in triazine-resistant biotypes of *Echinochloa crus-galli* and *Conyza canadensis*, *Weed Sci.*, 48, 176, 2000.

164. Browde, J. A., Pedigo, L. P., Owen, M. D. K., and Tylka, G. L., Soybean yield and pest management as influenced by nematodes, herbicides, and defoliating insects, *Agron. J.*, 86, 601, 1994.

165. Ghersa, C. M. and Roush, M. L., Searching for solutions to weed problems, *Bioscience*, 43(2), 104, 1993.

166. Holt, J. S. and Powles, S. B., Mechanisms and agronomic aspects of herbicide resistance, *Annu. Rev. Plant Physiol. Mol. Biol.*, 44, 203, 1993.

167. Rotteveel, T. J., Joost, W. F. M., de Goeij, F. M., and van Gemerden, A. F., Towards the construction of resistance risk evaluation scheme, *Pestic. Sci.*, 51, 407, 1997.

168. Ranson, J. K. and Ranson, C. V., Socio-economics and risk influence the adoption of integrated weed management, in *Proc. 3rd Int. Weed Sci. Congr.*, 3, Legere, A., Ed., International Weed Science Society, Corvallis, OR, 2000, 187.

169. Krummel, J. and Hough, J., Pesticides and controversies: benefits versus costs, in *Symp. Environmental, Socioeconomic, and Political Aspects of Pest Management Systems*, Am. Assoc. Adv. Sci. 43, Pimentel, D., Ed., Westview Press, Boulder, CO, 1980, 159.

170. Gressel, J., Synergizing weed control, in *Proc. 2nd Int. Weed Congr.*, 2, Brown, H., Cussans, G. W., Devine, M. D., Duke, S. O., Fernandez-Quintanilla, C., Helweg, A., Labrada, R. E., Landes, M., Kudsk, P., and Streibig, J., Eds., International Weed Science Society, Corvallis, OR, 1996, 1.

171. Matthews, J. M., Management of herbicide resistant weed populations, in *Herbicide Resistance in Plants: Biology and Biochemistry*, Powles, S. B. and Holtum, J. A. M., Eds., CRC Press, Boca Raton, FL, 1994, 317.

172. Menalled, F. D. and Gross, K. L., Long term influence of agricultural practices on weed community dynamics, in *Proc. Weed Sci. Soc. Am.*, 40, Wilcutt, J. W., Ed., Allen Press, Lawrence, KS, 2000, 72.

173. Shaner, D., Herbicide-resistant crops in resistant weed management: an industrial perspective, *Phytoprotection*, 75(Suppl.), 79, 1994.

174. Radosevich, S. R., Ghersa, C. M., and Comstock, G., Concerns a weed scientist might have about herbicide-tolerant crops, *Weed Technol.*, 6, 635, 1992.

175. Legere, A., Beckie, H. J., Stevenson, F. C., and Thomas, A. G., Survey of management practices affecting the occurrence of wild oat (*Avena fatua*) resistance to acetyl-coA carboxylase inhibitors, *Weed Technol.*, 14, 366, 2000.

176. Beckie, H. J., Thomas, A. G., Legere, A., Kelner, D. J., Van Acker, R. C., and Meers, S., Nature, occurrence, and cost of herbicide-resistant wild oat (*Avena fatua*) in small-grain production areas, *Weed Technol.*, 13, 612, 1999.

177. Thomas, A. G., Leeson, J. Y., Beckie, H. J., and Legere, A., Identification of farm management systems at risk for ACCase inhibitor resistant wild oat (*Avena fatua* L.), *Proc. Weed Sci. Soc. Am.*, 39, Wilcutt, J. W., Ed., Allen Press, Lawrence, KS, 1999, 33.

178. Stephenson, G. R., Dykstra, M. D., McLaren, R. D., and Hamill, A. S., Agronomic practices influencing triazine-resistant weed distribution in Ontario, *Weed Technol.*, 4, 199, 1990.

179. Goodwin, M., An extension program for ACCase inhibitor resistance in Manitoba: a case study, *Phytoprotection*, 75(Suppl.), 97, 1994.

180. Hartzler, R. G. and Owen, M. D. K., *Herbicide Resistance — Concerns and Management*, IPM-39, Iowa State University, Ames, 1994, 8 pp.

181. Jordan, N., Weed prevention: priority research for alternative weed management, *J. Prod. Agric.*, 9, 485, 1996.

182. Owen, M. D. K., The value of alternative strategies for weed management, in *Proc. 3rd Int. Weed Science Congr.*, 3, Legere, A., Ed., International Weed Science Society, Corvallis, OR, 2000, 50.

183. Coble, H. D., Future directions and needs for weed science research, *Weed Technol.*, 8, 410, 1994.

184. Murphy, S. D. and Swanton, C. J., Integrated farming systems increases weed species diversity but reduces weed pressure and weed seedbanks, in *Proc. Weed Sci. Soc. Am.*, 40, Wilcutt, J. W., Ed., Allen Press, Lawrence, KS, 2000, 72.

185. Mohler, C. L., Ecological bases for the cultural control of annual weeds, *J. Prod. Agric.*, 9(4), 468, 1996.

186. Powles, S. B., Success from adversity: herbicide resistance can drive changes to sustainable weed management systems, *Brit. Crop Prot. Conf.*, 3, 1119, 1997.

187. Owen, M. D. K., North American developments in herbicide tolerant crops, *Brit. Crop Prot. Conf.*, 3, 955, 1997.

188. Thill, D. C., O'Donovan, J. T., and Mallory-Smith, C. A., Integrating weed management strategies for delaying herbicide resistance in wild oats, *Phytoprotection*, 75(Suppl.), 61, 1994.

189. Goudy, H. J., Bennett, K. A., Brown, R. B., and Tardif, F. J., Evaluation of site-specific weed control in a corn and soybeans rotation, in *Proc. Weed Sci. Soc. Am.*, 40, Wilcutt, J. W., Ed., Allen Press, Lawrence, KS, 2000, 70.

190. Moss, S., Strategies for the prevention and control of herbicide resistance in annual grass weeds, in *Weed and Crop Resistance to Herbicides*, De Prado, R., Jorrín, J., and García-Torres, L., Eds., Kluwer Academic Publishers, London, 1996, 283.

191. Bellinder, R. R., Gummesson, G., and Karlsson, C., Percentage-driven government mandates for pesticide reduction: the Swedish model, *Weed Technol.*, 8, 350, 1994.

192. Gressel, J., Creeping resistances: the outcome of using marginally-effective or reduced rates of herbicides, *Brit. Crop Prot. Conf.*, 3, 587, 1993.

193. Boger, P. and Sandmann, G., *Target Sites of Herbicide Action*, CRC Press, Boca Raton, FL, 1989, 248.

194. Retzinger, E. J. and Mallory-Smith, C., Classification of herbicides by site of action for weed resistance management strategies, *Weed Technol.*, 11, 384, 1997.

195. Gressel, J. and Segel, L. A., Herbicide rotations and mixtures — effective strategies to delay resistance, in *Managing Resistance to Agrochemicals*, Green, M. B., LeBaron, H. M., and Moberg, W. K., Eds., American Chemical Society, Washington, D.C., 1990, 430.

196. Buhler, D. D., Gunsolus, J. L., and Ralston, D. F., Common cocklebur (*Xanthium strumarium*) control in soybean (*Glycine max*) with reduced bentazon rates and cultivation, *Weed Sci.*, 41, 447, 1993.

197. Renner, K. A. and Woods, J. J., Influence of cultural practices on weed management in soybeans, *J. Prod. Agric.*, 12(1), 48, 1999.

198. Zhang, J. and Cavers, P. B., Effect of herbicide application on fruit characters of *Xanthium strumarium* L. populations, *Weed Res.*, 34, 319, 1994.

199. Wrubel, R. P. and Gressel, J., Are herbicide mixtures useful for delaying the rapid evolution of resistance? A case study, *Weed Technol.*, 8, 635, 1994.

200. Lich, J. M. and Renner, K. A., Glyphosate tank mixtures for control of velvetleaf (*Abutilon theophrasti* Medik.) and common lambsquarters (*Chenopodium album* L.), in *Proc. Weed Sci. Soc. Am.*, 36, Murray, D. S., Ed., Weed Science Society of America, Champaign, IL, 1996, 10.

201. Hartzler, R. G., Buhler, D. D., and Stoltenberg, D. E., Emergence characteristics of four annual weed species, *Weed Sci.*, 47, 578, 1999.

202. Griffin, J. L., Reynolds, D. B., Vidrine, P. R., and Saxton, A. M., Common cocklebur (*Xanthium strumarium*) control with reduced rates of soil and foliar-applied imazaquin, *Weed Technol.*, 6, 847, 1992.

203. Buhler, D. D., Doll, J. D., Proost, R. T., and Visocky, M. R., Integrating mechanical weeding with reduced herbicide use in conservation tillage corn production systems, *Agron. J.*, 87, 507, 1995.

204. Steckel, L. E., DeFelice, M. S., and Sims, B. D., Integrating reduced rates of post-emergence herbicides and cultivation for broadleaf weed control in soybeans (*Glycine max*), *Weed Sci.*, 38, 541, 1990.

205. Mulder, T. A. and Doll, J. D., Integrating reduced herbicide use with mechanical weeding in corn (*Zea mays*), *Weed Technol.*, 7, 382, 1993.

206. Hartzler, R. G., *Reducing Herbicide Rates for Narrow-Row Soybeans*, IPM-511, Adcock, E., Ed., Iowa State University, Ames, 1996, 2 pp.

207. Eadie, A. G., Swanton, C. J., Shaw, J. E., and Anderson, G. W., Banded herbicide applications and cultivation in a modified no-till corn (*Zea mays*) system, *Weed Technol.*, 6, 535, 1992.

208. Poston, D. H., Murdock, E. C., and Toler, J. E., Cost-efficient weed control in soybean (*Glycine max*) with cultivation and banded herbicide applications, *Weed Technol.*, 6, 990, 1992.

209. Wallinga, J., Groeneveld, R. M. W., and Lotz, L. A. P., Measures that describe weed spatial patterns at different levels of resolution and their applications for patch spraying of weeds, *Weed Res.*, 38, 351, 1998.

210. Paice, M. E. R., Day, W., Rew, L. J., and Howard, A., A stochastic simulation model for evaluation the concept of patch spraying, *Weed Res.*, 38, 373, 1998.

211. Faechner, T. and Hall, L. M., Operational site specific spraying, in *Proc. Weed Sci. Soc. Am.*, 39, Wilcutt, J. W., Ed., Allen Press, Lawrence, KS, 1999, 34.

212. Kemp, M. S. and Caseley, J. C., Synergist to combat herbicide resistance, in *Herbicide Resistance in Weeds and Crops*, Caseley, J. C., Cussans, G. W., and Atkin, R. K., Eds., Butterworth-Heinemann, Oxford, 1991, 279.

213. Duke, S. O., Rimando, A. M., Dayan, F. E., Tellez, M. R., Canel, C., and Scheffler, B. E., Strategies for use of plant-derived natural products in weed management, in *Proc. 3rd Int. Weed Science Congr.*, 3, Legere, A., Ed., International Weed Science Society, Corvallis, OR, 2000, 102.

214. Gerwick, B. C., Graupner, P. R., Gray, J. A., Peacock, C. L., Hahn, D. R., Chapin, E. L., and Schmitzer, P. R., New herbicides from screening microbes, in *Proc. 3rd Int. Weed Science Congr.*, 3, Legere, A., Ed., International Weed Science Society, Corvallis, OR, 2000, 103.

215. Young, A. L., Genetically modified crops: the real issues hindering public acceptance, *Environ. Sci. Pollut. Res.*, 7, 1, 2000.

216. Re, D. B., Rogers, S. G., Stone, T. B., and Serdy, F. S., Herbicide-tolerant plants developed through biotechnology: regulatory considerations in the United States, in *Herbicide-Resistant Crops — Agricultural, Environmental, Economic, Regulatory, and Technical Aspects*, Duke, S. O., Ed., Lewis Publishers, Boca Raton, FL, 1996, 341.

217. Cetiom, A.M., Management of herbicide tolerant crops in Europe, *Brit. Crop Prot. Conf.*, 3, 947, 1997.
218. Wilcut, J. W., Coble, H. D., York, A. C., and Monks, D. W., The niche for herbicide-resistant crops in U. S. agriculture, in *Herbicide-Resistant Crops — Agricultural, Environmental, Economic, Regulatory, and Technical Aspects*, Duke, S. O., Ed., Lewis Publishers, Boca Raton, FL, 1996, 213.
219. Burnside, O. C., Rationale for developing herbicide-resistant crops, *Weed Technol.*, 6, 621, 1992.
220. Mazur, B. J. and Falco, S. C., The development of herbicide resistant crops, *Annu. Rev. Plant Physiol. Plant Mol. Biol.*, 40, 441, 1989.
221. Berzenyi, Z., Kopacsi, J., Arendas, T., Bonis, P., and Lap, D. Q., Three-years experiences on the efficacy and selectivity of glufosinate-ammonium in transgenic maize, *Z. Pflanzenkr. Pflanzenschutz*, 16, 391, 1998.
222. Dunwell, J. M., Time-scale for transgenic product development, *Field Crops Res.*, 45, 135, 1996.
223. Lenardon, A., Arregui, M. C., Maitre, M. I., Sanchez, D., Scotta, R., and Enrique, S., Glyphosate residues in transgenic RR soybean plants and grains, in *Proc. 3rd Int. Weed Science Congr.*, 3, Legere, A., Ed., International Weed Science Society, Corvallis, OR, 2000, 126.
224. Holliday, M. J. and Keen, N. T., The role of phytoalexins in the resistance of soybean leaves to bacteria: effect of glyphosate on glyceollin accumulation, *Phytopathology*, 72(11), 1470, 1982.
225. Egley, G. H. and Elmore, D., Germination and the potential persistence of weedy and domestic okra (*Abelmoschus esculentus*) seeds, *Weed Sci.*, 35, 45, 1987.
226. Buckelew, L. D., Effect of Weed Management Systems on Canopy-Inhabiting and Surface-Active Arthropods in Iowa Soybeans, Ph.D. dissertation, Iowa State University, Ames, 1999, 106 pp.
227. Nalewaja, J. D., Cultural practices for weed resistance management, *Weed Technol.*, 13, 643, 1999.
228. Johnson, G. A., Hoverstad, T. R., and Greenwald, R. E., Integrated weed management using narrow corn row spacing, herbicides, and cultivation, *Agron. J.*, 90, 40, 1998.
229. Naylor, J. M. and Jana, S., Genetic adaptation for seed dormancy in *Avena fatua*, *Can. J. Bot.*, 54, 306, 1976.
230. Buhler, D. D., Population dynamics and control of annual weeds in corn (*Zea mays*) as influenced by tillage systems, *Weed Sci.*, 40, 241, 1992.
231. Derksen, D. A., Lafond, G. P., Thomas, A. G., Loeppky, H. A., and Swanton, C. J., Impact of agronomic practice on weed communities: tillage systems, *Weed Sci.*, 41, 409, 1993.
232. Pareja, M. R., Staniforth, D. W., and Pareja, G. P., Distribution of weed seed among soil structural units, *Weed Sci.*, 33, 182, 1985.
233. Buhler, D. D., Tillage systems and weed population dynamics and management, in *Integrated Weed and Soil Management*, Hatfield, J. L., Buhler, D. D., and Stewart, B. A., Eds., Ann Arbor Press, Chelsea, MI, 1998, 223.
234. Buhler, D. D., Gunsolus, J. L., and Ralston, D. F., Integrated weed management techniques to reduce herbicide inputs in soybean, *Agron. J.*, 84, 973, 1992.
235. Mt. Pleasant, J., Burt, R. F., and Frisch, J. C., Integrating mechanical and chemical weed management in corn (*Zea mays*), *Weed Technol.*, 8, 217, 1994.
236. Mohler, C. L., Frisch, J. C., and Mt. Pleasant, J., Evaluation of mechanical weed management programs for corn (*Zea mays*), *Weed Technol.*, 11, 123, 1997.

237. Oriade, C. and Forcella, F., Maximizing efficacy and economics of mechanical weed control in row crops through forecasts of weed emergence, in *Expanding the Context of Weed Management*, Buhler, D. D., Ed., Haworth Press, Binghamton, NY, 1999, 189.

238. Gunsolus, J. L., Mechanical and cultural weed control in corn and soybeans, *Am. J. Alt. Agric.*, 5(3), 114, 1990.

239. Lovely, W. G., Weber, C. R., and Staniforth, D. W., Effectiveness of the rotary hoe for weed control in soybeans, *Agron. J.*, 50, 621, 1958.

240. Liebman, M. and Owen, M. D. K., Diversifying cropping systems for weed management — where to from here? in *Proc. 3rd Int. Weed Science Congr.*, 3, Legere, A., Ed., International Weed Science Society, Corvallis, OR, 2000, 60.

241. Thill, D. C. and Mallory-Smith, C. A., The nature and consequence of weed spread in cropping systems, *Weed Sci.*, 45, 337, 1997.

242. Brommer C. L. and Witt, W. W., Impact of cultural practices on common pokeweed (*Phytolacca americana*) control in Kentucky no-till soybean production, in *Proc. Weed Sci. Soc. Am.*, 40, Wilcutt, J. W., Ed., Allen Press, Lawrence, KS, 2000, 69.

243. Crookston, R. K., The rotation effect: what causes it to boost yields? *Crops and Soils*, 31, 12, 1984.

244. Crookston, R. K., Kurle, J. E., Copeland, P. J., Ford, J. H., and Lueschen, W. E., Rotational cropping sequence effects yields of corn and soybean, *Agron. J.*, 83, 108, 1991.

245. Liebman, M. and Davis, A. S., Integration of soil, crop and weed management in low-external-input farming systems, *Weed Res.*, 40, 27, 2000.

246. Brecke, B., *Senna obtusifolia* seed dynamics as affected by corn and soybean cropping systems, in *Proc. 3rd Int. Weed Science Congr.*, 3, Legere, A., Ed., International Weed Science Society, Corvallis, OR, 2000, 16.

247. Deuber, R. and Nunes Gerin, M. A., Seedbank reduction with rotations of *Arachis hypogaea, Mucuna aterrima* and *Zea mays*, in *Proc. 3rd Int. Weed Science Congr.*, 3, Legere, A., Ed., International Weed Science Society, Corvallis, OR, 2000, 14.

248. Hoffman, M. L., Buhler, D. D., and Owen, M. D. K., Multi-year evaluation of model-based weed control under variable crop and tillage conditions, in *Expanding the Context of Weed Management*, Buhler, D. D., Ed., Haworth Press, Binghamton, NY, 1999, 207.

249. Owen, M. D. K. and Pecinovsky, K. T., Effect of alternative management strategies on weed populations, in *Proc. North Cent. Weed Sci. Soc.*, 48, Ross, M. and Lembi, C., Eds., North Central Weed Control Conference, Champaign, IL, 1993, 97.

250. Whigham, K., Farnham, D., Lundvall, J., and Tranel, D., *Soybean Replant Decisions*, PM-1851, Edwards, E., Ed., Iowa State University, Ames, 2000, 8 pp.

251. Norwood, C. A. and Currie, R. S., Tillage, planting date, and plant population effects on dryland corn, *J. Prod. Agric.*, 9(1), 119, 1996.

252. Nafzinger, E. D., Corn planting date and plant population, *J. Prod. Agric.*, 7(1), 59, 1994.

253. Norsworthy, J. K. and Oliver, L. R., Competitive potential and return of a glyphosate-tolerant/conventional soybean mix, in *Proc. 3rd Int. Weed Science Congr.*, 3, Legere, A., Ed., International Weed Science Society, Corvallis, OR, 2000, 20.

254. Teasdale, J. R., Influence of narrow row/high population corn (*Zea mays*) on weed control and light transmittance, *Weed Technol.*, 9, 113, 1995.

255. Nieto, J. H. and Staniforth, D. W., Corn-foxtail competition under various production conditions, *Agron. J.*, 53, 1, 1961.

256. Bello, I. A., Owen, M. D. K., and Hatterman-Valenti, H. M., Effect of shade on velvetleaf (*Abutilon theophrasti*) growth, seed production and dormancy, *Weed Technol.*, 9, 452, 1995.

257. Forcella, F., Westgate, M. E., and Warnes, D. D., Effect of row width on herbicide and cultivation requirements in row crops, *Am. J. Alt. Agric.*, 7(4), 161, 1992.

258. Puricelli, E., Faccini, D., Orioli, G., and Sabbatini, M. R., Effect of glyphosate doses and soybean row spacing on *Anoda cristata* (L.) Schlecht biomass and seed production, in *Proc. 3rd Int. Weed Science Congr.*, 3, Legere, A., Ed., International Weed Science Society, Corvallis, OR, 2000, 31.

259. Quakenbush, L. S. and Andersen, R. N., Effect of soybean (*Glycine max*) interference on eastern black nightshade (*Solanum ptycanthum*), *Weed Sci.*, 32, 638, 1984.

260. Reddy, K., Ultra-narrow row planting as a weed management strategy for soybean in Miss. Delta, in *Proc. Weed Sci. Soc. Am.*, 40, Wilcutt, J. W., Ed., Allen Press, Lawrence, KS, 2000, 67.

261. Nurse, R. E. and Swanton, C. J., The effect of row width and soil tillage on predispersal weed seed predation, in *Proc. Weed Sci. Soc. Am.*, 40, Wilcutt, J. W., Ed., Allen Press, Lawrence, KS, 2000, 100.

262. Lindquist, J. L., Mortensen, D. A., and Johnson, B. E., Mechanisms for crop tolerance and velvetleaf suppressive ability, *Agron. J.*, 90, 787, 1998.

263. Teasdale, J. R., Contribution of cover crops to weed management in sustainable agricultural systems, *J. Prod. Agric.*, 9(4), 475, 1996.

264. Williams, M. M., II, Mortensen, D. A., and Doran, J. W., Assessment of weed and crop fitness in cover crop residues for integrated weed management, *Weed Sci.*, 46, 595, 1998.

265. Williams, M. M., II, Mortensen, D. A., and Doran, J. W., No-tillage soybean performance in cover crops for weed management in the western corn belt, *J. Soil Water Conserv.*, 55, 79, 2000.

266. Teasdale, J. R. and Mohler, C. L., The physical properties of mulches contributing to weed suppression, in *Proc. 3rd Int. Weed Science Congr.*, 3, Legere, A., Ed., International Weed Science Society, Corvallis, OR, 2000, 95.

267. Ekeleme, F., Akobundu, I. O., Isichei, A. O., and Chikoye, D., Planted fallow reduces weed seedbank in south-western Nigeria, in *Proc. 3rd Int. Weed Science Congr.*, 3, Legere, A., Ed., International Weed Science Society, Corvallis, OR, 2000, 13.

268. Dyck, E. and Liebman, M., Soil fertility management as a factor in weed control: the effect of crimson clover residues, synthetic nitrogen fertilizer, and their interaction on emergence and early growth of lambsquarters and sweet corn, *Plant and Soil*, 167, 227, 1994.

269. Weston, L. A., Utilization of allelopathy for weed management in agroecosystems, *Agron. J.*, 88, 860, 1996.

270. Hartzler, R. G., Velvetleaf (*Abutilon theophrasti*) population dynamics following a single year's seed rain, *Weed Technol.*, 10, 581, 1996.

271. Pareja, M. R. and Staniforth, D. W., Seed-soil microsite characteristics in relation to seed germination, *Weed Sci.*, 33, 190, 1985.

272. Ball, D. A., Weed seedbank response to tillage, herbicides and crop rotation sequence, *Weed Sci.*, 40, 654, 1992.

273. Hoffman, M. L., Owen, M. D. K., and Buhler, D. D., Effects of crop and weed management on density and vertical distribution of weed seeds in soil, *Agron. J.*, 90, 793, 1998.

274. Pitty, A., Staniforth, D. W., and Tiffany, L. H., Fungi associated with caryopses of *Setaria* species from field-harvested seeds and from soil under two tillage systems, *Weed Sci.*, 35, 319, 1987.

275. Eyherabide, J. and Calvino, P., Control of *Setaria viridis* (L.) Beauv. and *Digitaria sanguinalis* (L.) Scop in fallow during the autumn prior to no-till corn seeding, in *Proc. 3rd Int. Weed Science Congr.*, 3, Legere, A., Ed., International Weed Science Society, Corvallis, OR, 2000, 14.

276. Boutsalis, P. and Powles, S. B., Seedbank characteristics of herbicide-resistant and susceptible *Sisymbrium orientale*, *Weed Res.*, 38, 389, 1998.

277. Staniforth, D. W., Responses of soybean varieties to weed competition, *Agron. J.*, 54, 11, 1062.

278. Anderson, R. L., Tanaka, D. L., Black, A. L., and Schweizer, E. E., Weed community and species response to crop rotation, tillage, and nitrogen fertility, *Weed Technol.*, 12, 531, 1998.

279. Knezevic, S., Evans, S., Shapiro, C., and Lindquist, J., Effect of nitrogen on critical period of weed control in corn, in *Proc. 3rd Int. Weed Science Congr.*, 3, Legere, A., Ed., International Weed Science Society, Corvallis, 2000, 52.

280. Povey, R. D., Smith, H., and Watt, T. A., Predation of annual grass weed seeds in arable field margins, *Ann. Appl. Biol.*, 122(2), 323, 1993.

281. Kremer, R. J., Management of weed seedbanks with microorganisms, *Ecol. Appl.*, 3, 42, 1993.

282. Ennis, W. B., Biological weed control in weed management in the developing countries, *FAO Plant Prod. Prot.*, 31, 113, 1983.

283. Kenneday, A. C. and Kremer, R. J., Microorganisms in weed control strategies, *J. Prod. Agric.*, 9(4), 480, 1996.

284. Brust, G. E. and House, G. J., Weed seed destruction by arthropods and rodents in low-input soybean agroecosystems, *Am. J. Alt. Agric.*, 3(1), 19, 1988.

285. Hasan, S. and Ayes, P. G., The control of weeds through fungi: principles and prospects, *New Phytol.*, 115, 201, 1990.

286. Altman, J., Neate, S., and Rovira, A. D., Herbicide-pathogen interactions and myco-herbicides as alternative strategies for weed control, in *Microbes and Microbial Products as Herbicides*, Hoagland, R. E., Ed., American Chemical Society, Washington, D.C., 1990, 240.

287. Quimby, P. C. and Boyette, C. D., *Alternaria cassiae* can be integrated with selected herbicides, in *Proc. Weed Sci. Soc. Am.*, 26, Ahrens, J. F., Ed., Weed Science Society of America, Champaign, IL, 1986, 52.

288. Kennedy, A. C. and Smith, K. L., Soil microbial diversity and the sustainability of agricultural soils, *Plant and Soil*, 170, 75, 1995.

289. Rahe, J. E., Levesque, C. A., and Johal, G. S., Synergistic role of soil fungi in the herbicidal efficacy of glyphosate, in *Microbes and Microbial Products as Herbicides*, Hoagland, R. E., Ed., American Chemical Society, Washington, D.C., 1990, 260.

290. Mohler, C. L., A model of the effects of tillage on emergence of weed seedlings, *Ecol. Appl.*, 3(1), 53, 1993.

291. Marra, M. C. and Carlson, G. A., An economic threshold model for weeds in soybeans (*Glycine max*), *Weed Sci.*, 31, 604, 1983.

292. Coble, H. D. and Mortensen, D. A., The threshold concept and its application to weed science, *Weed Technol.*, 6, 191, 1992.

293. Cousens, R., Theory and reality of weed control thresholds, *Plant Prot. Q.*, 2, 13, 1987.

294. Bauer, T. A. and Mortensen, D. A., A comparison of economic and economic optimum thresholds for two annual weeds in soybeans, *Weed Technol.*, 6, 228, 1991.
295. Collender, R. N., Estimation risk in farm planning under uncertainty, *Am. J. Agric. Econ.*, 71, 996, 1989.
296. Cashman, C. M., Martin, M. A., and McCarl, B. A., Economic consequences of bans on corn (*Zea mays*) and soybean (*Glycine max*) herbicides commonly used on Indiana farms, *Weed Sci.*, 29, 323, 1981.
297. Hoffman, M. L., Buhler, D. D., and Owen, M. D. K., Weed population and crop yield response to recommendations from a weed control decision aid, *Agron. J.*, 91, 386, 1999.
298. Czapar, G. F., Curry, M. P., and Wax, L. M., Grower acceptance of economic thresholds for weed management in Illinois, *Weed Technol.*, 11, 828, 1997.
299. Maxwell, B. D. and Mortimer, A. M., Selection for herbicide resistance, in *Herbicide Resistance in Plants: Biology and Biochemistry*, Powles, S. B. and Holtum, J. A. M., Eds., CRC Press, Boca Raton, FL, 1994, 263.
300. Dyer, W. E., Herbicide-resistant crops: a weed scientist's perspective, *Phytoprotection*, 75(Suppl.), 71, 1994.
301. Foster, K. R., Vecchia, P., and Repacholi, M. H., Science and the precautionary principle, *Science*, 288, 979, 2000.

5 World Wheat and Herbicide Resistance

Donald C. Thill and Deirdre Lemerle

CONTENTS

5.1 INTRODUCTION

Herbicide resistance continues to escalate worldwide (see Chapter 1), causing significant yield losses and increasing the cost of food production.[1] In this chapter we outline the major wheat-growing (*Triticum aestivum* L. and durum Desf.) areas of the world, the factors implicated with the development of herbicide resistance, and the management factors commonly used in different regions to combat resistance once it has developed.

0-8493-2219-7/01/$0.00+$.50
© 2001 by CRC Press LLC

5.2   WHEAT AGROECOSYSTEMS

Wheat is grown worldwide; it is the most widely adapted of all the cereals.[2] It has the largest total production and total area, and it is the number one food grain consumed by humans.[3] Winter wheat is grown more extensively than spring wheat, with major production occurring in the United States, western Europe, the Balkans, southern Russia, and China. Spring wheat is often grown in cool, dry climates that are unfavorable for winter wheat production. These areas include the northern United States, the Canadian prairies, Argentina, and central and northern Russia. Spring wheat is also autumn-planted in milder climates, such as those of Mexico, Brazil, India, Australia, and the southwestern United States.[2] In 1998, over 224 million ha of wheat was harvested worldwide.[4] In this year, wheat was harvested on about 100.5 million ha in Asia, 59.8 million ha in Europe (including the former U.S.S.R.), 34.6 million ha in North America, 12.0 million ha in Latin America, 11.5 million ha in Oceania, and 9.9 million ha in Africa. Only 29% of the harvested area was in industrialized countries.[4]

Wheat is a cool-season crop with production concentrated between latitudes 30 and 60°N and 27 to 40°S.[5,6] According to Briggle,[7] the minimum temperature for wheat growth is about 3 to 4°C, while the optimal temperature is about 25°C. Briggle and Curtis[3] further state that wheat grows best on well-drained soils from sea level to about 3000 m above sea level, but it has been grown in Tibet at 4270 to 4570 m.[5] Wheat can be grown where annual precipitation ranges from 250 to 1750 mm. However, most wheat is grown in areas with 375 to 875 mm annual rainfall. Globally, wheat is grown in rotations with many different crops, including barley (*Hordeum vulgare* L.), oilseed rape or canola (*Brassica napus* L.), grain legumes, sorghum (*Sorghum vulgare* Pers.), potato (*Solanum tuberosum* L.), sugar beet (*Beta vulagris* L.), corn (*Zea maize* L.), oats (*Avena sativa* L.), and rice (*Oryza sativa* L.).

5.2.1   AFRICA

About 17.9 million t of wheat was produced on 9.9 million ha of land in Africa in 1998 (Table 5.1).[4] Major wheat-producing countries included Algeria (13%), Egypt (34%), Morocco (25%), South Africa (8%), and Tunisia (8%). Most of the wheat produced in Egypt is grown under irrigation. Wheat is produced under rainfed conditions throughout much of the rest of Africa.[3]

5.2.2   ASIA

China ranks first among developing countries in wheat production (Table 5.1).[4] In 1998, China produced 110 million t of wheat on 30 million ha. Wheat is grown throughout China, however, most is grown east of the 100th meridian.[3] Based on wheat growth habit, there are three main wheat production regions in China: the northern winter wheat region, the southern winter wheat region (spring and facultative types are planted in the autumn), and the spring wheat region. About 60% of the wheat production area is winter wheat and 40% is spring wheat. However, only 17% of the spring wheat is planted in the spring; the remainder is autumn planted. Semidwarf wheat varieties are commonly grown in China.[3] Production constraints

TABLE 5.1
Worldwide Wheat Production and Harvested Area in 1998

Area	Production (million t)	Harvested Area (million ha)
Africa	17.9	9.9
China	110.0	30.0
Oceania	22.1	11.5
Europe	183.0	59.8
Middle East	55.2	34.7
U.S.A.	69.4	23.9
Canada	24.4	10.8
Mexico	3.2	0.8
South America	15.1	7.2

vary across the regions and include spring droughts and hot dry winds during the grain-filling stage. Wheat is often double and triple cropped with rice in the southern and central wheat-growing regions of China.

North and South Korea and Mongolia produced 364,000 t of wheat on 428,000 ha of land in 1998.[4] In South Asia (India, Pakistan, Nepal, and Bangladesh), more than 87.5 million t of wheat was produced on 35.4 million ha of land in 1998 (Table 5.1).[4] India and Pakistan accounted for 75 and 21%, respectively, of the total production. Most wheat in India and Pakistan is grown under irrigation, while production in Nepal and Bangladesh is under rainfed conditions.[3] In India, wheat is grown on the northwestern plains in a wheat–rice rotation. Common white spring wheat is widely grown in the region, but it is planted in the autumn and harvested the following spring. The dramatic increase in wheat production in South Asia is attributed to availability of well-adapted semidwarf cultivars, which possess the attributes of increased yield and disease resistance. A coordinated training, research, and extension program has been effective.[8]

5.2.3 Australia

Oceania wheat growers produced 22.1 million t of wheat on 11.5 million ha of land in 1998 (Table 5.1).[4] Australian growers accounted for over 98% of the total wheat production from this region of the world. Cereals (wheat, barley, oat, and triticale) are grown intensively in rotation with canola, grain legumes (lupin [*Lupinus* spp.], field pea [*Pisum sativum* L.], and chickpea [*Cicer arientinum*]), mainly in winter from April to November under rainfed, dry land conditions.[9] These legume-based pastures are an essential component of the farming system for maintaining soil fertility and structure, providing greater flexibility for weed management, and their duration can vary from 2 to 8 years. A decline in the price of wool, combined with the availability of selective herbicides since the 1980s, has led to an intensification of crop production, with a consequent reduction in pasture and livestock. In the cropping areas of northern New South Wales and southern Queensland, summer rainfall enables crops such as sorghum to be included in rotation with cereals or grain legumes. Summer cropping is rarely an option in the winter rainfall areas of the southern wheat belt (southern New South Wales, Victoria, South Australia, and Western Australia) and the ground is fallowed over summer (Figure 5.1). Recently, some farmers have grown opportunistic summer crops, such as millet (*Panicum* spp.) for grain or silage production. In the southern areas, winter crops are grown continuously or in rotation with a pasture phase.

5.2.4   EUROPE

In Europe, under continuous cropping systems, winter and spring cereals, oilseed rape and sugar beet, are rotated with corn and soybean (*Glycine max* L.) in southern areas or potatoes in northern areas. The growing season for spring cereals is April to September, whereas winter cereals are sown in September and October, and are harvested the following autumn. In the last decade, there has been a trend toward growing winter cereals because of higher grain yields and profitability. European farmers produced about 183 million t of wheat on 59.8 million ha of land in 1998 (Table 5.1).[4] Countries with the highest wheat production included France (22%), Russian Federation (15%), Germany (11%), United Kingdom (8%), and Ukraine (8%).

5.2.5   MIDDLE EAST

The Middle East (Turkey, Iran, Afghanistan, Syria, Iraq, Israel, and other Eastern Mediterranean countries) produced 55.2 million t of wheat on 34.7 million ha of land in 1998 (Table 5.1).[4] Most production was in Turkey (17%), Kazakhstan (17%), and Iran (13%). Climatic conditions throughout the region are diverse and variable.[3]

FIGURE 5.1   Map of the Australian wheat belt.

Common white and durum wheat varieties are planted throughout much of the region. Except for Turkey, cultural practices have not been modernized, resulting in low grain yields.

5.2.6 NORTH AMERICA

In 1998, over 69 million t of wheat was harvested in the United States from about 23.9 million ha (Table 5.1).[4] Five market classes of wheat are grown in four regions in the United States. Hard red winter wheat is grown mostly in the Central and Southern Plains; soft red winter wheat is grown in the East; white winter and spring wheat are grown in the Pacific Northwest; some white winter wheat is grown in Michigan and New York; and most hard red spring and durum wheat is grown in the Northern Great Plains.[3] Most wheat is grown under rainfed conditions, with weather conditions ranging from dry and cold to warm and wet. Cropping systems within regions vary from monoculture wheat (wheat–wheat or wheat–fallow) to diverse annual crop rotations where wheat is grown every second to fourth year of the rotation. Rotational crops include barley, grain legumes, oilseed rape and canola, potatoes, sugar beets, and sorghum. Specific practices associated with wheat production within each wheat-producing region of the United States were reviewed in Donald.[10] Weed management in grain crops was reviewed by Donald and Easton,[11] and in no-till and reduced-tillage winter wheat production systems by Wicks.[12]

Wheat is the principal crop in Canada.[3] Canadian farmers produced 24.4 million t of wheat on 10.8 million ha of land in 1998.[4] About 95% of all wheat grown in Canada is produced in the southern regions of the Prairie Provinces of Manitoba, Saskatchewan, and Alberta. Long, cold winters and short, hot summers are common in this region of Canada. Precipitation is often limiting and periods of drought are common. Hard red spring wheat is most commonly grown in the Prairie Provinces. Severe winters throughout most of the region limit winter wheat production, although some winter wheat is grown in southern Alberta. Wheat is often grown after a fallow year or in rotation with canola, grain legumes, barley, and other spring-planted crops. Hunter et al.[13] reviewed cropping practices used to grow wheat in Canada.

Mexico and Guatemala are the only countries that produce a significant amount of wheat in Central America. In 1998, Mexican farmers produced 3.2 million t of wheat on 772,000 ha,[4] mostly under irrigation in the coastal valleys of the northwestern states of Sonora and Sinaloa.[3] About 100,000 ha of wheat is grown in the central plateau, called the Bajío, and some is grown in the rainfed highlands. Mexican wheat producers use high-yielding, semidwarf varieties and modern cultural practices.[3] In Guatemala, about 5000 t of wheat was produced on 3000 ha in highland valleys under rainfed conditions.[3,4]

5.2.7 SOUTH AMERICA

Peru, Bolivia, Colombia, and Ecuador produced about 350,000 t of wheat on 351,000 ha (Table 5.1).[4] Most wheat is produced under rainfed conditions, but annual rainfall is highly variable in both quantity and distribution.[3] Yields and production are relatively low, constrained by variable rainfall, diseases, and inadequate cultural practices.

Argentina, Brazil, Chile, Paraquay, and Uruguay produced 14.7 million t of wheat on 6.8 million ha in 1998, with almost 90% being produced in Brazil and Argentina.[4] Wheat is produced in a variety of environments in this Southern Cone region ranging from the highly productive humid plains of Argentina to the acid soils of Brazil.[3] Production constraints in Brazil include acid soils with high levels of soluble aluminum and strong phosphorus fixation, heavy disease incidence, variable rainfall (often too much, especially at harvest), and unseasonable frosts. Variable rainfall and an unstable supply of fertilizer are major wheat production constraints in Argentina.[3]

5.3 FACTORS AFFECTING WEED RESISTANCE TO HERBICIDES

The pattern of herbicide usage is closely related to the farming systems in different regions, levels of agronomic input, and socioeconomic factors and can directly influence the development of resistance. The distribution of weed species at the regional level and the biological characteristics of particular species will determine the rate of evolution of resistance in any particular region. In many areas, a change from multiple cultivation for weed control to reduced tillage to prevent soil erosion has led to greater dependence on herbicides. Other associated factors include the wide availability of a cheap herbicide and/or reliance on few herbicide modes of action. An increase in the cost of hand-weeding resulting from socioeconomic changes in developing countries may result in greater dependence on herbicides. Growers and advisers are sometimes slow to recognize or accept the existence of resistant weed populations and they may fail to act appropriately, resulting in rapid increases in densities of resistant populations. Finally, as resistance usually starts as a patch, this may not be noticed by the farmer until it is too late, especially as farmers have less time available for monitoring fields for weed control.

Herbicide-resistant weeds have been reported for wheat production systems in only 20 of the 123 wheat-producing countries in the world (Tables 5.2 and 5.3).[4,14] Of these, 20 countries, 15 are categorized as industrial or developed (Australia, Belgium-Luxembourg, Bulgaria, Canada, Denmark, France, Germany, Hungary, Israel, the Netherlands, South Africa, Spain, Switzerland, United Kingdom, and the United States) and 5 are classified as developing (Chile, China, India, Mexico, and Saudi Arabia).[4] The industrial or developed countries accounted for about 93% of the reported cases of herbicide-resistant weeds in wheat (cereal) production systems, while developing countries accounted for only 7% of the cases. The actual number of cases of herbicide-resistant weeds in developing and some developed countries may be greatly underestimated because thorough surveys have not been conducted. However, the incidence of herbicide resistance in these countries will likely increase in the future as monitoring and reporting procedures improve and as herbicide use increases. The risk of selecting for herbicide-resistant weed biotypes in both developing and developed countries could be very high if increased herbicide use is not coordinated with effective educational programs on management strategies to delay or prevent selection of resistant weeds.

The estimated U.S. dollar value of the global herbicide market for 1996 to 1997 was $16.5 billion, with wheat accounting for $1.7 billion, ranking fourth behind corn, soybean, and industrial/noncrop usage.[15] In North America this value was $450 million, in Europe $900 million, in Asia $150 million, and in the rest of the world (South America, Africa, and Oceania) $200 million.[15] Globally, the average amount spent on herbicides to control weeds in wheat in 1997 was $7.73/ha. The average amount spent in Asia was $1.72/ha, in Europe $12.68/ha, in North America $12.16/ha, and in the rest of the world $8.00/ha.[15] Thirteen of the fifteen industrial or developed countries reporting the occurrence of herbicide-resistant weeds in wheat production systems are in Europe or North America (see p. 170). These countries accounted for about 52% of the 1997 worldwide area of wheat production, 79% of the global herbicide market for wheat, and 91% of the reported cases of herbicide-resistant weeds in wheat production systems. In 1997, Asia accounted for almost 40% of the land from which wheat was harvested, but less than 9% of the global herbicide usage in wheat was in Asia.[15] Only three cases of herbicide-resistant weeds have been reported in Asian wheat production systems — two in China and one in India.[14]

Not surprisingly, the amount of herbicide used to control weeds in wheat is clearly correlated with the number of cases of herbicide-resistant weeds reported in different wheat-growing regions of the world.[16] There is a greater likelihood of selecting for herbicide-resistant weed biotypes in herbicide-intensive agricultural production systems (i.e., where selection pressure is greatest). Selection of a herbicide-resistant weed biotype is usually associated with monoculture or short rotation cropping systems (e.g., wheat–rotational crop or fallow–wheat) where the same or similar herbicides are applied repeatedly over a number of years to the same area of land.[16,17] The number of herbicide applications required to select a herbicide-resistant weed biotype depends on herbicide chemical properties (e.g., target site and soil persistence), the weed species, and the specific agronomic practices.[18]

Acetyl coenzyme A carboxylase (ACCase)-inhibiting herbicides (e.g., diclofop-methyl) and acetolactate synthase (ALS)-inhibiting herbicides (e.g., chlorsulfuron) account for well over 75% of the reported cases of herbicide resistance in wheat production systems.[14,19] These herbicides have been used extensively for many years because they effectively and affordably control economically important weeds. In many cases, a herbicide-resistant weed biotype was selected following as few as three annual applications of one of these herbicides.[20,21] Herbicide-resistant weed biotypes in wheat-based production systems have also been reported for many other herbicide groups (Tables 5.2 and 5.3).[14]

ACCase-inhibiting herbicides were introduced for use in cereals in the late 1970s,[20] and a resistant biotype was first found in an Australian wheat field in 1980.[22] Resistance may be caused by several mechanisms, including altered target site and metabolism as reviewed by Hall et al.[23] In many cases, resistance is conferred by a single, semidominant nuclear trait.[20] However, resistance is phenotypically dominant at typical herbicide field use rates.[24] The unpredictable pattern of cross-resistance between herbicide families in ACCase-herbicide-resistant weeds indicates that more than one mutant allele of the ACCase gene confers for resistance.[24,25] Little or no differences in fitness have been observed between ACCase-herbicide-resistant and -susceptible weed biotypes.[20] Many of the reported grass species with ACCase-herbicide-resistant biotypes are cross-pollinated (*Lolium rigidum* Gaud. [rigid ryegrass], *Alopecurus myosuroides* Huds. [blackgrass], *L. multiflorum* Lam [Italian ryegrass]), are prolific seed producers (*Lolium* spp., *A. myosuroides*, and *Avena fatua* L. [wild oat]), and have short to long seed dormancy (*A. myosuroides* vs. *A. fatua*, respectively). Grass seeds are usually dispersed naturally only a short distance from the mother plant, but they may be dispersed long distances by human and animal activity.

ALS-inhibiting herbicides were introduced for use in wheat in 1982[21] and a resistant weed biotype was discovered in a U.S. wheat field in 1987.[26] With few exceptions (e.g., *L. rigidum, A. myosuroides, A. fatua*), resistance has been due to an altered form of ALS with decreased sensitivity to ALS-inhibiting herbicides rather than to differences in herbicide uptake, translocation, or metabolism.[27–29] Resistance is usually conferred by a single point mutation of the ALS protein,[21,30] which may alter regulation of branched-chain amino acid synthesis differently in resistant and susceptible biotypes.[30] Target-site cross-resistance among ALS-inhibitor herbicides within an ALS-herbicide-resistant weed biotype is not predictable and likely depends on the site of the point mutation on the ALS gene.[31] Fitness studies have shown that the relative competitiveness, seed output, and seed longevity are nearly equal between ALS-herbicide-resistant and -susceptible weed biotypes or become more competitive.[21] Seeds of ALS-herbicide-resistant biotypes often germinate faster than susceptible biotypes, especially at cooler temperatures, and thus may be more susceptible to early season tillage or herbicide application.[21] Many ALS-herbicide-resistant weeds are cross-pollinated, which can spread the trait within a local weed population (e.g., *Kochia scoparia* [L.] Schrad. [kochia], *L. rigidum, L. multiforum*, and *Salsola iberica* Sennen & Pau [Russian thistle]).[21,31] Other weeds have special seed dispersal mechanisms that can distribute seed long distances from the mother plant (*Lactuca serriola* L. [prickly lettuce], *K. scoparia*, and *S. iberica*). Most of the resistant biotypes are prolific seed producers; however, seeds do not tend to persist in the soil seedbank for long periods.[33–35] In a field situation, once ALS-inhibiting herbicide use was discontinued, the proportion of resistant *L. serriola* in the population inexplicably decreased, but its range increased, probably through wind-dispersed seed.[36] However, it was postulated that resistant biotypes may persist, at some level, in infested fields for many years, even in the absence of any additional selection pressure. Persistence will depend on effectiveness of specific control tactics (to prevent continued seed production and promote removal), seed longevity in soil, fitness, and immigration of herbicide-resistant seeds into a field.

5.3.1 AFRICA

Diclofop-methyl-resistance in *A. fatua* and *L. multiflorum* in South Africa and in *L. rigidum* in Saudi Arabia are the only cases of herbicide resistance reported in Africa.[14] In South Africa, populations of both weeds are increasing within current resistant sites and new sites continue to be identified.[14] *L. rigidum* infests less than 4000 ha in Saudi Arabia. Although no information is available on the crop management system(s) under which selection occurred, continued use of the herbicide to the exclusion of other weed management tactics is likely.

5.3.2 Asia

Chlorotoluron-resistant (Group 7) *Beckmannia syzigachne* (Steud.) (American sloughgrass) and *Alopecurus japonicus* (Steud.) (Japanese foxtail) in China are the only reported cases of herbicide resistance in cereals in East Asia (Table 5.2).[14] These herbicide-resistant weed biotypes infest less than 8000 ha. However, herbicide usage is expected to increase rapidly in China as the country becomes more economically stable and as herbicide technology is adopted.

Isoproturon-resistant (Group 7) *Phalaris minor* (Retz.) (littleseed canarygrass) in India is the only reported case of herbicide resistance in wheat in South Asia (Table 5.2).[14] The resistant populations infest less than 4000 ha, but new sites continue to be identified. Malik et al.[37] reviewed the development of resistance in *P. minor* and its management. In summary, the introduction of poorly competitive semidwarf wheat cultivars in the mid-1960s "green revolution" increased farmer dependence on herbicides for *P. minor* control in the 1970s. The continuous use of isoproturon for 15 years led to resistance in the mid-1980s.

In India, *P. minor* is a widely distributed weed that occurs at very high densities (more than 1000 plants/m^2) with prolific seed production, high seed dormancy, and a staggered germination pattern.[37] The weed is favored by high surface moisture for seedling emergence (wheat is irrigated in India), a phenology ideally suited to the climatic conditions and adoption of late sowing (December rather than November). The weed spreads between farms mainly as a seed contaminant and in floodwater, and once resistance develops, the rate of spread increases rapidly because of little competition from other weeds. In addition, reduced herbicide efficacy due to stubble burning, poor application, or late application has exacerbated development of resistance. This allows weed survival, seed production, and a buildup of weed numbers in subsequent seasons, therefore increasing the likelihood of resistance developing.

5.3.3 Australia

The main postemergence herbicides used in cereals in Australia are ACCase and ALS inhibitors and it is to these groups that major resistance problems have developed, as is the case in Europe and North America. The pre-emergence herbicides, trifluralin and pendimethalin, are now used more widely in response to resistance to ACCase- and ALS-inhibiting herbicides, in particular for control of *Lolium rigidum*.

In canola, a crop often grown in rotation with wheat, the ACCase herbicides are used for grass weed control, while most broadleaf weeds (and some grass weeds) are controlled pre-emergence with trifluralin or postemergence with clopyralid. Nonselective herbicides such as glyphosate and paraquat are also used as preseeding treatments before both cereals and canola. Triazines are used in triazine-resistant canola. The first triazine-resistant cultivar, cv. Siren, was introduced in 1993, followed by cv. Karoo in 1995.[38] Adoption of triazine-resistant cultivars was most rapid in Western Australia from 1995 onward. By 1998, 95% of canola in Western Australia

TABLE 5.2
Herbicide-Resistant Weeds Reported in Small Grain Cereal Production Systems Worldwide (except for the United States)[14,53]

Country	Weed Species	Herbicide Group	Cases	Year
Australia	*Avena fatua* (L.)	1 (A)	11–50	1985
	A. sterilis L. ssp. *Ludoviciana* (Durien) Nyman	1 (A)	101–500	1989
	Brassica tournefortii Gouan	2 (B)	2–5	1992
	Echium plantagineum L.	2 (B)	2–5	1997
	Fallopia convulvulus (L.) A. Löve	2 (B)	2–5	1993
	Fumaria densiflora DC	3 (K$_1$)	2–5	1999
	Hordeum glaucum Steud.	22 (D)	11–50	1982
	Lactuca serriola L.	2 (B)	11–50	1994
	Lolium rigidum Gaudin	1 (A)	501–1000	1982
		2 (B)	501–1000	1984
		3 (K$_1$)	6–10	1984
		5 (C$_1$)	11–50	1984
		9 (G)	2–5	1996
		15 (K$_3$)	1–5	1984
	Phalaris paradoxa L.	1 (A)	2–5	1997
	Raphanus raphanistrum L.	2 (B)	51–100	1997
		5 (C$_1$)	2–5	1999
	Rapistrum rugosum (L.) All.	2 (B)	2–5	1996
	Sinapis arvensis L.	2 (B)	2–5	1996
	Sisymbrium orientale L.	2 (B)	6–10	1990
	S. thellungii O. Schulz	2 (B)	1	1996
	Sonchus oleraceus L.	2 (B)	2–5	1990
	Urochloa panicoides P. Beauv.	5 (C$_1$)	2–5	1996
Belgium	*Alopecurus myosuroides* Huds.	1 (A)	11–50	1996
		2 (B)	11–50	1997
		7 (C$_2$)	Unknown	1996
	Avena fatua (L.)	1 (A)	1	1996
Bulgaria	*Amaranthus retroflexus* L.	5 (C$_1$)	2–5	1984
		7 (C$_2$)	2–5	1984
	Chenopodium album L.	5 (C$_1$)	2–5	1989
Canada	*Amaranthus powellii* S. Wats.	2 (B)	2–5	1998
	A. retroflexus L.	2 (B)	1	1998
	Avena fatua L.	1 (A)	1001–10000	1990
		2 (B)	501–1000	1994
		8 (N,Z)	501–1000	1989
		25 (Z)	501–1000	1994
	Galeopsis tetrahit L.	2 (B)	1	1995
		4 (O)	1	1998
	Galium spurium L.	2 (B)	Unknown	1997
		4 (O)	Unknown	1997
	Kochia scoparia (L.) Schrad.	2 (B)	11–50	1988

TABLE 5.2 (CONTINUED)
Herbicide-Resistant Weeds Reported in Small Grain Cereal Production Systems Worldwide (except for the United States)[14,53]

Country	Weed Species	Herbicide Group	Cases	Year
	Neslia paniculata (L.) Desv.	2 (B)	1	1998
	Setaria viridis (L.) Beauv.	1 (A)	501–1000	1992
		3 (K₁)	501–1000	1988
	Sinapis arvensis L.	2 (B)	2–5	1992
		3 (K₁)	2–5	1992
		5 (C₁)	1	1994
		4 (O)	6–10	1991
	Sonchus asper (L.) Hill	2 (B)	2–5	1996
	Stellaria media (L.) Vill.	2 (B)	Unknown	1992
Chile	*Avena fatua* L.	1 (A)	Unknown	1995
	Lolium rigidum Gaudin	1 (A)	Unknown	1995
China	*Alopecurus japonicus* Steudel.	7 (C₂)	2–5	1990
	Beckmannia syzigachne Steudel.	7 (C₂)	101–500	1993
Denmark	*Stellaria media* L.	2 (B)	1	1993
France	*Alopecurus myosuroides* Huds.	1 (A)	101–500	1993
		2 (B)	Unknown	1993
	Avena fatua L.	1 (A)	11–50	1996
	A. sterilis L.	1 (A)	11–50	1996
	Lolium multiflorum Lam.	1 (A)	Unknown	1993
	L. rigidum Gaudin	1 (A)	Unknown	1993
	Matricaria perforata Merat.	O (4)	2–5	1975
Germany	*Alopecurus myosuroides* Huds.	2 (B)	11–50	1997
		7 (C₂)	51–100	1983
	Apera spic-venti L	7 (C₂)	Unknown	Unknown
Greece	*Lolium rigidum* Gaudin	1 (A)	1	1997
		2 (B)	1	1998
	Papaver rhoeas L.	2 (B)	1	1998
India	*Phalaris minor* Retz.	7 (C₂)	10001–100000	1991
Ireland	*Chrysanthemum segetum* L.	2 (B)	1	1997
	Stellaria media L	2 (B)	1	1996
Israel	*Phalaris minor* Retz.	1 (A)	1	1993
Italy	*Avena sterilis* L.	1 (A)	11–50	1992
	Lolium multiflorum Lam.	1 (A)	2–5	1995
		2 (B)	2–5	1995
	Papaver rhoeas L.	2 (B)	2–5	1998
Mexico	*Phalaris minor* Retz.	1 (A)	501–1000	1996
	P. paradoxa L.	1 (A)	501–1000	1996
New Zealand	*Stellaria media* L.	2 (B)	Unknown	1995
Norway	*Chenopodium album* L.	5 (C₁)	1	1994
Saudi Arabia	*Lolium rigidum* Gaudin	1 (A)	101–500	1992
South Africa	*Avena fatua* L.	1 (A)	2–5	1986
	Lolium multiflorum Lam.	1 (A)	Unknown	1993

TABLE 5.2 (CONTINUED)
Herbicide-Resistant Weeds Reported in Small Grain Cereal Production Systems Worldwide (except for the United States)[14,53]

Country	Weed Species	Herbicide Group	Cases	Year
Spain	*Alopecurus myosuroides* Huds.	2 (B)	2–5	1997
		7 (C$_2$)	1	1991
	Bromus tectorum L.	7 (C$_2$)	2–5	1990
	Lolium rigidum Gaudin	1 (A)	11–50	1992
		2 (B)	1	1997
		7 (C$_2$)	11–50	1992
	Papaver rhoeas L.	2 (B)	11–50	1993
		4 (O)	2–5	1993
Sweden	*Chrysanthemum segetum* L.	2 (B)	1	1997
	Cirsium arvense (L.) Scop.	4 (O)	Unknown	1979
	Stellaria media L.	2 (B)	1	1995
Switzerland	*Alopecurus myosuroides* Huds.	7 (C$_2$)	2–5	1997
	Apera spic-venti L.	7 (C$_2$)	1	1994
The Netherlands	*Alopecurus myosuroides* Huds.	2 (B)	11–50	1998
		7 (C$_2$)	2–5	1989
United Kingdom	*Alopecurus myosuroides* Huds.	1 (A)	501–1000	1982
		2 (B)	Unknown	1984
		7 (C$_2$)	501–1000	1982
		3 (K$_1$)	Unknown	1987
	Avena fatua L.	1 (A)	2–5	1994
	A. sterilis L.	1 (A)	2–5	1993
	Chenopodium album L.	5 (C$_1$)	Unknown	1989
	Lolium multiflorum Lam.	1 (A)	51–100	1990
	Matricaria perforata Merat.	4 (O)	2–5	1975
	Stellaria media L.	4 (O)	11–50	1985

was triazine-resistant, whereas in the eastern states of Australia adoption was less than 30% of the total canola crop[39] due to greater resistance in *L. rigidum* and/or *Raphanus raphanistrum*.

The first report of resistance in *L. rigidum* was to diclofop-methyl.[22] This resulted from the availability and widespread use of diclofop, followed by an increasing number of ACCase herbicides from 1978 onward. The availability of many selective ACCase and ALS herbicides has enabled simplification of cropping rotations, which has led to increased reliance on herbicides. Currently, at least 20 weed species of wheat cropping systems are resistant to herbicides from seven modes of action (Table 5.2).[14] Fourteen weed species have developed resistance since 1990. The most widespread resistance is to ACCase and ALS herbicides, reflecting their use patterns.

L. rigidum is the weed species with the highest level of resistance, extending to multiple-resistance across many herbicide groups (Table 5.2). It is a desirable species of the pasture phase, and it is therefore widespread and abundant and has biological characteristics that favor the rapid development of resistance.[40] *L. rigidum* is out-crossing, genetically diverse, and has prolific seed production with low seed dormancy.[41] Resistance has developed in the field after as few as three herbicide applications.[42] Multiple-resistance in many *L. rigidum* populations occurs simultaneously across many separate chemical groups and may include all selective herbicides that normally provide control.[40,43] The mechanisms of resistance include herbicide-resistant target sites and enhanced metabolism. Multiple-resistance is due to the accumulation of multiple mechanisms controlled by individual genes.[43] As *L. rigidum* is genetically diverse, a population can respond rapidly to any new selection pressure. This, together with the varying patterns of multiple-resistance in *L. rigidum*, makes recommendations of reliable alternative herbicides for farmers virtually impossible.[43] The widespread development of multiple-resistance in *L. rigidum* in Australia has forced the implementation of alternative nonchemical control options for this weed to supplement chemical control.

The use of ALS herbicides in the 1980s for the management of *L. rigidum* with ACCase resistance and for control of broadleaf weeds led to ALS resistance in *L. rigidum* and many Brassicaceae weeds, including *Rapistrum rugosum* (L.) All. (turnip weed), *Raphanus raphanistrum* L. (wild radish), and *Sisymbrium* spp. (Table 5.2). The subsequent evolution of resistance to ALS herbicides resulted in reliance on other herbicide groups, and this has led to resistance to trifluralin (Group 3) in *Fumaria densiflora* DC (dense-flowered Fumitory) and diflufenican (Group 12) resistance in *R. raphanistrum*.[14] Resistance to these groups is expected to increase. It is generally accepted that *L. rigidum* will develop resistance more quickly than *Avena* spp. or *R. raphanistrum*, because these latter species have longer seed dormancy, lower total numbers, and less genetic diversity.[41,44,45] However, resistance in *Avena* spp. and *R. raphanistrum* is expected to become more widespread in the future.

Resistance to the triazine (Group 5) herbicides was first reported in *L. rigidum* in the late 1980s (Table 5.2). Triazine herbicides have traditionally been used in lupins, and in pastures prior to cropping, for control of a wide range of grass weed species, such as *Vulpia* spp. The recent rapid adoption of triazine-tolerant canola, particularly in Western Australia, is likely to exacerbate triazine resistance in *L. rigidum* and *R. raphanistrum*, and resistance may evolve in other species such as *Vulpia* spp.[39]

Resistance to glyphosate (Group 9) was first reported in a crop field where glyphosate had been used for 15 years,[46] and then in an orchard.[47] The mechanism of glyphosate resistance in *L. rigidum* is unclear at this stage.[47,48] Glyphosate resistance in *L. rigidum* is of considerable concern worldwide as glyphosate is the most important nonselective herbicide for weed control in reduced-tillage systems. Resistance in *Hordeum glaucum* (wall barley) to paraquat and diquat (Group 22), used to a lesser degree than glyphosate in reduced-tillage systems, was reported following 10 to 15 years of application.[49] Other species with limited reports of paraquat resistance are *Arctotheca calendula* (L.) Levyns (capeweed), *H. leporinum* Link (barley grass), and *Vulpia bromoides* Godr. (silvergrass) (Table 5.2). If farmers revert to paraquat and diquat for managing glyphosate-resistant populations, resistance to these herbicides may increase in the future.

5.3.4   EUROPE

The onset and the levels of herbicide resistance in weeds of Europe vary with country (Table 5.2).[14] Most of the herbicides currently used in western Europe in cereals and rapeseed are ACCase- and ALS-inhibiting herbicides; it is to these herbicides that most resistance has developed in recent years. Prior to this, the Groups 5 and 7 herbicides (triazine and phenylureas) were widely used. Problems with soil persistence have led to deregistration of atrazine, and now resistance to these herbicides is less important in western regions, but it may persist in the east. Likewise, phenoxy herbicides (Group 4) have largely been replaced and the first record of resistance in 1979 in *Cirsium arvense* (L.) Scop. (Canada thistle) has not spread. Resistance is currently more of a problem in northwestern Europe than in the eastern or Mediterranean areas. Older herbicides such as trifluralin are now being used to help combat resistance to ACCase- and ALS-inhibiting herbicides in cereals. In oilseed rape, clopyralid, metazachlor, propyzamide, and trifluralin are used for broadleaf control and ACCase-inhibiting herbicides for grass control. Nonselective herbicides such as glyphosate and paraquat are also used as preplant or pre-emergence treatments in all crops.

Avena myosuroides is the weed species with the most widespread resistance in Europe. Currently there is documented evidence of resistance in four groups of herbicides (Table 5.2). *A. myosuroides* is cross-pollinated, a prolific seed producer, and the seed has only short-term dormancy.[50] Resistance of *A. myosuroides* to a range of ACCase-inhibiting herbicides was first reported at two sites in the U.K. and later in other countries.[51] The most common resistance mechanism is enhanced metabolism giving various levels of cross-resistance to many selective herbicides.[50] The first report of resistance in *A. myosuroides* to ALS-inhibiting herbicides was in 1991 in the U.K., and later in other northern European countries (Table 5.2). Most ALS resistance is metabolic although some grass species show multiple-resistance.[52] The more recent reports are of resistance to ALS-inhibiting herbicides with target-site resistance and enhanced metabolism mechanisms.[53] In *A. myosuroides*, resistance was quicker to develop to the highly active, foliar-absorbed Group 1 herbicides than to the phenylureas, whose variable performance is more affected by environmental conditions.[52]

ACCase resistance occurs in *Avena* spp. and *L. multiflorum* in the northern regions, while *L. rigidum* resistance is becoming widespread in the Mediterranean areas (Table 5.2). The rate of development of herbicide resistance to ALS-inhibiting herbicides in weeds of Europe was slower than in North America.[53] *L. rigidum* resistance in Spain and Greece is expanding, with multiple-resistance to ACCase- and ALS-inhibiting herbicides reported in the late 1990s following use of diclofop-methyl since the mid-1970s and chlorsulfuron since the late 1980s.[54] For most of the metabolic resistance reported in grasses, ALS-inhibiting herbicides were not the selecting agent.[53] ALS resistance in *Papaver rhoeas* L. (corn poppy) in cereals has been found in Spain, Italy, and Greece and is suspected to be target-site resistance.[53] This weed is widely distributed in Europe, is cross-pollinated, has protracted germination, produces a large number of persistent seeds, and is spreading rapidly. In contrast, ALS resistance in *Stellaria media* was the first case recorded in Europe in the early 1990s, but it seems to have stabilized with only two other cases identified.[14] There were only two reports of ALS resistance in *Chrysanthemum segetum* L., in 1997 in Sweden and Ireland.[53]

Current levels of resistance in Europe are unclear as recent surveys have not been undertaken in many countries; however, resistance is thought to be increasing, especially to the ALS-inhibiting herbicides (P. Niemann and M. Sattin, personal communication, 1999). Little herbicide resistance is reported currently in eastern Europe, probably because of the longer cropping rotations and a greater use of low risk herbicides (triazines and phenoxys).[14] This situation is likely to change in the near future as herbicide use expands, cropping rotations become shorter, species diversity decreases, and weed densities increase (P. Niemann, personal communication, 1999). European Union crop subsidies often dictate cropping rotations rather than optimal, sustainable production systems, so it is often difficult to predict patterns of herbicide use by farmers.

5.3.5 MIDDLE EAST

Fenoxaprop-*p*-ethyl-resistant *Phalaris minor* in Israel is the only reported case of herbicide resistance in wheat in the Middle East (Table 5.2).[14] Resistant biotypes infest less than 4000 ha and apparently are not increasing.

5.3.6 NORTH AMERICA

Some 43 cases of herbicide resistance in eight weed species in over 4000 fields have been reported since 1987 in small grain cereal production systems in 18 U.S. states.[14] Five states (Idaho, Montana, North Dakota, Oregon, and Washington) accounted for about 65% of the reported cases of herbicide resistance. All cases of herbicide resistance in these five states occurred in dry land, monoculture wheat production systems (wheat–wheat or wheat–fallow–wheat) where the same herbicide was applied repeatedly over several growing seasons.[21] A total of 21 cases of herbicide resistance were to ALS-inhibitor herbicides (Group 2); 16 were to ACCase-inhibitor herbicides (Group 1); 3 were to lipid synthesis-inhibitor herbicides with unknown target sites (Group 8); 2 were to synthetic auxin herbicides (Group 4); and 1 was to a microtubule assembly inhibitor herbicide (Group 3) (Table 5.3).

In the United States, the main herbicide-resistant weeds are *Kochia scoparia*, *Lolium* sp., *A. fatua*, and *S. iberica*, accounting for over 80% of the reported cases of herbicide resistance (Table 5.3). *K. scoparia* and *S. iberica* are widespread, competitive weeds infesting much of the low-rainfall (<400 mm annually), wheat-producing areas of the United States. Control of these two weeds with older Groups 4 and 5 herbicides was poor to fair, thus use of sulfonylurea herbicides was adopted quickly throughout much of the dry land wheat-producing areas of the United States. Many growers relied on sulfonylurea herbicides to control these weeds, even during the fallow year of the production system, resulting in selection of resistance within 5 years. Some *K. scoparia* biotypes have multiple-resistance to sulfonylurea and triazine herbicides and at least some populations contained triazine-resistant biotypes prior to the selection of sulfonylurea resistance.[55] About 70% of the sampled

TABLE 5.3
Herbicide-Resistant Weeds Reported in Small Grain Cereal Production Systems in the United States by State[14]

State	Weed Species	Herbicide Group	Cases	Year
Arkansas	*Lolium multiflorum* Lam.	1 (A)	Unknown	1995
California	*Avena fatua* L.	8 (N, Z)	11–50	1996
Colorado	*Avena fatua* L.	1 (A)	6–10	1997
	Kochia scoparia (L.) Schrad.	2 (B)	501–1000	1989
Georgia	*Lolium multiflorum* Lam.	1 (A)	11–50	1995
Idaho	*Anthemis cotula* L.	2 (B)	1	1997
	Avena fatua L.	1 (A)	11–50	1992
		8 (N, Z)	51–100	1993
	Kochia scoparia (L.) Schrad.	2 (B)	501–1000	1989
	Lactuca serriola L.	2 (B)	501–1000	1987
	Lolium multiflorum Lam.	1 (A)	101–500	1992
	Salsola iberica Sennen and Pau	2 (B)	101–500	1990
Kansas	*Kochia scoparia* (L.) Schrad.	2 (B)	501–1000	1987
Maryland	*Lolium multiflorum* Lam.	1 (A)	Unknown	1998
Michigan	*Daucus carota* L.	4 (O)	6–10	1993
Minnesota	*Avena fatua* L.	1 (A)	51–100	1991
	Kochia scoparia (L.) Schrad.	2 (B)	11–50	1994
Montana	*Avena fatua* L.	1 (A)	11–50	1990
		2 (B)	2–5	1996
		8 (N)	501–1000	1990
	Kochia scoparia (L.) Schrad.	2 (B)	501–1000	1989
		4 (O)	101–500	1995
	Lolium persicum Boiss. and Hohen. ex Boiss.	1 (A)	1	1993
	Salsola iberica Sennen and Pau	2 (B)	2–5	1987
North Carolina	*Lolium multiflorum* Lam.	1 (A)	501–1000	1990
North Dakota	*Avena fatua* L.	1 (A)	101–500	1991
		2 (B)	2–5	1996
	Kochia scoparia (L.) Schrad.	2 (B)	501–1000	1987
		4 (O)	6–10	1995
	Setaria viridis (L.) Beauv.	3 (K$_1$)	501–1000	1989
Oklahoma	*Kochia scoparia* (L.) Schrad.	2 (B)	501–1000	1992
Oregon	*Avena fatua* L.	1 (A)	51–100	1990
		15 (K$_3$)	51–100	1990
	Bromus tectorum L.	2 (B)	2–5	1997
	Kochia scoparia (L.) Schrad.	2 (B)	51–100	1993
	Lactuca serriola L.	2 (B)	101–500	1993
	Lolium multiflorum Lam.	1 (A)	101–500	1987
	Salsola iberica Sennen and Pau	2 (B)	501–1000	1993
South Carolina	*Lolium multiflorum* Lam.	1 (A)	101–500	1990
South Dakota	*Kochia scoparia* (L.) Schrad.	2 (B)	501–1000	1998
Virginia	*Lolium multiflorum* Lam.	1 (A)	501–1000	1993

TABLE 5.3 (CONTINUED)
Herbicide-Resistant Weeds Reported in Small Grain Cereal Production Systems in the United States by State[14]

State	Weed Species	Herbicide Group	Cases	Year
Washington	*Avena fatua* L.	1 (A)	51–100	1991
	Kochia scoparia (L.) Schrad.	2 (B)	501–1000	1989
	Lactuca serriola L.	2 (B)	11–50	1993
	Lolium multiflorum Lam.	1 (A)	6–10	1991
	Salsola iberica Sennen & Pau	2 (B)	501–1000	1987
Wyoming	*Kochia scoparia* (L.) Schrad.	2 (B)	2–5	1996

area in the dry land wheat-producing area of eastern Washington State was infested with sulfonylurea herbicide-resistant *S. iberica* by 1992, which was directly related to herbicide use patterns.[56] Also, selection of resistant plants at most sites was likely independent of other sites. Over 50% of the areas sampled in Idaho were infested with sulfonylurea-resistant *K. scoparia* by 1993, which was not associated with sulfonylurea herbicide use in croplands (D. Thill, unpublished data). However, weed resistance was closely related to ALS-inhibitor herbicide use on roadside rights-of-way. *K. scoparia* is a tumbleweed and resistant plants along roadsides likely tumbled onto adjacent croplands (D. Thill, unpublished data).[34] This clearly demonstrates that resistance management for some weeds is a landscape issue, rather than confined to individual farms.

Currently, *Bromus tectorum* L. (downy brome) resistance to primisulfuron (Group 2) is of concern in the Pacific Northwest region of the United States. Primisulfuron-resistant *B. tectorum* was selected in a Kentucky bluegrass (*Poa pratensis* L.) seed production field following two annual applications of the herbicide over 3 years.[57] Sulfosulfuron and imazamox effectively control *B. tectorum* in winter wheat,[58–60] which is grown in rotation with Kentucky bluegrass. However, these ALS-inhibiting herbicides do not control primisulfuron-resistant *B. tectorum*.[61] Without stringent weed management plans (see Section 5.3), ALS-resistant *B. tectorum* populations will likely increase quickly in winter wheat production systems, resulting in the loss of these highly effective herbicides shortly after market introduction.

A similar trend occurred for *A. fatua* and *L. multiflorum* in the higher rainfall (>400 mm) wheat-producing areas of the United States. ACCase herbicides were used to control these weeds in wheat and many crops grown in rotation with wheat. For example, at one time diclofop-methyl was used to control *A. fatua* and *L. multiflorum* in wheat, barley, pea, and lentil crops grown in the Palouse region of the Pacific Northwest. A typical crop rotation was winter wheat–spring barley–grain legume. Unfortunately, diclofop-methyl was sometimes applied to every crop grown in this 3-year rotation, which resulted in the selection of diclofop-methyl-resistant weed biotypes.

Densities of herbicide-resistant weeds are increasing within infested sites in about 80% of the reported cases and new sites continue to be identified. Population increases are expected with many of the herbicide-resistant weeds currently infesting wheat fields. Given that resistance is often a single, semidominant nuclear trait, cross-pollination can occur between resistant and susceptible biotypes in many species. In addition several species have specialized, long-distance seed dispersal mechanisms.[62] However, it is estimated that only about 1 to 2% of the lands used to produce wheat in the United States is infested with herbicide-resistant weeds.[4,14]

In Canadian cereal production systems, the first reported case of herbicide resistance was *Setaria viridis* resistant to ethalfluralin in 1988.[14] Currently, eight weed species are reported with resistance to one or more herbicides used to control weeds in small grain cereals (Table 5.2). Herbicide-resistant weed biotypes have been reported to herbicide Groups 1, 2, 3, 4, 5, 8, and 25, with resistance most common to ACCase- and ALS-inhibiting herbicide groups (1 and 2) for the same basic reason as in the United States, Europe, and Australia. ACCase-inhibiting herbicides have been used extensively to control grass weeds such as *A. fatua* and *S. viridis* in cereal production systems. ALS-inhibiting herbicides are used to control many broadleaf weeds in cereals. Resistance is most widespread in *A. fatua* with some populations resistant to clodinafop-propargyl, diclofop-methyl, fenoxaprop-*p*-ethyl, sethoxydim, clethodim, and tralkoxydim; some resistant to fenoxaprop-*p*-ethyl, imazamethabenz, rimsulfuron, triallate, and flamprop-methyl; and others resistant to triallate and difenzoquat.[14] Herbicide-resistant *A. fatua* and *S. viridis* populations are increasing within infested sites and new sites continue to be identified. Herbicide-resistant broadleaf weeds in cereal production systems include *Galeopsis tetrahit* L. (common hempnettle), *Galium spurium* L. (false cleavers) (biotypes resistant to Group 2 or 4 herbicides),[63] *Neslia paniculata* (L.) Desv. (ball mustard), *B. kaber*, *Sonchas asper* (L.) Hill (spiny sowthistle), and *S. media* (resistant to Group 2 herbicides) (Table 5.2).[14] Based on current reports, as much as 15% of Canadian wheat lands may be infested with herbicide-resistant weeds.[4,14] Fairly widespread herbicide resistance in Canada has likely contributed greatly to the quick adoption of glyphosate-, glufosinate-, and imazamox-resistant canola into cereal/canola-based production systems. Unfortunately, volunteer canola plants with multiple-resistance to glyphosate, glufosinate, and imazamox have been identified.[64] Pollen transfer, rather than mutation, is implicated.

Fenoxaprop-*p*-ethyl-resistant *Phalaris paradoxa* L. (hood canarygrass) and *P. minor* are the only cases of herbicide resistance reported for Mexican wheat production systems (Table 5.2).[14] The populations of both species are increasing within infested sites and new sites continue to be identified. However, less than 8000 ha of wheat-producing lands are infested with herbicide-resistant weeds.[14]

5.3.7 SOUTH AMERICA

Chile has the only reported cases of herbicide-resistant weeds in small grain cereals in South America (Table 5.2).[14] Diclofop-methyl-resistant *A. fatua* and *L. rigidum* were reported in 1995.

5.4   MANAGEMENT OF HERBICIDE RESISTANCE

Successful management of resistance relies on reducing the seedbank and limiting the dispersal of resistant weed seed. The longevity of weed seed in the seedbank varies with species, soil type, climate, and farming system. Integrated weed management (IWM) is defined as the combination of chemical, cultural, and biological control tactics for weed management. More recently, a preferred term is best practice management (BPM). A systems approach to weed management relies on the adoption of tactics that are often more expensive, less convenient, and less effective than the herbicides that were used prior to the development of weed resistance. Herbicides remain the primary method of weed control in cereal crops because they are very cost effective. This is why farmers are reluctant to adopt IWM until they have a serious herbicide-resistance weed problem. The IWM strategies available to farmers in different regions depend on the level of resistance, the weed species, the cropping or pasture options available, and socioeconomic considerations. When resistance exists to few herbicides, farmers generally initially change to other modes of herbicide action (including herbicide-resistant crops); however, when resistance extends to many herbicides, farmers are forced to adopt more radical nonchemical or cultural control options.

5.4.1   Australia

Multiple herbicide resistance in *L. rigidum* has spearheaded a considerable research and extension effort to develop control tactics for managing this weed.[65,66] Herbicide resistance in *L. rigidum* can now be successfully managed in continuous cropping systems by eliminating return of weed seed to the soil, combined with techniques to accelerate seedbank decline and reduce weed population densities.[67] Control strategies have been reviewed by Gill[65] and Matthews and Powles,[66] and include the rotation of herbicide groups with increased use of alternative "low risk" herbicides such as trifluralin, combined with cultural practices such as strategic tillage, stubble burning, or grazing. A number of important cultural and chemical control tactics are applied at various stages of the weed life cycle. These include delayed seeding combined with a shallow tillage or a nonselective preplant herbicide such as glyphosate, cutting a crop for green manure or hay, enhancing crop competitive ability by choice of cultivar or higher crop seeding rates, and capture of weed seeds at harvest.[68–71] A considerable research effort is in progress to improve the competitive ability of wheat by manipulating crop agronomy. In the longer term, breeding for cultivars with enhanced competitive ability, while maintaining yield potential, is necessary.[72]

A number of innovative practices have been developed to stop grass weed seed production while having no impact on the grain yield of the crop. In pulse crops, low rates of paraquat are applied late in the season, a practice known as "crop-topping."[65,66] Another technique is "selective spray-topping," where a selective herbicide is used to kill, for example, wild oat seed in wheat.[73] This concept of the "double knock" is to always follow up any weed control strategy with at least one other to minimize residual weed densities, seed production, and seedbank replenishment. This is a dramatic shift from the concept of weed control thresholds, where herbicides were used to simply reduce yield loss from weed competition and low weed densities were acceptable.

Utilization of a pasture phase in the farm rotation can play an important role in preventing seed input either through grazing, use of nonselective chemicals, or cutting for hay at or before flowering and seed set.[68,74] The controversial role of crop stubble burning is useful for destroying seeds of some weed species at or near the soil surface.[68] The benefits of rotating crops for weed management is well recognized as providing opportunities for a wider range of both chemical and cultural management options. In the northern wheat belt, rotation of chemical groups in-crop is less effective than rotation to winter fallow, pasture, or summer crops for managing herbicide-resistant weeds.[75]

The pros and cons of herbicide-resistant crops (HRC) for weed management have been summarized by Powles et al.[1] The main benefit of HRC is to increase the range of herbicide options available to farmers; however, caution is needed in their use to avoid selecting for herbicide resistance in weeds. Bowran et al.[76] argue the need for longer cropping rotations, which allow the HRC and herbicide to remain effective over a longer time period, combined with alternative nonchemical control measures, to minimize selection pressure on weed populations. In Australia, a number of HRC (imidazolinones, glufosinate, glyphosate, and bromoxynil) are being developed and are likely to be released in the next 5 years.[77] Glufosinate will be the only new mode of action, while the others provide new uses for existing modes of action. The introduction of a bromoxynil- or glufosinate-resistant crop is unlikely to lead to major resistance in the short term because neither herbicide is widely used in cropping systems. However, the introduction of the imidazolinone-resistant crops is likely to exacerbate the existing high levels of ALS resistance.[77]

Given the importance of glyphosate in minimum tillage systems and the associated increase in selection pressure, the likely introduction of glyphosate-resistant crops represents a significant risk for future resistance problems.[78] Also, management of volunteer canola with multiple-resistance is a major concern (see Section 5.3.6, North America). Herbicide-resistant crops have an important role to play in managing herbicide-resistant weeds, but they must be used only as one part of a system (IWM). Farmers with herbicide resistance have generally accepted the need for IWM; however, the release of HRC could again result in greater reliance on herbicides. The development of resistance to the nonselective herbicides, which the zero-tillage systems are based upon, has paradoxically forced a return in some situations to the environmentally unsustainable practices of periodic cultivation and stubble burning. The challenge is to continue to develop alternative strategies. Adoption of HRC may be dictated by market forces and community concerns about transgenic crops.

The range of management tactics required to implement IWM for *L. rigidum*, *Avena* spp., and *R. raphanistrum* varies depending on the cropping options, climate, soil type, resource availability, and preferences of farmers.[45,68,79] Mandatory labeling of herbicide containers with herbicide group information has played an important role in educating farmers about rotating herbicide groups.

5.4.2 Europe

Traditionally in Europe, weeds in cropping systems were controlled by plowing or shallow tillage (following straw burning), and crop species were rotated to provide a wide range of weed management options. However, very high densities of weeds resulted from trends to shorter cropping rotations (including continuous winter wheat) for higher grain yields, earlier planting in mid-September rather than in mid-October, minimum tillage, and the banning of stubble burning.[52] As a result, farmers became dependent on selective herbicides for weed control, and thus considerable selection pressure was applied, leading to herbicide resistance in the 1980s and 1990s. Where resistance developed, farmers have been forced to use traditional practices, such as annual plowing to bury weed seed, crop rotation (including grass ley or "set-aside") and a move to spring sown crops, and later seeding with the associated risks of poor establishment and yield loss.[80]

The essential components of managing herbicide-resistant weed populations in Europe, as elsewhere, include the use of nonselective herbicides (glyphosate or paraquat), preseeding combined with low risk pre-emergence herbicides such as triallate or trifluralin and mixtures with isoproturon, strongly competitive crops and cultivars, in-crop cultivation, and correct herbicide timing and rate of application.[52,80] It is well recognized that a combination of cultural and chemical options, including herbicide-resistant crops, is needed to slow down resistance development in some weeds.[81] A management system designed to retard, reduce the risk of, or prevent resistance by avoiding early seeding and adopting annual plowing may be more expensive, but in the long term, resistance management will be economical for farmers.[80] HRC have not yet been widely grown.

5.4.3 India

Alternative herbicide groups have recently been explored for control of isoproturon-resistant *P. minor*; however, there are problems of crop phytotoxicity with trifluralin, but chlorotoluron and tralkoxydim show promise.[82] The usefulness of herbicide rotation depends on the level of cross-resistance to other herbicide groups. Delayed seeding or later seeded rotational crops combined with tillage or preseeding non-selective herbicides shows promise for resistance management.[83] Recently, Mahajan et al.[84] found that while *P. minor* was more vigorous in wheat under conventional compared with zero tillage, delaying seeding by 10 days suppressed the weed by nearly 50% in the conventional tillage systems. This benefit was less under zero tillage and was also offset by a reduced grain yield of 10 to 23%.

Rotational crops are an important tool for reducing resistance increase. Crops such as sunflower (*Helianthus annuus* L.) and sugarcane (*Saccharum officinarum* L.) provide a noncereal phase (December to January), which can be used to stimulate the germination of *P. minor* and allow the use of tillage or a nonselective herbicide. A survey in 1993 showed that weed occurrence was 67% under a rice–wheat cropping system, compared with only 8 to 16% where wheat was rotated with sugarcane/vegetables, pigeon pea, clover (*Trifolium* spp.), or sunflower.[85] Alternative crops provide opportunities for different herbicide groups and fodder crops. However, adoption of alternative crops has proven difficult because of low profitability. For example, sugarcane adoption was hampered by price instability and uncertain delivery schedules to the sugar mills.

Manual weeding is less practical and is becoming expensive because of socio-economic changes in India, particularly toward a greater emphasis on education for woman and children. Hand weeding is also difficult because crops are irrigated and weeds remain within rows.[83] Mechanical weeding, direct-drilling combined with preplant, nonselective herbicides, early-seeded wheat, strongly competitive wheat cultivars at higher seeding rates, and narrow-row spacing and optimal fertilizer rates all show potential for resistance management.[82,83]

The need for the integration of cultural practices and herbicides for *P. minor* control in India is well recognized.[82] The potential of herbicide-resistant crops is seen as a possibility but would only be adopted if economically viable.[82] Greater resources must be allocated to educate farmers about the long-term benefits of rotation of crops and herbicides and the use of other cultural weed control tactics to minimize the spread of resistance in the future.

5.4.4   NORTH AMERICA

Wheat producers have been dealing with weed species shifts for nearly as long as they have been growing wheat. Chancellor[86] points out that weed problems in wheat will change with changing crop management practices, including herbicide use. Tillage practices, herbicide use, and the number and kind of crops grown in rotation with wheat are important elements in determining which weeds will infest wheat fields and at what densities. For example, Freyman et al.[87] reported on weed problems that limited yield of spring wheat in a long-term (1912 to 1970) study near Lethbridge, Canada. Prior to the use of 2,4-D in 1950, broadleaf weeds, such as *S. iberica*, *C. album*, *A. retroflexus*, and *B. kaber*, reduced wheat production. Controlling these broadleaf weeds has allowed grass weeds (*A. fatua* and *S. viridis*) to develop as problem weeds in wheat. Introduction of triallate for wild oat control in 1961 resulted in improved wheat yields. Selection of herbicide-resistant weed biotypes represents a specific type of weed shift in a cropping system.

The occurrence of herbicide-resistant weeds in North American cereal production systems has resulted in a proliferation of management strategies to deal with the problem. These include IWM strategies to prevent (or at least slow) the onset of resistance, such as rotating herbicide groups, using short-residual herbicides, employing more diverse crop rotations, using cultivation where and when possible, planting weed-free crop seed, and keeping accurate herbicide application records. Strategies to deal with situations where herbicide-resistant weeds infest a field include monitoring fields for escapes, preventing the resistant weed biotype from spreading, and changing crop, herbicide, and tillage systems. These recommendations are effective only if adopted by the farming community and the agricultural chemical industry as a whole. Unfortunately, in many wheat-growing regions of the United States (and likely also in Canada), these practices have not been adopted. Record low farm commodity prices, including wheat and the crops grown in rotation with wheat, do not facilitate implementation of long-term weed management strategies to control existing populations of herbicide-resistant weeds or to use weed management practices to prevent selection of new resistant weed biotypes. Rather, many farmers use herbicides that are effective and affordable until they encounter a herbicide-resistant weed problem. Then they simply change from one herbicide to another. Manufacturers of agricultural chemicals have not added herbicide group number information to U.S. herbicide labels, although this is done voluntarily in Canada and mandated in Australia. A recent survey of Australian farmers showed that mode of action labeling was somewhat beneficial.[88] However, the simplicity of the system did result in some confusion regarding herbicide grouping and herbicide mixtures with more than one mode of action. It was suggested that mode of action labeling should be considered only as one part of a herbicide-resistant weed educational program.

Use of herbicide-resistant crops has been quickly adopted by farmers in the major wheat-producing regions of North America. This trend likely will continue as new HRC enter the market place. For example, herbicide-resistant canola (imidazolinone, glyphosate, and glufosinate) was planted on about 3 million ha in 1998 in Canada and is projected to reach nearly 4 million ha by 2003.[89] Glyphosate-resistant canola was introduced in the United States in 1999 and was planted on about 81,000 ha (S. Halter, personal communication, January 2000). Imidazolinone-resistant canola was available to U.S. canola growers by spring 2000. Herbicide-resistant wheat (glyphosate and imidazolinone) will be available to U.S. and Canadian farmers within 2 to 3 years. Some concerns have been expressed regarding the need for glyphosate-resistant wheat, especially in direct-drill (no-till) seeding systems. Glyphosate currently is the only effective preplant herbicide for controlling weeds and volunteer crop plants in direct-seed systems in much of Canada and the United States. This powerful weed control technology, if used judiciously, will allow farmers to control many problem weeds, including herbicide-resistant weeds. Overuse will result in weed species shifts and selection of herbicide-resistant weeds.[1] The key to success is to use this technology as part of an IWM system. Many of the techniques developed in Australia to manage multiple herbicide resistance will need to be explored in both North America and Europe, and adapted to local conditions as necessary, to provide more opportunities for farmers to manage herbicide resistance in weeds.

5.5 CONCLUSIONS

Herbicide resistance in weeds is continuing to increase at a rapid rate, and there is little sign that this will do anything but accelerate in the near future, particularly in the intensive cropping systems of Europe and North America. Herbicide resistance in developing countries, such as China, is likely to increase as cropping rotations are simplified and herbicide use expands. Whether resistance in these countries can be retarded depends on many factors. Considerable resources will be required to continue to educate farmers about the importance of IWM in both developed and developing countries. Greater cooperation between the agricultural chemical industry and government or farmer extension agencies will enhance farmer adoption of IWM.

The patterns and rates of herbicide resistance development are now well understood and quite predictable. The main factors leading to major herbicide-resistant weed problems are (1) simple cropping rotations that favor few dominant weed species; (2) weed is present at high densities, is widely distributed, is genetically variable, and is a prolific seed producer; and (3) multiple applications of single or similar mode-of-action herbicides are used. In some situations, resistance has developed rapidly, due to the characteristics of a particular weed species and the herbicide mode of action, while resistance in other weed/herbicide associations has been much slower to develop. Despite the experience of widespread multiple-resistance of *L. rigidum* in Australia, where alternative herbicide recommendations for resistance management are not feasible and farmers must use less profitable cultural control tactics, farmers in both Australia and other countries continue not to practice IWM. However, it is possible that the development of glyphosate resistance in weeds may be a sufficient threat to lead to significant changes in the future.

Short-term economic gains have dictated the responses of both farmers and the agrochemical industry to the development of herbicide resistance in weeds. IWM is well known to reduce herbicide reliance and reduce plant densities in weed populations. However, farmers will generally not adopt IWM until they have a significant resistance problem, due to short-term profit imperatives. Therefore, despite considerable research and extension to develop strategies for IWM in the face of spreading resistance, farmer adoption is a problem worldwide. Decision support models may be valuable tools to demonstrate the long-term benefits of IWM.

There is some evidence that adoption of IWM can be improved by working with farmer groups, and by emphasizing the need for long-term weed population management. This involves many techniques for reducing the weed seedbank prior to seeding, maximizing the competitive ability of the crop, preventing seed return to the seedbank, spreading the risk over the whole farm, and enforcing strong quarantine procedures between fields on the farm and between farms. The benefits and risks of HRC for managing herbicide-resistant weeds are obvious. It remains to be seen whether farmers will use them exclusively as just another mode of action, or as part of IWM. The management of HRC is often more complicated than for conventional crops and the short-term benefits may be outweighed by longer-term disadvantages.

REFERENCES

1. Powles, S. B., Preston, C., Bryan, I. B., and Jutsum, A. R., Herbicide resistance: impact and management, *Adv. Agron.*, 58, 57, 1997.
2. Allen, R. E., Wheat, *Principles of Cultivar Development*, Vol. 2, *Crop Species*, Fehr, W. R., Ed., Macmillan, New York, 1987, chap. 18.
3. Briggle, L. W. and Curtis, B. C., Wheat worldwide, in *Wheat and Wheat Improvement*, 2nd ed., Agronomy Monograph No. 13, Heyne, E. G., Ed., American Society of Agronomy, Madison, WI, 1987, chap. 1.
4. Food and Agriculture Organization Web site, http://www.fao.org/, Area harvested in 1998 sorted by geographic region, 1999.

5. Percival, J., *The Wheat Plant, Monograph*, E. P. Dutton, New York, 1921.
6. Nuttonson, M. Y., *Wheat-Climatic Relationships and the Use of Phenology in Ascertaining the Thermal and Photo-Thermal Requirements of Wheat*, American Institute of Crop Ecology, Washington, D.C., 1955.
7. Briggle, L. W., Origin and botany of wheat, in *Wheat*, Häfliger, E., Ed., Documenta Ciba-Geigy, Basel, Switzerland, 1980, 6.
8. Curtis, B. C., Potential for a yield increase in wheat, in *Proc. National Wheat Research Conf., Beltsville, Maryland, October 26–28*, National Association of Wheat Growers Foundation, Washington, D.C., 1982, 5.
9. Kelleher, F. M., Climate and crop production, in *Principles of Field Crop Production*, Pratley, J. E., Ed., Sydney University Press, Sydney, 1994, chap. 2.
10. Donald, W. W., Ed., *Systems of Weed Control in Wheat in North America*, Weed Science Society of America, Champaign, IL, 1990, 488.
11. Donald, W. W. and Easton, E. F., Weed management systems for grain crops, in *Handbook of Weed Management Systems*, Smith, A. E., Ed., Marcel Dekker, New York, 1995, 401.
12. Wicks, G. A., Weed control in conservation tillage systems: small grains, in *Weed Control in Limited-Tillage Systems*, Wiese, A. F., Ed., Weed Science Society of America, Champaign, IL, 1985, 72.
13. Hunter, J. H., Morrison, I. N., and Rourke, D., The Canadian prairie provinces, in *Systems of Weed Control in Wheat in North America*, Donald, W. W., Ed., Weed Science Society of America, Champaign, IL, 1990, 51.
14. Heap, I. M., International survey of herbicide-resistant weeds, Weed Science Society of America and the Herbicide Action Committee, http://www.weedscience.com/, 2000.
15. Hopkins, W. L., *Global Herbicide Directory*, 2nd ed., Ag Chem Information Services, Indianapolis, IN, 1997, 207.
16. Rubin, R., Herbicide resistance outside North America and Europe: cause and significance, in *Weed and Crop Resistance to Herbicides*, De Prado, R., Jorrín, J., and García-Torres, L., Eds., Kluwer Academic Publishers, Dordrecht, the Netherlands, 1997, chap. 4.
17. Shaner, D. L., Herbicide resistance in North America: history, circumstances of development, and current situation, in *Weed and Crop Resistance to Herbicides*, De Prado, R., Jorrín, J., and García-Torres, L., Eds., Kluwer Academic Publishers, Dordrecht, the Netherlands, 1997, chap. 3.
18. Mallory-Smith, C. A., Thill, D. C., and Morishita, D. W., *Herbicide-Resistant Weeds and Their Management*, Pacific Northwest Extension Publication 437, 1999, 6.
19. Retzinger, E. J., Jr. and Mallory-Smith, C. A., Classification of herbicides by site of action for weed resistance management, *Weed Technol.*, 11, 384, 1997.
20. Devine, M. D. and Shimabukuro, R. H., Resistance to acetyl coenzyme A carboxylase inhibiting herbicides, in *Herbicide Resistance in Plants, Biology and Biochemistry*, Powles, S. B. and Holtum, J. A. M., Eds., Lewis Publishers, Boca Raton, FL, 1994, chap. 5.
21. Saari, L. L., Cotterman, J. C., and Thill, D. C., Resistance to acetolactate synthase inhibiting herbicides, in *Herbicide Resistance in Plants, Biology and Biochemistry*, Powles, S. B. and Holtum, J. A. M., Eds., Lewis Publishers, Boca Raton, FL, 1994, chap. 4.
22. Heap, I. M. and Knight, R., A population of ryegrass tolerant to the herbicide diclofop-methyl, *J. Aust. Inst. Agric. Sci.*, 48, 156, 1982.

23. Hall, L. M., Holtum, J. A. M., and Powles, S. B., Mechanism responsible for cross resistance and multiple resistance, in *Herbicide Resistance in Plants, Biology and Biochemistry*, Powles, S. B. and Holtum, J. A. M., Eds., Lewis Publishers, Boca Raton, FL, 1994, chap. 9.

24. Seefeldt, S. S., Hoffman, D. L., Gealy, D. R., and Fuerst, E. P., Inheritance of diclofop resistance in wild oat (*Avena fatua*) biotypes from the Willamette Valley of Oregon, *Weed Sci.*, 46, 170, 1998.

25. Bourgeois, L., Kenkel, N. C., and Morrison, I. N., Characterization of cross-resistance patterns in acetyl-CoA carboxylase inhibitor resistant wild oat (*Avena fatua*), *Weed Sci.*, 45, 750, 1997.

26. Mallory-Smith, C. A., Thill, D. C., and Dial, M. J., Identification of sulfonylurea herbicide-resistant prickly lettuce (*Lactuca serriola*), *Weed Technol.*, 4, 163, 1990.

27. Mallory-Smith, C. A., Identification and Inheritance of Sulfonylurea Herbicide-Resistance in Prickly Lettuce (*Lactuca serriola* L.), Ph.D. dissertation, University of Idaho, Moscow, 1990, 58.

28. Moss, S. R. and Clarke, J. H., Guidelines for the prevention and control of herbicide-resistant black-grass (*Alopecurus myosuroides* Huds), *Crop Prot.*, 13, 230, 1994.

29. Saari, L. L., Cotterman, J. C., and Primiani, M. M., Mechanisms of sulfonylurea herbicide resistance in the broadleaf weed, *Kochia scoparia*, *Plant Physiol.*, 93, 55, 1990.

30. Eberlein, C. V., Guttieri, M. J., Berger, P. H., Fellman, J. K., Mallory-Smith, C. A., Thill, D. C., Baerg, R. J., and Belknap, W. R., Physiological consequences of mutation for ALS-inhibitor resistance, *Weed Sci.*, 47, 383, 1999.

31. Saari, L. L. and Maxwell, C. A., Target-site resistance for acetolactate synthase inhibitor herbicides, in *Weed and Crop Resistance to Herbicides*, De Prado, R., Jorrín, J., and García-Torres, L., Eds., Kluwer Academic Publishers, Dordrecht, the Netherlands, 1997, chap. 8.

32. Guttieri, M. J., Eberlein, C. V., and Souza, E. J., Inbreeding coefficients of field populations of *Kochia scoparia* using chlorsulfuron resistance as a phenotypic marker, *Weed Sci.*, 46, 521, 1998.

33. Alcocer-Ruthling, M., Thill, D. C., and Shafii, B., Seed biology of sulfonylurea-resistant and susceptible biotypes of prickly lettuce (*Latuca serriola*), *Weed Technol.*, 6, 858, 1992.

34. Stallings, G. P., Thill, D. C., Mallory-Smith, C. A., and Lass, L. W., Plant movement and seed dispersal of Russian thistle (*Salsola iberica*), *Weed Sci.*, 43, 63, 1995.

35. Stallings, G. P., Thill, D. C., Mallory-Smith, C. A., and Sahfii, B., Pollen-mediated gene flow in sulfonylurea-resistant kochia (*Kochia scoparia*), *Weed Sci.*, 43, 95, 1995.

36. Alcocer-Ruthling, M., Thill, D. C., and Mallory-Smith, C. A., Monitoring the occurrence of sulfonylurea-resistant prickly lettuce (*Lactuca serriola*), *Weed Technol.*, 6, 437, 1992.

37. Malik, R. K., Gill, G., and Hobbs, P., Herbicide resistance — a major issue for sustaining wheat productivity in rice-wheat cropping systems in the Indo-Gangetic plains, in *Rice-Wheat Consortium Paper,* Series 3, Rice-Wheat Consortium for the Indo-Gangetic Plains, New Delhi, India, 1998.

38. Salisbury, P. A. and Wratten, N., Canola in Australia: the first 30 years, in *Brassica Napus Breeding*, Salisbury, P. A., Potter, T. D., McDonald, G., and Green, A. G., Eds., 10th Int. Rapeseed Congr. Organising Committee, Canberra, 1999, 29.

39. Lemerle, D., Blackshaw, R. E., Potter, T., Marcroft, S., and Barrett-Lennard, R., Incidence of weeds in canola crops across southern Australia, in *Proc. 10th Int. Rapeseed Congress* [CD-ROM], Wratten, N. and Salisbury, P. A., Eds., IRC Organising Committee, Canberra, 1999, 4.

40. Powles, S. B. and Matthews, J. M., Multiple herbicide resistance in annual ryegrass (*Lolium rigidum*): the driving force for the adoption of integrated weed management, in *Achievements and Developments in Combating Pest Resistance*, Denholm, I., Devonshire, A. L., and Hollomon, D. W., Eds., Elsevier Science Publishers, London, 1992, 75.

41. Monaghan, N. M., The biology and control of Lolium as a weed of wheat, *Weed Res.*, 20, 117, 1980.

42. Gill, G. S., Development of herbicide resistance in annual ryegrass (*Lolium rigidum*) populations in the cropping belt of Western Australia, *Aust. J. Exp. Agric.*, 35, 67, 1995.

43. Preston, C., Tardif, F. J., and Powles, S. B., Multiple mechanisms and multiple herbicide resistance in *Lolium rigidum*, in *Molecular Genetics and Evolution of Pesticide Resistance*, Brown, T. M., Ed., ACS Symp. Ser. 645, American Chemical Society, Washington, D.C., 1996, 117.

44. Medd, R. W., Ecology of wild oats, *Plant Prot. Q.*, 11, 185, 1996.

45. Cheam, A. H. and Code, G. R., *Raphanus raphanistrum* L, *Biology of Australian Weeds*, Vol. 2, Panetta, F. D., Groves, R. H., and Shepherd, R. C. H., Eds., R. G. and F. J. Richardson, Meredith, Australia, 1998, 207.

46. Pratley, J. E., Baines, P., Eberbach, P., Incerti, M., and Broster, J., Glyphosate resistance in annual ryegrass, in *Proc. 11th Annu. Conf. Grasslands Soc. NSW*, Virgona, J. and Michalk, D., Eds., The Grasslands Society of New South Wales, Wagga Wagga, 1996, 126.

47. Powles, S. B., Lorraine-Colwill, D. F., Dellow, J. J., and Preston, C., Evolved resistance to the herbicide glyphosate in rigid ryegrass (*Lolium rigidum*) in Australia, *Weed Sci.*, 46, 604, 1998.

48. Pratley, J., Urwin, N., Stanton, R., Baines, P., Broster, J., Cullis, K., Schafer, D., Bohn, J., and Krueger, R., Resistance to glyphosate in *Lolium rigidum, Weed Sci.*, 47, 405, 1999.

49. Alizadeh, H. M., Preston, C., and Powles, S. B., Paraquat-resistant biotypes of *Hordeum glaucum* Sted. from zero-till wheat, *Weed Res.*, 38, 139, 1998.

50. Moss, S. R., Herbicide cross-resistance in slender foxtail (*Alopecurus myosuroides*), *Weed Sci.*, 38, 492, 1990.

51. Moss, S. R. and Cussans, G. W., Variability in the susceptibility of *Alopecurus myosuroides* (black-grass) to chlorotoluron and isoproturon, *Asp. Appl. Biol.*, 9, 91, 1985.

52. Moss, S. R. and Clarke, J. H., Guidelines for the prevention and control of herbicide-resistant black-grass (*Alopecurus myosuroides* Huds.), *Crop Prot.*, 13, 230, 1994.

53. Claude, J. P. and Cornes, D., Status of ALS resistance in Europe, in *Abstr. Proc. 11th European Weed Research Symposium*, Basel, 1999, 156.

54. Kotoulo-Syka, E., Avi, T., and Rubin, B., Multiple resistance to ACCase inhibitors and chlorsulfuron in annual ryegrass (*Lolium rigidum*) from northern Greece, in *Abstr. Proc. 11th European Weed Research Symposium*, Basel, 1999, 151.

55. Thompson, C. R., Thill, D. C., Mallory-Smith, C. A., and Sahfii, B., Characterization of chlorsulfuron resistant and susceptible kochia (*Kochia scoparia*), *Weed Technol.*, 8, 470, 1994.

56. Stallings, G. P., Thill, D. C., and Mallory-Smith, C. A., Sulfonlyurea-resistant Russian thistle (*Salsola iberica*) survey in Washington state, *Weed Technol.*, 8, 258, 1994.

57. Mallory-Smith, C. A., Hendrickson, P., and Mueller-Warrrant, G., Cross-resistance of primisulfuron-resistant *Bromus tectorum* L. (downy brome) to sulfosufuron, *Weed Sci.*, 47, 256, 1999.

58. Hanson, B. D., Brewster, B. B., and Mallory-Smith, C. A., Downy brome control in winter wheat with imazamox, *West. Soc. Weed Sci. Res. Prog. Rep.*, 183, 2000.

59. Rainbolt, C. R., Thill, D. C., and Ball, D. A., Response of rotational crops to BAY MKH 6561, *Weed Technol.*, in press.

60. Shinn, S. L., Thill, D. C., Price, W. J., and Ball, D. A., Response of downy brome (*Bromus tectorum*) and rotational crops to MON 37500, *Weed Technol.*, 12, 690, 1998.

61. Ball, D. A. and Mallory-Smith, C. A., Sulfonylurea herbicide resistance in downy brome, in *Proceedings of the Western Society of Weed Science*, in press.

62. Thill, D. C. and Mallory-Smith, C. A., The nature and consequence of weed spread in cropping systems, *Weed Sci.*, 45, 337, 1997.

63. Hall, L. M., Stromme, K. M., Horsman, G. P., and Devine, M. D., Resistance to acetolactate synthase inhibitors and qunclorac in a biotype of false cleavers (*Galium spurium*), *Weed Sci.*, 46, 390, 1998.

64. Hall, L. M., Topinka, K., Huffman, J., Davis, L., and Good, A., Pollen flow between herbicide resistant canola (*Brassica napus*) is the cause of multiple resistant *B. napus* volunteers, *Weed Sci.*, 48, 688, 2000.

65. Gill, G. S., Management of herbicide resistant ryegrass in Western Australia — research and its adoption, in *Proc. 11th Aust. Weeds Conf.*, Shepherd, R. C. H., Ed., Weed Science Society of Victoria, Melbourne, 1996, 542.

66. Matthews, J. M. and Powles, S. B., Managing herbicide resistant annual ryegrass, southern Australia research, in *Proc. 11th Aust. Weeds Conf.*, Shepherd, R. C. H., Ed., Weed Science Society of Victoria, Melbourne, 1996, 537.

67. Roy, W., A systems approach to the control of herbicide resistant ryegrass, in *Proc. 12th Aust. Weeds Conf.*, Bishop, A. C., Boersma, M., and Barnes, C. D., Eds., Tasmanian Weed Society, Hobart, 1999, 226.

68. Gill, G. S. and Holmes, J. E., Efficacy of cultural control methods for combating herbicide-resistant *Lolium rigidum*, *Pestic. Sci.*, 51, 352, 1997.

69. Lemerle, D., Verbeek, B., Cousens, R. D., and Coombes, N., The potential for selecting wheat varieties strongly competitive against weeds, *Weed Res.*, 36, 505, 1996.

70. Powles, S. B. and Matthews, J. M., Integrated weed management for the control of herbicide resistant annual ryegrass (*Lolium rigidum*), in *Proc. 2nd Int. Weed Control Congress*, Brown, H., Cussans, G. W., Devine, M. D., Duke, S. O., Fernandez-Quintanilla, C., Helweg, A., Labrada, R. E., Landes, M., Kunsk, P., and Streibig, J., Eds., Department of Weed Control and Pesticide Ecology, Flakkebjerg, 1996, 407.

71. Walker, S. R., Robinson, G. R., and Medd, R. W., Management of wild oats and paradoxa grass with reduced dependence on herbicides, in *Proc. 9th Aust. Agronomy Conf.*, Michalk, D. L. and Pratley, J. E., Eds., Australian Society of Agronomy, Wagga Wagga, 1998, 572.

72. Lemerle, D., Verbeek, B., and Martin, P., Breeding wheat cultivars more competitive against weeds, in *Proc. 2nd Int. Weed Control Congress*, Brown, H., Cussans, G. W., Devine, M. D., Duke, S. O., Fernandez-Quintanilla, C., Helweg, A., Labrada, R. E., Landes, M., Kunsk, P., and Streibig, J., Eds., Department of Weed Control and Pesticide Ecology, Flakkebjerg, 1996, 1323.

73. Medd, R. W., McMillan, M. G., and Cook, A. S., Spray-topping of wild oats with selective herbicides, *Plant Prot. Q.*, 7(2), 62, 1992.
74. Reeves, T. G. and Smith, I. S., Pasture management and cultural methods for the control of annual ryegrass (*Lolium rigidum*) in wheat, *Aust. J. Exp. Agric. Anim. Husb.*, 5, 527, 1975.
75. Robinson, G. R., Willis, D. A., Walker, S. R., and Adkins, S. W., An appraisal for alternative strategies for managing herbicide-resistant weeds, in *Proc. 12th Aust. Weeds Conf.*, Bishop, A. C., Boersma, M., and Barnes, C. D., Eds., Tasmanian Weed Society, Hobart, 1999, 227.
76. Bowran, D. G., Hamblin, J., and Powles, S. B., Incorporation of transgenic herbicide resistant crops into integrated weed management systems, in *Commercialisation of Transgenic Crops: Risk, Benefit and Trade Considerations*, McLean, G. D., Waterhouse, P. M., Evans, G., and Gibbs, M. J., Eds., Bureau of Resource Sciences, Canberra, 1997, 301.
77. Preston, C., Roush, R. T., and Powles, S. B., Herbicide resistance in weeds of southern Australia: why are we the worst in the world? in *Proc. 12th Aust. Weeds Conf.*, Bishop, A. C., Boersma, M., and Barnes, C. D., Eds., Tasmanian Weed Society, Hobart, 1999, 454.
78. Prately, J. E., Lemerle, D., Luckett, D. J., Brennan, J. P., and Cornish, P. S., The role of herbicide-resistant crops in the winter rainfall farming systems of south-eastern Australia, in *Herbicide-Resistant Crops and Pastures in Australian Farming Systems*, McLean, G. D. and Evans, G., Eds., Bureau of Resource Science, Canberra, 1995, 127.
79. Nietschke, B. S., Medd, R. W., Matthews, J. M., Reeves, T. G., and Powles, S. B., Managing herbicide-resistant wild oats — options and adoption, in *Proc. 11th Aust. Weeds Conf.*, Shepherd, R. C. H., Ed., Weed Science Society of Victoria, Frankston, Australia, 1996, 546.
80. Orson, J. H. and Harris, D., The technical and financial impact of herbicide resistant black-grass (*Alopecurus myosuroides*) on individual farm businesses in England, in *Proc. Brighton Crop Protection Conference — Weeds*, British Crop Protection Council, Surrey, 1997, 1127.
81. Read, M. A., Palmer, J. J., and Howard, S., An integrated strategy for the successful management of herbicide-resistant *Alopecurus myosuroides* (black-grass) in UK, in *Proc. Brighton Crop Protection Conference — Weeds*, British Crop Protection Council, Surrey, 1997, 343.
82. Singh, S., Kirkwood, R. C., and Marshall, G., New management approaches for isoproturon-resistant *Phalaris minor* in India, in *Proc. Brighton Crop Protection Conference — Weeds*, British Crop Protection Council, Surrey, 1997, 357.
83. Malik, R. K. and Singh, S., Evolving strategies for herbicide use in wheat: resistance and integrated weed management, in *Proc. Int. Symp. on Integrated Weed Management for Sustainable Agriculture*, Indian Society of Weed Science, Hisar, 1993, 225.
84. Mahajan, G., Brar, L. S., and Sardana, V., The effect of tillage and time of sowing on the efficacy of herbicides against *Phalaris minor* in wheat, in *Proceedings of the 17th Asian-Pacific Weed Science Society Conference*, 1999, 193.
85. Malik, R. K. and Singh, S., Liverseed canarygrass (*Phalaris minor*) resistance to isoproturon in India, *Weed Technol.*, 9, 419, 1995.
86. Chancellor, R. J., The long-term effects of herbicides on weed populations, *Ann. Appl. Biol.*, 91, 141, 1979.
87. Freyman, S., Palmer, C. J., Hobbs, E. H., Dormaar, J. F., Schaalje, G. B., and Moyer, J. R., Yield trends on long-term dryland wheat rotations in Lethbridge, *Can. J. Plant Sci.*, 62, 609, 1982.

88. Shaner, D. L. and Howard, S., Effectiveness of mode of action labelling for resistance management: a survey of Australian farmers, in *Proc. 1999 Brighton Conference — Weeds*, 3, 797, 1999.
89. Wood, Mackenzie, Kintore House, 74-77 Queen St., Edinburgh, EH2 4NS, November 1998.

6 World Rice and Herbicide Resistance

Bernal E. Valverde and Kazuyuki Itoh

CONTENTS

0-8493-2219-7/01/$0.00+$.50

6.1   INTRODUCTION

Rice is the most important food crop in the world. Globally, it provides 23 and 16% of human per capita energy and protein, respectively.[1] About 153 million ha of rice was planted in 1999 for a total estimated production of 589 million t.[2] However, by the year 2025, rice production must increase by 40% over current levels if the growing demands for food are to be satisfied,[3] although low world market prices threaten producer income and future grain supply.[4] Overcoming constraints to rice production and increased yields, including weed management, is vital to satisfy future needs for this grain.

6.2   THE RICE AGROECOSYSTEMS
AND RICE-PRODUCING AREAS OF THE WORLD

Rice is produced under a wide variety of climatic conditions throughout the world, ranging from the wettest areas to deserts. Production systems vary from very small plots requiring an enormous amount of human labor to sophisticated, large holdings such as those in Australia and the United States.[1] The rice environment, however, compared to that of other major crops, is unique as it is dominated by surface flooding patterns, making surface hydrology the key factor in classification of rice ecosystems.[5] Khush[6] originally defined the most widely used classification of rice ecosystems and subecosystems based on hydrological characteristics, topography, and soil types. Rice ecosystems are not strictly defined and there may be overlap in their description, especially in relation to rainfed lowland rice.[7] The area distribution among rice ecosystems worldwide is presented in Table 6.1.

TABLE 6.1
Distribution of Rice Area According to Major Ecosystems and Regions of the World[1]

Region	Rice Area (1000 ha)				
	Irrigated	Rainfed Lowland	Upland	Floodprone	Total
South Asia	24,210	18,241	7,218	7,551	57,220
Southeast Asia	15,033	13,849	4,157	3,828	36,867
East Asia	34,304	1,975	677	0	36,956
Latin America	1,744	388	3,317	23	5,472
Africa	1,170	1,268	2,507	1,062	6,007
Other countries	2,878	656	1,084	330	4,951
Total	79,339	36,377	18,960	12,797	147,473
Percent	53.8	24.7	12.9	8.6	100.0

Source: Adapted from Reference 1.

6.2.1 PRINCIPAL RICE ECOSYSTEMS

6.2.1.1 Irrigated Rice

Irrigated rice is grown in bunded, puddle fields. Irrigation allows for one or more crops per year.[1] Depending on the rainfall, two subecosystems may be recognized: irrigated wet season, in which irrigation water is used to supplement that provided by natural rainfall during the wet season, and irrigated dry season, where rice cannot be grown in the dry season unless irrigation water is provided. Worldwide, about 79 million ha of rice is grown with irrigation; 93% of this production occurs in Asia, with yields varying between 3 and 9 t ha^{-1}. More than 75% of the world's rice is produced in irrigated rice lands.[1]

The most important yield-limiting factors in irrigated rice production are deficient input management, losses from pests, inadequate and inefficient water use, poor drainage, and environmental stresses.[1] Among pests, weeds play a major role in reducing yields, especially in areas where inadequate land leveling and water management are conducive to high weed infestations.

6.2.1.2 Rainfed Lowland Rice

In rainfed lowlands, rice fields are bunded and usually level to slightly sloping with noncontinuous flooding of variable depth and duration[8] and without access to irrigation water.[9] The crop may experience fluctuating hydrological conditions from complete submergence through drought during the growing season. Alternation between aerobic and anaerobic soil conditions plays a significant role in root growth, nutrient availability, and, especially, weed interference.[10] About one fourth of the world's total rice land is rainfed[1] (Table 6.1). The amount and timing of the water supply is the most important constraint to rice production in rainfed lowlands.[8,11]

There are a variety of rainfed lowland subecosystems depending on water depth and availability: favorable rainfed lowland, drought-prone, submergence-prone, drought- and submergence-prone, and medium-deep water. In these subecosystems, planted rice varieties vary in height, sensitivity to photoperiod, and length of life cycle.[12] There is also ample local hydrological variation, which is influenced by the surrounding landscape. Thus, on a single farm some fields may be drought-prone, while others may be flood- and submergence-prone in the same season.[1] These characteristics at the local level make weed floras and control practices vary from field to field, a factor that should be considered when designing and implementing herbicide-resistance management tactics.

Weeds are more economically damaging and important in high-input rainfed lowland systems, whereas diseases and insect pests are more relevant in low-input systems.[13] Weed problems will increase, especially in Asian rainfed lowlands, with the current trend to transform rice production from the predominant transplant system to direct seeding.[14] Rising wages and decreased labor availability are responsible for this change.[15] Direct seeding can reduce substantially the labor requirement for crop establishment. In Thailand, transplanting a hectare of rice takes 25 to 30 person-days, but just one person can broadcast the same area within a day.[16] Direct seeding also contributes to early planting since water requirements for sowing are much lower than for transplanting, thus reducing the risk of crop failure associated with high variability of rainfall at time of planting.[15] However, weed problems and high costs to control them could be a major constraint to widespread adoption of direct seeding, especially dry seeding.[15,16] Inevitably this will force rice growers to be more dependent on herbicides for weed control, with the subsequent risks of herbicide resistance.

6.2.1.3   Upland Rice

Upland rice is grown under diverse landforms varying from low-lying valley floors to undulating and steep slopes with high runoff and lateral water movement.[1] Upland rice accounts for about 13% of the world rice area but contributes only 4% to total rice production, since it is mostly a subsistence crop. Soils are usually dry-prepared and rice is direct-seeded in unbunded fields. Yields are generally very low (about 1 t ha^{-1}), except in some large farms in Latin America that produce 2 to 3 t ha^{-1}.[1]

Weeds are the most important biological constraint to upland rice production, causing yield losses from 30 to 100%. In some areas, small farmers lack resources for weed control and are forced to abandon their fields after a few years of shift cultivation.[1]

6.2.1.4 Flood-Prone Rice

The flood-prone ecosystem is quite diverse, and rice types must be able to tolerate conditions ranging from temporary (1 to 10 days) submergence to long periods (1 to 5 months) of stagnant water, or daily tidal fluctuations.[1] Flash floods after heavy rains may completely submerge the crop for several days. The rice crop is generally dry-seeded although some farmers also transplant tall seedlings.[17] Rice grown under these conditions is common in South and Southeast Asia and in particular areas in Africa (Guinea, Nigeria, and Sierra Leone). During the initial dry period after planting, typical rice weeds infest the crop but, after flooding, tall grasses such as *Echinochloa stagnina* and *Leersia hexandra* elongate their culms to survive as emergent plants, and weedy species adapted to rising water become predominant.[17]

6.2.2 GEOGRAPHICAL AREAS OF RICE PRODUCTION

The most important rice-producing countries are located in South, Southeast, and East Asia (Table 6.1). The two largest world producers are China (200 million t) and India (128 million t), with planted areas of 32 and 43 million ha, respectively. Eight Asian countries contribute 80% of total world rice production.

6.2.2.1 Asia

In Asia, the irrigated ecosystem represents about 56% of the 131 million ha of rice. In East Asia, more than 34 million ha of rice (93% of the area planted) is irrigated, which constitutes 43% of the world's rice irrigated area. About 41% of rice areas in the tropics of South Asia and Southeast Asia is also irrigated.[1] Highest yields in these areas are obtained in China, Japan, Indonesia, Vietnam, and Korea.

Rainfed lowlands in Asia comprise about 36.4 million ha, which represents about 25% of the total rice cultivated area, with an average yield of 2.3 t ha^{-1}.[1] The rainfed lowlands are mostly in the warm subhumid and humid tropics of South and Southeast Asia.[12]

Almost 10% of the rice area in Asia is planted in upland ecosystems (Table 6.1). In South and Southeast Asia most upland rice is grown on rolling and mountainous lands.

Currently it is estimated that about 29 million ha of rice is direct seeded in Asia, including upland and submergence-prone environments where opportunities for transplanting are limited. If only rainfed lowlands and irrigated ecosystems are considered, the rice area that is direct-seeded is 15 million ha.[15]

In China, rice production is concentrated in the region south of the Qinling Mountains and the Huaihe River.[1] In the 1990s an estimated 13 million ha of rice, representing about 40% of the planted area in China, was produced using herbicides.[18] In the subtropical province of Jiangsu, the total of about 3 million ha of rice is treated with herbicides. Predominant weed species are *Echinochloa crus-galli*, *Cyperus difformis*, *Scirpus planiculmis*, *Monochoria vaginalis*, *Eclipta alba*, *Rotala indica*, and *Ludwigia prostrata*, causing yield reductions of 8 to 15%. Thiobencarb, molinate, and quinclorac are widely used, especially against grasses and bensulfuron-methyl for broadleaf weeds. If preemergence herbicides fail, high doses of molinate or quinclorac are applied later in the season. Herbicide mixtures are commonly used, including those of propanil plus either bentazon or MCPA, oxadiazon plus butachlor, and the three-way mixture of bensulfuron, metsulfuron-methyl, and acetochlor. The average cost of a herbicide treatment in China is only U.S. $3.50 ha^{-1}.[19] Post-emergence herbicides manufactured in China do not contain adjuvants; thus, some sulfonylureas and ACCase herbicides are applied at higher rates than normal to compensate for the lower activity in their absence.[20] Obviously, the consequences are increased costs, risk of carryover with residual compounds, and increased selection pressure by those herbicides that also exhibit residual activity.

In East Asia, rice production systems are input-intensive and it is not expected that a major shift to direct seeding will occur in this region. Similarly, in South Asia economic incentives for a shift to direct seeding are weak since population density is high and overall economic growth is slow.[15,21] Where direct seeding is most likely to expand is in Southeast Asia, especially in countries with low population densities and where labor costs are escalating. In densely populated areas (Java, western China, and the Red River Delta of Vietnam) transplanting is expected to remain the dominant production system.[15]

6.2.2.2 The Americas

Upland rice predominates in Latin America, comprising about 60% of the total rice area. Only about 32% of the rice land in the most important rice-growing countries in Latin America is irrigated.[1] About 7% is grown in rainfed lowlands. On larger farms in tropical Latin America and the Caribbean, production is predominantly mechanized, direct-seeded rice.

In Brazil, the major rice producer in the Americas, the predominant upland rice is mostly grown on level to gently rolling land, mostly under mechanized cultivation, in the central and northeastern parts of the country. In other areas of Brazil, similar to many other forested areas of Latin America, rice is grown under shifting cultivation. About 27% of the rice area in Brazil is under irrigation, mostly concentrated in the southern states of Rio Grande do Sul and Santa Catarina, where rice is predominantly produced in 2 consecutive years followed by a rotation with pastures for 3 years. Rainfed lowland rice is produced in Minas Gerais, Rio de Janeiro, and Espiritu Santo.[1] In Colombia, the second largest rice producer in Latin America, rice is grown under favorable moisture regimes. Two thirds of the area is irrigated and the crop is predominantly broadcast-seeded into dry soil. In areas infested with red rice, the stale seedbed preparation is widely adopted under which weed germination is promoted by both tillage and irrigation flushes followed by broadcast applications of glyphosate or oxyfluorfen or both.[22] This practice is also becoming more common in Central America, both for red rice control and propanil-resistant *Echinochloa colona* management.

The United States produces almost as much rice as Brazil but in less than half the area. Rice is planted in three main areas: the Grand Prairie and Mississippi River

Delta of Arkansas, Louisiana, Mississippi, and Missouri; the Gulf Coast of Florida, Louisiana, and Texas; and the Sacramento and San Joaquin valleys of California. In all production areas, rice is direct-seeded; predominantly dry-seeded in the Mississippi Delta; both dry- and water-seeded in the Gulf Coast; and water-seeded in California.[23] Water-seeding has been a key component of red rice management under monoculture. For water culture, fields are laser-leveled and pregerminated rice seed is normally broadcast by airplane onto preflooded fields. A continuous water coverage is maintained until close to harvest. This surface water serves as an important weed suppressor. Prevalent weeds in each system vary: grass weeds dominate dry-seeded rice, whereas both aquatic weeds and large-seeded *Echinochloa* spp. infest water-seeded rice.

6.2.2.3 Africa

In Africa, the proportion of rice land under irrigation is low (about 12%), except for the entire 462,000 ha (mostly transplanted) in Egypt that is irrigated.[1] In the Near East, hand weeding is the most common weed control method, especially in Iran, Morocco, and Egypt.[24] In West Africa, upland rice predominates, growing on hills in the humid zone and on flatland in the drought-prone and moist forest zones. Production systems range from shifting cultivation to relatively intensive operations, using hand, animal, or mechanized tillage and rotations with other crops. Grain yields are generally low, about 1 t ha^{-1}. In the humid forests, weed invasion is a principal factor for abandoning the land; hand weeding is the most common control method. Typical rice grass weeds are predominant including *E. colona*.[25] About 21% of the West African rice area is located in rainfed lowlands, where the crop is either direct-sown or transplanted. The most serious weeds include several sedges, four *Echinochloa* grass weeds (*E. crus-galli*, *E. crus-pavonis*, *E. glaberescens*, and *E. pyramidalis*), wild *Oryza* species, and some broadleaf weeds.[25] About 200,000 ha of irrigated rice is planted in the Sahel, where Asian technologies for rice production are being adapted to local biophysical and economic conditions.[26]

6.2.2.4 Australia

About 140,000 ha of rice is planted in Australia, yielding an average 10 t ha^{-1}, the highest yield in the world. Rice productivity is attributed to high radiation levels, absence of major diseases and insect pests, locally bred high-yielding varieties, adequate decision support systems for farmers, and effective weed control.[27] Rice growing is fully mechanized and geographically confined to irrigated areas in the Murrumbidgee and Murray River valleys and Coleambally in New South Wales.[28] Rice is often rotated with legume pastures and dryland crops. Almost the entire area is sown by aircraft,[1] a planting system that increased largely as a result of the introduction of bensulfuron for weed control.[27,29] In this system, annual aquatic weeds proliferate, especially *C. difformis*, *Damasonium minus*, *Sagittaria montevidensis*, and *Alisma lanceolatum*.[30] Bensulfuron resistance is now becoming widespread in this ecosystem.

6.2.2.5   Europe

Europe contributes less than 1% of the world rice production, which is insufficient to satisfy the local market,[31] yet intensity of production and herbicide use in some areas have resulted in eight weed species becoming resistant to rice herbicides. Five countries in the European Union produce rice: Italy (the largest producer with 1.4 million t), Spain, France, Portugal, and Greece. Among other European rice-producing countries, the Russian Federation is the largest, contributing 450,000 t harvested from 146,000 ha.[2]

6.3   CURRENT STATUS OF HERBICIDE RESISTANCE IN RICE WEEDS WORLDWIDE

6.3.1   WEEDS AS A CONSTRAINT TO RICE PRODUCTION

Overall, weeds are the most important biological constraint to rice production.[32–34] Uncontrolled weeds interfering with rice throughout the growing season can cause severe losses depending on local conditions and weed infestation. Under commercial cropping conditions, weeds are by far more important than other biological factors in determining yields.[34] The same weed species can have differential impact depending on the rice cropping system. For example, the same relative *Echinochloa* spp. density causing a 20% yield loss in transplanted rice caused a 70% loss in direct-seeded rice.[23]

Usually a diverse weed flora is associated with rice production. A survey conducted along the irrigation canals at a representative farm in the irrigated rice area in Costa Rica registered 131 and 144 weed species belonging to 45 botanical families in the dry and rainy seasons, respectively.[35] Obviously, a large, diverse weed flora has the potential to invade the crop if suitable niches are created by agricultural practice. Different species also occupy different sections of the canals (levee, internal and external slopes, and bottom). Interestingly, the most important weeds associated with rice are fairly similar within ecological and production systems across geographical areas of the world. Weed floras associated with rice production and the surrounding landscape will respond to any selective control practice by changing their composition and the relative importance of certain species. When herbicides are used, weed shifts usually occur before resistance evolves. Changes in planting method and production system also have a dramatic impact on the composition of weed populations.

The long-term, persistent use of herbicides has led to infestations of rice fields with perennial weeds in Japan[36] and shifts in weed flora for different parts of Asia.[37] Changes in predominance of weed species have been well documented in Korea. Until the mid-1960s there was only limited use of herbicides, mostly 2,4-D and later PCP (pentachlorophenol). In the 1970s nitrofen became popular and, later in the decade, butachlor and alachlor. Under this herbicide history, populations of *E. crus-galli* and *R. indica* were reduced, but *M. vaginalis* became dominant, especially in the 1980s when the use of carbamate herbicides became predominant. Decreased use of butachlor in the 1990s resulted in a resurgence of *E. crus-galli*. In the last decade, herbicide mixtures, most of them including ALS-inhibiting herbicides, especially bensulfuron and pyrazosulfuron-ethyl, have dominated rice weed control. This change to ALS-inhibiting sulfonylureas has resulted in *Eleocharis kuroguwai* and *Sagittaria trifolia* becoming the most troublesome weeds in transplanted rice in Korea, as these two species are not satisfactorily controlled by sulfonylureas applied soon after rice transplanting.[38]

The change from transplanting to direct seeding favors a shift from less competitive broadleaf weeds to grasses. One of the most serious threats to rice production with the spread of direct seeding is the likely increase in weedy *Oryza* species infestation. In Malaysia, some farmers have been forced to switch back to transplanting to overcome weedy *Oryza* in direct-seeded plots.[39] Four years after widescale adoption of direct seeding onto saturated puddled soil in traditionally transplanted irrigated fields in Malaysia, 21 new weed species were incorporated into the weed community and 15 of the original 46 common weeds were excluded. The weed flora shifted from predominantly semi-aquatic and aquatic species, represented by *Sagittaria guyanensis*, *M. vaginalis*, and *Limnocharis flava*, to heavy infestations of *Echinochloa* spp. Thus, within 4 years of adopting direct seeding, *E. crus-galli* that had been extremely rare under a transplanting system became the predominant weed under direct seeding.[40]

Weed control represents an important proportion of total rice production costs. The total cost of weed control in Japan was about U.S. $445 ha^{-1} in 1990, or 4.6% of the total production cost of U.S. $9,714 ha^{-1}, most of which is machinery and labor.[41] In Colombia, cost of rice weed control ranges from 16 to 18% of the total production cost, without considering that about 24 to 28% of the production cost is dedicated to soil preparation and planting.[42]

6.3.2 WEEDS ASSOCIATED WITH RICE THAT HAVE EVOLVED RESISTANCE TO HERBICIDES

Thirty weed species have evolved resistance to herbicides in rice (Table 6.2) in all growing areas of the world, except Africa. Resistance has occurred under all agroecosystems, except in very low input growing systems such as slash and burn upland rice. Resistance to sulfonylureas is most frequent although this group of ALS-inhibiting herbicides is relatively new to rice. Twenty weed species are known to be resistant to ALS-inhibiting herbicides, especially bensulfuron. Other economically important cases of resistance are those of *Echinochloa* spp. that evolved resistance to propanil, molinate, butachlor, thiobencarb, and quinclorac and, more recently, multiple-resistance to several rice herbicides (Table 6.2).

6.3.2.1 Resistance to ALS-Inhibiting Herbicides (Group 2)

Several weed species have evolved resistance to ALS-inhibiting herbicides in various crops within a few years after their commercial introduction (reviewed by Saari et al.[93]). In rice, resistance to this group of herbicides is predominant in Asia, especially in Japan, where currently nine species, most of them aquatic annuals, are

TABLE 6.2
Weeds Associated with Rice That Have Evolved Resistance to Herbicides[a]

Species	Botanical Family	Life Cycle	Herbicide to Which Resistance Evolved	Country	References
1 Alisma plantago-aquatica	Alismataceae	Perennial	Bensulfuron, cinosulfuron	Portugal	43, 44
				Italy	45
			Bensulfuron	Spain	46
2 Ammania auriculata	Lythraceae	Annual	Bensulfuron	U.S.A.	47
3 A. coccinea	Lythraceae	Annual	Bensulfuron	U.S.A.	47
4 Bacopa rotundifolia	Scrophulariaceae	Annual	Sulfonylureas	Malaysia	48
5 Cyperus difformis	Cyperaceae	Annual	Bensulfuron	Australia	494
				U.S.A.	50
				Spain	46
6 Damasonium minus	Alismataceae	Perennial	Bensulfuron	Australia	51
7 Echinochloa colona	Poaceae	Annual	Propanil	Central America	52
				Colombia	53
				Mexico	54
				U.S.A.	55
				Venezuela	67
			Fenoxaprop	Costa Rica	57
				Nicaragua	Valverde et al., unpublished
			Propanil, fenoxaprop, azimsulfuron	Colombia	50, Valverde et al., unpublished
				Costa Rica	59, 60
			Quinclorac	Colombia	60

No.	Species	Family	Life cycle	Herbicide	Country	Reference
8	*E. crus-galli*	Poaceae	Annual	Propanil	Greece	61
					U.S.A.	62
					Sri Lanka	63
					Thailand	64
				Butachlor, thiobencarb	China	18, 65
				Butachlor	Thailand	66
				Butachlor, propanil	Thailand	66
				Quinclorac	U.S.A.	67, 68
					Spain	69
					Brazil	70, 71
9	*E. crus-pavonis*	Poaceae	Annual	Quinclorac	Brazil	70
10	*E. hispidula*	Poaceae	Annual	Quinclorac	Spain	69, 72
11	*E. oryzicola*	Poaceae	Annual	Quinclorac	Spain	69, 72
12	*E. oryzoides*	Poaceae	Annual	Molinate, thiobencarb	U.S.A.	73
				Quinclorac	Spain	69, 72
13	*E. phyllopogon*	Poaceae	Annual	Fenoxaprop, molinate thiobencarb, bispyribac	U.S.A.	73
14	*Elatine triandra* var. *pedicellata*	Elatineceae	Annual	Sulfonylureas	Japan	74
15	*Fimbristylis miliacea*	Cyperaceae	Annual	2,4-D	Malaysia	75
16	*Ischaemum rugosum*	Poaceae	Annual	Fenoxaprop	Colombia	58
17	*Limnocharis flava*	Butomaceae	Perennial	2,4-D	Indonesia	76
				2,4-D and bensulfuron	Malaysia	77
18	*Limnophila sessiliflora*	Scrophulariaceae	Perennial	Sulfonylureas	Japan	78
19	*Lindernia dubia* sub spp. *dubia*, *L. dubia* var. *major*[b]	Scrophulariaceae	Annual	Sulfonylureas	Japan	78
					Malaysia	48
20	*L. micrantha*	Scrophulariaceae	Annual	Sulfonylureas	Japan	78
21	*L. procumbens* (= *L. pyxidaria*)	Scrophulariaceae	Annual	Sulfonylureas	Japan	78
22	*Monochoria korsakoii*	Pontederiaceae	Annual	Sulfonylureas	Japan	80, 81
					Korea	82, 83
23	*M. vaginalis*	Pontederiaceae	Annual	Sulfonylureas	Japan	84
					Korea	83, 85

TABLE 6.2 (CONTINUED)
Weeds Associated with Rice That Have Evolved Resistance to Herbicides[a]

Species	Botanical Family	Life Cycle	Herbicide to Which Resistance Evolved	Country	References
24 *Rotala indica* var. *uliginosa*	Lythraceae	Annual	Sulfonylureas	Japan	86
25 *Sagittaria guyanensis*	Alismataceae	Annual	Sulfonylureas	Malaysia	87
26 *S. montevidensis*	Alismataceae	Perennial	Bensulfuron	Australia	49
			Bensulfuron	U.S.A.	50
			Bensulfuron, bispyribac	Brazil	70
27 *Scirpus juncoides* var. *ohwianus*[c]	Cyperaceae	Perennial	Sulfonylureas	Japan	88
28 *S. maritimus*	Cyperaceae	Perennial	Sulfonylureas	Spain	46
29 *S. mucronatus*	Cyperaceae	Perennial	Bensulfuron, cinosulfuron	Italy	89
			Bensulfuron	U.S.A.	47
30 *Sphenoclea zeylandica*	Sphenocleaceae	Annual	2,4-D	Malaysia	90
				Philippines	91, 92
				Thailand	Itoh, unpublished

[a] Only confirmed cases are listed.
[b] Both subspecies, introduced from North America to Asia, are distinct. Subspecies *major* was introduced between 50 to 80 years ago to Japanese paddy rice fields; subspecies *dubia*, probably 25 to 35 years ago.
[c] Variety should be correctly identified as *ohwianus*, not *juncoides*.

Source: Compiled from data presented by Heap[76] and indicated references.

reported resistant. In Japan, almost the entire rice area is mechanically transplanted and typically receives two herbicide applications, one each in pre- and postemergence under puddled conditions. Granular "one-shot" formulations are preferred; they were used in 84% of paddy fields in 1991.[41] Most of these one-shot formulations contain ALS-inhibiting herbicides, mainly bensulfuron and pyrazosulfuron. In 1991, 636,000 and 774,000 ha were applied with one-shot formulations containing ALS-inhibiting herbicides as pre- and postemergence applications, respectively.[41] More than half of Japan's rice fields were sprayed with a mixture of bensulfuron plus mefenacet in 1993; the second most popular treatment included these two herbicides plus thiobencarb.[36] Currently, in addition to these mixtures, one-shot formulations containing bensulfuron plus either esprocarb or dimepiperate, and pyrazosulfuron plus mefenacet, are widely used. Bensulfuron is a component of more than 40 one-shot formulations.[94] This is an example of how heavy reliance on a herbicide such as bensulfuron for weed control in rice translates into a generalized herbicide-resistance problem.

The first ALS-inhibiting herbicide-resistant species (*M. korsakowii*) in Japan was discovered in 1995, after only 5 years of sulfonylurea herbicide use. This weed no longer can be controlled with ALS herbicides, including pyrazosulfuron, imazosulfuron, and cyclosulfamuron.[81] The second confirmed case of ALS-herbicide resistance in Japan was that of the aquatic weed *L. micrantha* in a monoculture rice field treated for 7 consecutive years.[95] Subsequently, biotypes of *L. dubia* subsp. *dubia*, *L. dubia* subsp. *major*, *L. procumbens*, and *L. sessiliflora* were also found to be resistant to ALS-inhibiting herbicides.[78,80,81,96–99] Recently, four more ALS-herbicide-resistant species have been found: *S. juncoides* var. *ohwianus*[88,100,101] after more than 5 years of intensive sulfonylurea use, *R. indica* var. *uliginosa*,[86] *E. triandra* var. *pedicellata*,[74] and *M. vaginalis*.[84] These nine species (one of them, *L. dubia*, with two resistant varieties) are becoming increasingly widespread throughout the rice-producing area of Japan.

An ALS-herbicide-resistant biotype of *M. korsakowii* was also recently found in reclaimed rice fields of the west coastal area of Korea after being exposed to sulfonylureas for about 9 years.[82] *L. flava* multiple-resistant to ALS-inhibiting herbicides and 2,4-D has been documented in peninsular Malaysia.[77] Populations of *Lindernia* sp. and *B. rotundifolia* have also evolved resistance to ALS-inhibiting herbicides. *S. guyanensis* is the most recently confirmed case of this type of resistance in peninsular Malaysia.[87]

Other species resistant to ALS-inhibiting herbicides in rice fields are *A. plantago-aquatica* in Portugal, Italy and Spain; *A. auriculata* and *A. coccinea* in the United States; *C. difformis* in Australia, the United States, and Spain; *D. minus* in Australia; *S. montevidensis* in Australia, the United States, and Brazil; *S. mucronatus* in Italy and the United States; and *S. maritimus* in Spain (Table 6.2). *A. plantago lanceolata*, *S. montevidensis*, and *S. mucronatus* are listed among the important weeds associated with rice in Chile[102] and there is indication that these three species have also evolved resistance to ALS-inhibiting herbicides.[103]

Although normally not invasive in rice fields, the grass weed *Ixophorus unisetus* evolved resistance to ALS-inhibiting herbicides in Costa Rica, after about 5 years of treatment with imazapyr along irrigation and drainage ditchbanks in rice fields.[104]

Populations of *E. phyllopogon* that exhibit multiple-resistance in California (U.S.A.) are not controlled with the ALS-herbicide bispyribac-sodium.[73] Similarly, a population of *E. colona* from Costa Rica formerly identified as multiple-resistant to propanil and fenoxaprop-ethyl is also resistant to azimsulfuron.[60]

A common characteristic of the evolution of resistance to ALS-inhibiting herbicides is the persistent use of the same herbicides, especially bensulfuron, for controlling aquatic weeds and sedges. Thus, a close association between the herbicide use patterns and the distribution of resistant biotypes of *Lindernia* spp., mostly *L. dubia* var. *dubia*, was observed at several farms in the Yamagata Prefecture in Japan.[96] Aoki et al.[105] and Sugimoto et al.[106] made similar observations in relation to the distribution of resistant *L. micrantha*. In Australia, *C. difformis, D. minus*, and *S. montevidensis* evolved resistance to bensulfuron in areas where this herbicide had been used extensively (more than 90% of the rice crop treated for 10 seasons). Consequently, the possibilities of continued chemical weed control is substantially limited, as the only alternative herbicide for these species has been MCPA.[30,49,107]

The current situation in California (U.S.A.) illustrates how serious and widespread the ALS-herbicide-resistance problem can be. From the mid-1940s until 1990, phenoxy herbicides were widely used in rice production for broadleaf weed control, and in the early 1960s propanil was introduced to control grass weeds. From 1989, bensulfuron was rapidly adopted on nearly all Californian rice areas to replace previous broadleaf herbicides because it allowed control at earlier stages with more selectivity to the crop than the phenoxy herbicides.[23] Additionally, environmental concerns forced changes and limitations in the use of alternative herbicides in California. Propanil use was banned in the Sacramento Valley in 1969 because of drift and potential injury to prune and other deciduous orchard crops[108–110] being allowed only in about 10% of the rice area. In 1989, bentazon use in rice was suspended and its registration was cancelled the following year because of contamination of well water.[110,111] Recent use patterns of important rice herbicides in California are illustrated in Figure 6.1. In 1992, 3 years after introduction of bensulfuron, two populations of *C. difformis* and two of *S. montevidensis* were confirmed as resistant to this herbicide in separate counties (total of four fields in two counties). Three years later, resistance had been confirmed in 4118 fields in 13 counties that represented 95% of the rice-growing area. At least half of the fields had more than one resistant species and ALS-resistant *S. montevidensis* was present in over 60% of California's rice fields. Of the species that evolved resistance, *S. montevidensis* originally had been the most susceptible to bensulfuron.[47,112] Widespread ALS-herbicide resistance has forced a new shift in herbicide use patterns in California (Figure 6.1). Although not as well documented, a similar pattern of rapid and almost total bensulfuron adoption occurred in Australian rice areas, with subsequent widespread resistance.

In Italy, it is estimated that about 15,000 ha of rice (6% of the total area) is infested with ALS-herbicide-resistant weeds; the area affected by *A. plantago-aquatica* has stabilized but *S. mucronatus* is increasing.[45] Resistance has appeared in fields subjected to rice monoculture for several years and treated with either bensulfuron or cinosulfuron for more than 3 years, in some cases with sequential half-dose applications. In Portugal, *A. plantago-aquatica* became resistant to ALS-inhibiting

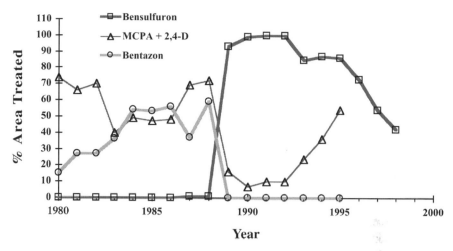

FIGURE 6.1 Use patterns of important broadleaf and sedge herbicides in California rice fields (1980–1998). (Adapted from data (1980–1995) provided by Carriere[112] and Hill[47] and from the State of California Environmental Protection Agency, Department of Pesticide Regulation, Pesticide Use Reports.[113–116])

herbicides shortly after bensulfuron was commercially introduced in rice.[43] Resistant populations are now prevalent in over 1700 ha in the Tagus, Sorraia, Sado, Caia, and Mondego valleys. Resistance is more frequent in fields considered high risk due to the higher selection pressure (more than 3 years of bensulfuron use) than in fields where the sulfonylurea herbicide had been rotated with herbicides of alternate modes of action.[44] Resistant populations of *A. plantago-aquatica* from Portugal are cross-resistant to other sulfonylureas but not to imazapyr or to herbicides with an alternative mode of action.[117]

It is common in weeds that have evolved resistance to ALS-inhibiting herbicides that the resistant biotypes exhibit different patterns of target-site cross-resistance to herbicides sharing the ALS mode of action.[93] Resistance indexes (RI), customarily calculated as the ratio of the herbicide doses required to reduce growth by 50% (GR50) in resistant and susceptible plants, in Japanese *Scrophulariaceae* weeds range from about 100 to 14,000 times. Itoh et al.[95] reported that a pyrazosulfuron-resistant population of *L. micrantha* (RI = 141) was also resistant to bensulfuron (RI = 282) and imazosulfuron (RI = 81). A population of *L. sessiliflora* had RI values of 896, 334, and 655 for bensulfuron, pyrazosulfuron, and imazosulfuron, respectively.[118] Based on a petri dish assay, Uchino et al.[119] determined RI values for bensulfuron of 60, 10,200 and 14,100 in *L. procumbens*, *L. dubia* var. *major*, and *L. dubia* var. *dubia*, respectively. Resistance indexes for pyrazosulfuron, imazosulfuron, and bensulfuron in *S. juncoides* were 40 to 80, 75 to 90, and 55 to 140, respectively.[88] In Italy, there are biotypes of both *S. maritimus* and *A. plantago-aquatica* resistant to bensulfuron that are cross-resistant to other sulfonylureas and to the triazolopyrimidine herbicide metosulam; one resistant population, however, is only marginally cross-resistant to metosulam. These biotypes had not been exposed previously to metosulam.[89] Populations of *S. montevidensis* from Brazil resistant to pyrazosulfuron are cross-resistant to bispyribac, metsulfuron, ethoxysulfuron, and cyclosulfamuron, leaving only the non-ALS herbicide bentazon as an effective alternative.[70] *I. unisetus* resistant to imazapyr is cross-resistant to other imidazolinone herbicides but not to triasulfuron.[104]

6.3.2.2   Resistance to the PS II-Inhibiting Herbicide Propanil (Group 7)

Biotypes of two species, *E. colona* and *E. crus-galli*, have evolved resistance to propanil in many areas. Propanil-resistant populations of *E. colona* were first found in the central Pacific region (Puntarenas province) of Costa Rica in 1989, during an initial survey prompted by growers' complaints about a lack of control with propanil.[120] Rice fields in that area had been treated with propanil up to three times per season for up to 15 years, with one or two crop seasons per year. The most resistant population came from the field with the strongest selection pressure and the propanil GR_{50} was determined to be 8.5 times higher in the resistant biotype. The observed levels of resistance closely corresponded with the propanil exposure history, except for two populations collected in fields where rice had been planted for 15 years (one crop per year). Interestingly, these two populations, which were barely 1.5 times more resistant than the control, had been regularly treated with pendimethalin in addition to the customary propanil applications, which probably delayed the evolution of resistance. Similar complaints about poor performance and a need to use higher doses of propanil to achieve adequate efficacy led to the discovery of propanil resistance in Colombia.[53] Colombian populations exhibited similar levels of resistance to those originally found in Costa Rica.[53]

A wider range of responses to propanil than that originally reported in Costa Rica and Colombia was found in a preliminary survey of the most important rice-producing areas in Central America.[52] Populations 70 times more resistant than the control susceptible population were found in areas with a long history of rice production. Here the crop had been planted twice a year and propanil used for more than 25 years. Currently about 80% of the populations sampled in Central America have been diagnosed as propanil resistant (Table 6.3).

It is estimated that at least 60% of the 250,000 ha of rice planted in the Central America area has propanil-resistance problems. Farmers still use propanil with little effect on resistant populations and an estimated U.S. $18 million per year (1998) is spent on propanil under such conditions. Assessment of the economic impact of propanil resistance should take into consideration that additional and more expensive herbicides, such as fenoxaprop or cyhalofop-butyl that are applied later in the cropping season, are required to control resistant populations. For example, in Costa Rica an estimated 24,000 ha per year were treated with the ACCase herbicides fenoxaprop and sethoxydim between 1990 and 1997 to control propanil-resistant *E. colona*. Additionally, the lack of efficacious control early in the season results in yield losses due to weed–crop competition.[121]

In Mexico, about 96% of 46 *E. colona* accessions collected in the major rice-growing areas of Tres Valles and Tierra Blanca (state of Veracruz), Zacatepec (Morelos), Tuxtepec (Oaxaca), and Palizada (Campeche) were slightly to moderately

TABLE 6.3

Summary of Survey to Determine the Response to Propanil of Selected Populations of *Echinochloa colona* from Central and South America (1996–1999)[a]

Country	Total Tested No.	Susceptible No.	Susceptible %	Intermediate Resistance[b] No.	%	Highly Resistant[c] No.	%	Total Resistant No.	%
Belize	2	1	50.0	1	50.0	0	0.0	1	50.0
Bolivia	3	3	100.0	0	0.0	0	0.0	0	0.0
Colombia	4	0	0.0	0	0.0	4	100.0	4	100.0
Costa Rica	31	3	9.7	5	16.1	23	74.2	28	90.3
El Salvador	21	4	19.0	3	14.3	14	66.7	17	81.0
Guatemala	4	2	50.0	1	25.0	1	25.0	2	50.0
Honduras	10	7	70.0	2	20.0	1	10.0	3	30.0
Nicaragua	19	4	21.1	1	5.3	14	73.7	15	78.9
Panama	7	0	0.0	0	0.0	7	100.0	7	100.0
Total	101	24	23.8	13	12.9	64	63.4	77	76.2

[a] Partial data from survey of Valverde et al., unpublished.
[b] Populations with resistance indexes between 2.0 and 4.0.
[c] Populations with resistance indexes greater than 4.0.

resistant to propanil; those with highest resistance levels came from Veracruz and Campeche.[54,122] Resistance is also suspected in the states of Tabasco and Colima.[121]

Propanil resistance in *E. colona* in Venezuela has now been confirmed in Portuguesa State.[56] The situation in Venezuela is likely to be similar to that of other regions in Mesoamerica in that different levels of resistance to propanil are found among *E. colona* populations and farms in the same region and even among fields within a farm. Thus it is not surprising that in a recent efficacy trial at one location in Portuguesa State, propanil at 3.6 kg ha[-1] controlled above 98% of grass weeds, primarily *E. colona*.[123] Propanil-resistant *E. colona* also occurs in the United States[55] and it is suspected in the Sahelian West Africa.[26] There are unconfirmed reports of *E. colona* resistance to propanil in Ecuador but no resistant populations were found in Bolivia (Valverde et al., unpublished).

Resistance to propanil in *E. crus-galli* has been documented in Greece,[61] the United States,[62] Sri Lanka,[63] and, more recently, in Thailand.[64] Of these, the best-studied case is that affecting rice production in the United States. Propanil and molinate were introduced and rapidly adopted in the early 1960s but propanil continues to be the backbone of chemical weed control in the southern United States; molinate has been the principal grass herbicide in California.[23] Failure to control *E. crus-galli* with propanil was identified in Arkansas in 1989 and Baltazar and Smith[124] confirmed resistance. A survey conducted in 1991 and 1992 identified resistant populations in most sampled areas among 19 rice-producing counties.[55] Resistance was common in fields where rice had been grown for several years; 70% of the fields surveyed had been planted with rice for more than 15 years. Resistant populations have been confirmed in another three states (Louisiana, Mississippi, and Texas) where propanil also has been widely used.[55] In California, legal limitations precluded widespread use of propanil for several years; thus, resistance has not evolved there. However, resistance to ALS-inhibiting herbicides recently forced re-registration of propanil although with severe restrictions on aerial application[47] and allowing only water-based flowable formulations.[111] In Greece, propanil resistance in barnyardgrass evolved in areas where rice had been grown for over a decade and propanil was used repeatedly.[61] In Sri Lanka, resistance to propanil also evolved under similar conditions.[63]

6.3.2.3   Resistance to Herbicides That Inhibit Lipid Synthesis (Thiocarbamates, Group 8, and Chloroacetamides, Group 15)

Resistance to the thiocarbamate herbicides thiobencarb and molinate and to the chloroacetamide herbicide butachlor, all considered as low risk herbicides for resistance evolution,[125] has been documented in *Echinochloa* spp.

Huang and Lin[65] initially reported resistance to butachlor in *E. crus-galli* in China in areas where the herbicide had been used for more than 8 years. Resistance to thiobencarb was also found in China.[18] Resistance level to both herbicides was determined to be directly related to the intensity of rice cropping (single vs. double cropping) and herbicide use history.[18,65] Populations monitored for 6 years in areas of South China, where rice is planted twice per year, exhibited a steady increase in resistance indexes to butachlor from 6.4 in 1991 to 15.1 in 1996.[18] Huang and Gressel[126] estimated that *E. crus-galli* populations resistant to butachlor and thiobencarb in China infest 2 million ha.

More recently, *E. crus-galli* resistant to butachlor[66] and *E. oryzoides* resistant to molinate and thiobencarb[73] were also found in Thailand and California (U.S.A.), respectively. At least one *E. crus-galli* population in Thailand is multiple-resistant to both butachlor and propanil.[66]

6.3.2.4 Resistance to Other Herbicides

Biotypes of three weed species, *F. miliacea*, *L. flava*, and *S. zeylandica*, are resistant to 2,4-D in Asia.[75,77,91,92] *F. miliacea* resistant to 2,4-D was identified in 1989 in a field in Malaysia where the herbicide had been used seasonally in double rice cropping since 1975. The resistant biotype required 32 times the recommended dose of 2,4-D for a 50% reduction in growth.[75] *L. flava* resistance was initially reported in Indonesia and more recently in Malaysia.[76,77] A population multiple-resistant to both 2,4-D (RI > 28) and bensulfuron (RI > 10) was found at an experimental site in Malaysia.[77] *S. zeylandica* resistant to 2,4-D was first documented in the Philippines[91,92] and later in Malaysia.[90]

Failure to control *E. colona* with propanil prompted farmers in Costa Rica to adopt fenoxaprop and sethoxydim as alternative herbicides, usually applied much later in the cropping season (about 30 to 40 days after planting). As expected, persistent dependency on ACCase herbicides within a few years resulted in fenoxaprop resistance in Costa Rica[57] and in Colombia and Nicaragua (Valverde, unpublished). Areas where *E. colona* has become resistant to fenoxaprop are increasing rapidly in Colombia as several ACCase herbicides are used for postemergence grass control.[58] In Costa Rica, there are populations resistant to either propanil or fenoxaprop or both.[59] Fenoxaprop-resistant *E. colona* populations from Costa Rica are also cross-resistant to cyhalofop, fluazifop-*p*-butyl, quizalofop-ethyl, clodinafop-propargyl, cycloxydim, and sethoxydim.[57,59,127]

Resistance to fenoxaprop has been confirmed in *Ischaemum rugosum* in Colombia. Most severely affected are the rice-producing areas of Granada and Cucuta in the departments of Meta and Norte de Santander, respectively, where about 70% of the populations tested were resistant.[58]

Echinochloa spp. also have evolved resistance to quinclorac. Quinclorac resistance was initially reported in Spain but there was confusion as to whether resistance evolved because of quinclorac-imposed selection pressure during 3 to 5 years of use, or from cross-resistance from an atrazine-resistant biotype, or was due to differential herbicide response among the species involved.[72,128–130] Recently, a survey covering the five *Echinochloa* species present in Spain (*E. crus-galli*, *E. hispidula*, *E. oryzoides*, *E. oryzicola*, and *E. colona*) confirmed resistance in both *E. crus-galli* and *E. hispidula*.[69] Since previous reports actually referred to resistance in *E. oryzoides* and *E. oryzicola*,[72] the only *Echinochloa* species that has not evolved quinclorac resistance in Spain is *E. colona*. However, there are populations of *E. colona* already resistant to quinclorac in Colombia.[60] Additionally, there are *E. crus-galli* populations resistant to quinclorac in two U.S. states: Louisiana[67] and

Mississippi,[68] and in Brazil. The Brazilian populations remain susceptible to herbicides with alternative modes of action.[70,71] *E. crus-pavonis* was also confirmed resistant to quinclorac in Brazil.[70]

6.3.2.5   Multiple Herbicide Resistance

Cases of multiple herbicide resistance in rice weeds are beginning to emerge but thus far do not pose the management challenges of *Lolium rigidum* in Australia. An *E. colona* biotype from Costa Rica already multiple-resistant to propanil and fenoxaprop exhibited reduced sensitivity to azimsulfuron in greenhouse trials with this sulfonylurea herbicide.[60] Most worrisome is the case of *E. phyllopogon* multiple-resistant to fenoxaprop, thiobencarb, molinate, and bispyribac in California.[73] The extent of the California rice area affected by multiple-resistance is not known. Multiple-resistance and cross-resistance patterns in *Echinochloa* species highlight the risk of mismanaging herbicides in rice production. In this respect, recommendations such as those suggesting doubling the dose of bispyribac to control these populations[131] should be avoided.

6.3.3   Genetics and Mechanisms of Resistance

6.3.3.1   Resistance to ALS-Inhibiting Herbicides (Group 2)

Several ALS-inhibiting herbicides are used or are being developed for rice to control broadleaf weeds and sedges (bensulfuron, cinosulfuron, cyclosulfamuron, ethoxysulfuron, halosulfuron-methyl, imazosulfuron, iodosulfuron, metsulfuron, pyrazosulfuron). A few also control grasses, including *Echinochloa* spp. (azimsulfuron). Three pyrimidinyl (oxy)benzoate ALS herbicides effectively control *Echinochloa* spp. in rice: bispyribac, pyribenzoxim, and pyriminobac-methyl. Imidazolinone and triazolopyrimidine ALS-inhibiting herbicides are not selective in rice; however, imidazolinone-resistant rice varieties (IMI-rice) soon will be introduced, especially aimed at control of red rice.[132] Although chemically dissimilar, all of these herbicides have the same ALS mode of action. Most cases of resistance to ALS-inhibitors in weeds are due to a target mutation (resistant ALS). In weeds that have evolved resistance to ALS herbicides in crops other than rice, a single gene or allele confers resistance. Resistance can be inherited as a dominant, incompletely dominant, or additive trait depending on the species or biotype.[133–135] Proline substitutions in the Domain A of the ALS enzyme frequently confer resistance.[93] Indications that resistance in the Japanese species *L. micrantha* is caused by a target-site mutation[136] were confirmed by Uchino and Watanabe,[137] who found that pro in Domain A of all susceptible biotypes of *Lindernia* spp. had been substituted by other amino acids in resistant biotypes: gly, ser, and ala in *L. micrantha*, *L. procumbens*, and *L. dubia* var. *major*, respectively. Shibuya et al.[138] also determined a pro to leu substitution in ALS-resistant *S. juncoides*. In Europe, resistance to bensulfuron in *A. plantago-aquatica*, *C. difformis*, *S. maritimus*, and *S. mucronatus* is ALS target-site based.[46]

The inheritance of resistance to sulfonylureas in *L. micrantha* was studied by Itoh et al.[95] F_1 seedlings from controlled crosses between susceptible and resistant plants were treated with pyrazosulfuron and the surviving plants were back-crossed with susceptible pollen and seed collected. Segregation of the seedlings of the progeny of the back-crosses based on their response to bensulfuron indicated that a dominant nuclear gene controlled resistance. *L. micrantha* is predominantly cross-pollinated[139] but the role of outcrossing in the spread of resistance is unknown.

Similar studies with *M. korsakowii* also point to a single dominant nuclear gene as the basis for resistance to ALS herbicides.[118] The flowers of *M. korsakowii* exhibit somatic enantiostyly and also stamen dimorphism to facilitate insect cross-pollination, although flowers are self-compatible and capable of setting seed in absence of pollinators.[140] The rate of outcrossing in *M. korsakowii* was experimentally estimated in 37 to 80% in presence of pollinators and nil in their absence. *M. vaginalis* is self-pollinated as its flowers are cleistogamous.[118] There is no available information on the role of mode of reproduction on the spread of resistance or on the inheritance of ALS resistance in other species or to other herbicides.

6.3.3.2 Resistance to the PS II-Inhibiting Herbicide Propanil (Group 7)

Propanil is an inhibitor of photosynthesis at Photosystem II (PS II). PS II herbicides bind to the D1 protein that is located in the chloroplast and prevent the normal flow of electrons rendering the photosynthetic apparatus subject to excess energy and photo-oxidative damage. Rice hydrolyzes propanil (Figure 6.2) to 3,4-dichloro-aniline (DCA) and propionic acid by an aryl acylamidase (AAA).[141–143] Further metabolism of propanil after its initial hydrolysis may also occur, before it is inactivated by conjugation with sugars[144–146] or incorporation with lignin.[143] Both carbamate and organophosphorous insecticides, especially the oxidized analogs of the latter, act as strong inhibitors of the rice AAA,[141,147,148] an important characteristic that led to the development of synergists for management of propanil resistance.[149] Selectivity of propanil to rice is achieved because most weeds have no AAA or, if present, the enzyme has a much lower activity compared to that of rice.[150] For many years, rice growers have been able to take advantage of this difference to control *Echinochloa* spp. and other grasses with propanil.[151]

Resistance to PS II triazine herbicides almost always has been shown to be endowed by target-site mutations (reviewed by Gronwald[153]). Resistance to propanil, however, is not target-site based. Propanil resistance in *E. colona* is due to increased

FIGURE 6.2 Propanil hydrolysis in *Echniochloa colona*. (Adapted from Reference 152.)

AAA activity.[148,154] Susceptible and resistant biotypes absorb similar amounts of propanil when applied to the leaves. Older plants absorb less propanil than young seedlings.[154] Carey et al.[155] also confirmed similarity in absorption and translocation of propanil in propanil-resistant and -susceptible *E. crus-galli*. Resistant *E. colona* biotypes metabolize propanil more rapidly and to a greater extent than susceptible biotypes because of their higher AAA activity.[154] Metabolism is also implicated in propanil resistance in *E. crus-galli*.[156] Propanil resistance is maintained or even increased in older *E. colona* plants even though AAA and propanil metabolism decrease with plant age.[154] Probably, in older plants, mechanisms other than AAA activity also contribute to propanil resistance.

6.3.3.3   Resistance to Herbicides That Inhibit Lipid Synthesis (Thiocarbamates, Group 8, and Chloroacetamides, Group 15)

Chloroacetamides, which include the rice herbicides butachlor and pretilachlor, are considered mitotic inhibitors but their mechanism of action and selectivity is still not completely elucidated. Rice tolerates pretilachlor because it is able to metabolize the herbicide more rapidly than *Echinochloa* spp., especially in the presence of the safener fenclorim, which is included in the formulation.[157,158] The resistance mechanism of *E. crus-galli* to butachlor is not completely known. Indirect measurements indicate that there is two to four times increased α-amilase or amide hydrolase activity in resistant plants compared to the susceptible.[18,65] There is also limited information about the mode of action and selectivity to rice of the thiocarbamate herbicides molinate and thiobencarb. Molinate is transformed to polar metabolites in rice but little appears to be conjugated with glycosides.[159] The mechanism of resistance to thiobencarb in *Echinochloa* spp. is unknown but the indirect evidence suggests enhanced metabolism.

6.3.3.4   Resistance to Other Herbicides

The mechanism of resistance to 2,4-D in rice weeds is also unknown. Auxinic herbicides cause numerous physiological and biochemical responses and their mode of action has not been elucidated.[160] It is probable that the basis for auxin-herbicide resistance is also diverse although it could be associated with an altered auxin binding site yet to be described.[161]

Some ACCase herbicides are used in rice especially to control *Echinochloa* spp. and other important grasses such as *Leptochloa* spp. and *I. rugosum*. These include clodinafop, cyhalofop, fenoxaprop, butroxydim, clefoxydim, and sethoxydim. The mechanism of resistance to fenoxaprop in *E. colona* from Central America and Colombia is still unknown, but resistance is not associated with an elevated metabolism or altered ACCase.[121] Resistance in most grass weeds to ACCase graminicides is usually due to a modification of the target site (ACCase), but in many *L. rigidum* and *A. myosuroides* biotypes there can be enhanced metabolism.[162,163]

The mechanism of action of quinclorac was recently elucidated.[164] Plants readily absorb quinclorac and translocate it both basipetally and acropetally. The effect of quinclorac in broadleaf weeds is similar to an auxin overdose stimulating the synthesis of ethylene and accumulation of abscisic acid. In *Echinochloa* spp. quinclorac leads to the accumulation of toxic cyanide. Cyanide does not accumulate in tolerant rice or in quinclorac-resistant biotypes of *E. hispidula* and *E. crus-galli* from Spain and Mississippi, respectively.[164,165]

6.3.3.5 Multiple Herbicide Resistance

The mechanisms that endow multiple herbicide resistance in *E. colona* from Costa Rica that is resistant to propanil, fenoxaprop, and azimsulfuron are unknown.[121] Initial studies on multiple-resistance in *E. phyllopogon* biotypes resistant to fenoxaprop, molinate, thiobencarb, and bispyribac from California suggest at least partial involvement of enhanced herbicide degradation.[166] Multiple-resistance in other grass weeds can involve multiple mechanisms[167] and therefore this also likely will be documented in these multiple-resistant rice weeds.

6.4 HERBICIDE-RESISTANCE MANAGEMENT

As for other crops and agroecosystems, two strategies should be considered in relation to herbicide-resistance management in rice. If herbicide resistance has not yet evolved, designing and implementing appropriate control measures should help in minimizing its occurrence. If resistance has already occurred, integrated management offers the best approach to minimize the extent of resistance and to control resistant populations. Rice farmers usually integrate different tactics to control weeds, thereby unconsciously practicing integrated weed management. An example is the standard practice in Asia of combining herbicide and manual weed control with production practices suitable for the local environment and production system.[168] In areas where manual and physical control are still the main methods used to combat weeds, there is little chance that herbicide resistance will evolve, as herbicides are seldom used. As more intensive production schemes are implemented, dependency on herbicides will result in weed shifts and herbicide resistance. Asia has a long tradition of hand-weeding in rice[169] but it is increasingly more expensive to hand-weed as labor becomes scarce. Inter-row cultivation using hand-pushed or engine-powered rotary weeders is a preferred weed control method in most of Asia.[168] Manual or rotary weeding accounts for more than half of the weed management practices in South and Southeast Asia,[168] but it is not commonly used in Japan, where herbicides are the control method of choice, especially the "one-shot" combinations containing two or three herbicides.[36] As more fields are transformed to direct seeding, dependence on herbicides will increase. About three times more labor is required to hand-weed direct-seeded rice compared to transplanted rice, making hand-weeding unprofitable in direct-seeded rice.[169] Direct-seeding is being rapidly adopted in South and Southeast Asia, including parts of Thailand, Malaysia, the Philippines, and Vietnam; after direct-seeding is adopted by a few farmers, it increases rapidly and uniformly.[169] The key for success in direct-seeded rice is efficient weed control,[15] and integrated weed management offers the best opportunity to achieve it and to minimize herbicide resistance.

6.4.1 Minimizing the Evolution of Herbicide Resistance

The main preventative tactic is obviously to decrease the selection pressure imposed by herbicides. The dose, efficacy, and frequency of herbicide use largely determines selection pressure. Typically, resistance in rice weeds has evolved in monocropping systems heavily dependent on a few herbicides for weed control. Highly effective herbicides used persistently impose high selection pressure that can result in herbicide-resistant populations evolving in just a few generations. Two relatively new groups of herbicides are considered especially prone for resistance evolution: the ALS- and ACCase-inhibiting herbicides.

Repeated application of a herbicide in the same season can increase selection pressure, particularly when there are weed species that complete more than one generation per season. Such herbicide regimes are responsible for propanil resistance in *E. colona* in Central America. A common recommendation is to apply mixtures or to rotate herbicides to reduce selection pressure. For resistance management, both herbicides in the mixture must be at full dose and effective on the target weed species, and possess similar persistence but different mechanisms of action and/or degradation pathways in the plant.[170] These criteria are different from the conventional practice of mixing herbicides, i.e., for broadening the weed control spectrum. Pendimethalin, for example, fulfills most of these requirements and has been considered an excellent partner for propanil both to minimize propanil-resistance evolution in *E. colona* and as an alternative product when propanil resistance has already evolved.[52, 57, 120] In Japan, where herbicide rotations were used, no resistance to ALS-inhibiting herbicides in *L. micrantha* was observed.[95]

Use of certified rice seed free of weed seeds and avoidance of contaminated farm equipment should help in preventing the introduction of resistant material to new areas or fields. Unfortunately, in many tropical areas, farmers save seed from previous harvests to lower production costs,[22,168] increasing the risk of disseminating resistant weeds. The importance of machinery cleaning in preventing the spread of resistant weeds was illustrated by Itoh et al.[96,97] A farmer in Japan selected a resistant population of *L. dubia* var. *major* after 5 years of using bensulfuron plus mefenacet in a rice field. Movement of farm machinery (transplanting and combine harvester) infested a completely separated upland field with resistant individuals in a pattern following equipment movement from the field entrance onward. While weed problems in general, and herbicide resistance in particular, can be spread to other sites, it is important to emphasize that herbicide selection pressure is the main driving force in the appearance of resistant populations. Indeed, studies using molecular markers in *L. micrantha* established that resistance to ALS herbicides evolved as multiple events and that it was not the result of a founder population being spread over several locations either by pollen or plant introductions in contaminated equipment.[136]

The accumulated knowledge from 30 years of herbicide resistance should enable us to anticipate herbicide problems. However, experience also demonstrates that preventative measures are rarely practiced to avoid selection of herbicide-resistant populations. For example, scientists warned about the risk of rice weeds evolving resistance to bensulfuron and becoming a problem in Australia.[27] This was particularly evident where up to three times the recommended dose was being used to control the hard-to-kill species *A. lanceolatum*.[171] The attractiveness (efficacy, selectivity, and profitability) of bensulfuron, however, made its use widespread and soon resulted in the selection of resistant populations.

6.4.2 MANAGEMENT OF RESISTANT POPULATIONS

Once resistant populations have evolved, appropriate tactics should be integrated to manage the problem so as to avoid major crop losses. We now describe and discuss some of the tactics suitable for rice production that could be integrated in managing resistant populations.

6.4.2.1 Agronomic Management of Resistant Populations

Good agricultural practice, based on a solid knowledge of the biology and ecology of the weeds associated with rice, must be the basis for integrated weed management.

6.4.2.1.1 Seed set prevention after harvest

At rice harvest, it is common to find weeds still in the vegetative stage. These plants either survived herbicides used in the crop or emerged late. If the survivers are left to mature, their seeds will enrich the soil seedbank. Ploughing down the rice stubble after harvesting, if conditions are suitable, could prevent seed setting on surviving weeds. Stubble incorporation, however, had no effect on *E. colona* density in the following cropping season or within the span of a series of field trials conducted for up to 3 years in Costa Rica.[172]

Nonselective herbicides such as glyphosate, glufosinate, and paraquat or rotary weeders are used in Japan and Korea to control winter weeds before land preparation or during the autumn after rice harvest. In Taiwan and China, farmers burn rice straw after harvest to control remnant weeds and to minimize soil seedbanks.[168] In Japan, perennial weeds that have been selected after continuous use of one-shot herbicide treatments are controlled by ploughing the fields in late autumn. This also exposes underground propagules that will die in the cold, dry winter air.[41]

6.4.2.1.2 Puddling

Puddling and transplanting, perfected in China, are ancient practices associated with the domestication of rice.[173] Puddling, however, consumes a large proportion of the estimated 5000 L of water required to produce 1 kg of rice. In rainfed systems, farmers are forced to delay planting because they rely on heavy rains to flood their fields for puddling, increasing the risk of the crop suffering from drought at the reproductive stage.[21] If the heavy rains do not come, the land may remain fallow, as occurs in northeastern Thailand, where 30 to 40% of the rainfed lowland area is not planted at all or is planted but the crop fails completely after transplanting.[16]

6.4.2.1.3 Delayed planting for preseeding weed control

In weed management, a rule of thumb is that in conventional tillage the better the seedbed is prepared the better the weed control will be. Discing or ploughing at intervals before rice planting achieves control of initial weed populations that otherwise would emerge with the dry-seeded crop. A smooth seedbed also improves the efficacy of preemergence herbicides. As mentioned before, weed management by delayed seeding with several cultivations for weed control is sometimes referred to as the stale seedbed system. Delayed planting to allow weeds to emerge and be eliminated with herbicides is a common practice in Latin America, especially in Colombia, for red rice and *E. colona* management.[22] *E. colona* can emerge in several flushes, especially early in the rice-cropping season.[121] In rainfed rice, the first tillage operation, usually a heavy disk harrowing, prior to planting can be timed to kill the first weed flush promoted by initial rains. A new flush will be stimulated by this tillage practice and the remaining soil moisture or further rains. Growers normally finish seedbed preparation by two additional passes of a light disk harrow, which if properly timed, can eliminate the second flush of *E. colona* emergence. If instead of planting immediately, *E. colona* is controlled with a nonselective herbicide such as glyphosate, future in-crop infestations are considerably reduced.[172] Mechanical control could substitute for the preplant glyphosate in reducing in-crop *E. colona* densities provided that weather allows use of machinery. For example, in an experiment conducted in Costa Rica, the initial *E. colona* population emerging with the crop was reduced from 623 plants m^{-2} in conventionally planted plots to 67 and 56 plants m^{-2} when controlled by shallow tillage or glyphosate in delay-planted plots, respectively.[121] Exceptionally, for water seeding, as practiced in the southern United States and California, the seedbed is left relatively rough, with intact clods. Breaking up of these clods upon flooding covers the seed and avoids the seed drifting during strong winds.[174]

In Italy, in rice fields where contamination with red rice forced growers to use stale seedbed preparation, and to apply oxadiazon before planting, resistance to ALS herbicides has not been found. Where, in addition to ALS herbicide, oxadiazon (or pretilachlor) has been applied but at low doses or later than recommended, ALS resistance has evolved.[89]

No-tillage planting of rice can also be an option in managing herbicide-resistant weeds. In Brazil, approximately 300,000 ha of rice are planted under minimum or no-till systems to control red rice. After the soil is prepared (minimum tillage) or before planting (no-till) glyphosate is applied to remove emerged red rice (25 to 30 cm tall), whose germination had been promoted by irrigation. As the soil is not inverted, excellent control is achieved before planting the rice crop.[175] Minimum or no-till also lowers farm investments in heavy farm machinery and labor for tillage operations.[176] Similar minimum tillage systems are being used in Colombia and in Central America. No-tillage systems can result in less weed emergence and therefore the lower initial weed densities should reduce herbicide use, thereby decreasing the selection intensity.

6.4.2.1.4   Crop rotation

There is usually a typical weed flora associated with rice in each unique ecosystem and production scheme. Intensive rice production in monoculture selects for weed floras highly compatible with the agricultural systems, which usually are also the most competitive and hard to kill. Moody[39] reports that in Pyinmana, Myanmar, a shift in production from transplanted rice rotated with an upland crop to rice mono-culture (transplanted rice followed by wet-seeded rice) increased weed problems in transplanted rice. Rice monoculture decreased the time available for land preparation and imposed a change from long-duration, competitive rice varieties to short-cycle, less competitive varieties. Before the introduction of the wet-seeded rice, the predominant weeds in transplanted rice were *Commelina diffusa* and *Echinochloa* spp., but since then *Isachne globosa* has become more important. Fields infested with hard-to-kill weeds are more likely to be persistently treated with herbicides.

Crop rotation, with the corresponding interruption of the weed–crop association, is useful in controlling these weeds and diminishing their seedbank by allowing other control practices to be implemented, including the use of herbicides with different modes of action. Crop rotation is practiced in rainfed areas in tropical Asia[168] but not in Latin America.[22] In the southern United States, rice is grown in 1 year and then rotated with soybean for 3 years, especially to control recalcitrant weeds such as *Althernanthera philoxeroides*, red rice, and *Leptochloa* spp. Conversely, species difficult to control in soybean (*Sorghum halepense* and *Xanthium strumarium*) are more easily controlled under the flooded rice system.[174] In Arkansas (U.S.A.), the incidence and level of propanil resistance in *E. crus-galli* were less pronounced in rice fields subjected to crop rotation.[55] Similar observations have been made in Central America[52] for *E. colona*. Rotations, however, should be carefully planned. In Bolivia, where rice is rotated with soybeans, ACCase-inhibiting herbicides effective on grass weeds (mostly fluazifop) are frequently used in soybeans especially to control *Eriochloa punctata,* a weed that is becoming prevalent in rice. This weed already has evolved resistance to ACCase-inhibiting herbicides.[177]

In Australia, management practices in the past favored long rotations rather than monoculture.[29] However, low returns from animal production and changes in water-use policy resulted in more intensive rice cropping.[28] As expected, weed floras varied and perennial and aquatic weeds became more prevalent. A long pasture rotation aids in the control of *A. lanceolatum* by reducing the viability of buried corms and could regain importance as a tactic to manage herbicide resistance.[28]

6.4.2.1.5   *Competitive varieties and increased rice seeding rate*

In the face of herbicide resistance, increased attention is being paid to rice varieties with ability to suppress weeds. Garrity et al.[178] evaluated 25 cultivars for their competitive ability against weeds. Weed suppressing ability was associated with rice plant height; fewer weeds were present when taller cultivars (>1.20 m) were planted compared to intermediate (1.0 to 1.15 m) and semidwarf cultivars (<1.0 m). The most competitive cultivar suppressed weed dry weight by up to 75%. Increased crop dry matter and leaf area index but not tiller number were also associated with weed suppression ability. An acceptable compromise could be achieved by improving intermediate varieties with combined weed suppression ability and superior yields. Tall varieties have the drawback of being susceptible to lodging, have a lower tillering capacity, and a relatively large leaf area index that results in mutual shading of the leaves. All these characters are normally related to low yield potential.[178] Conversely, early maturing varieties tend to be short and poor weed competitors in rainfed conditions; they are sensitive to delayed transplanting and mature during periods of heavy rainfall, making them unsuitable for the rainfed ecosystem. On the other hand, early maturing varieties selected for irrigated areas, with vigorous initial vegetative growth, are more competitive with weeds and use less irrigation water.[179] Another important trait is tolerance to anaerobiosis, which would allow pregerminated seeds to be sown into standing water.[169]

Fischer et al.[180] found that under severe weed pressure in irrigated conditions in Colombia the rice variety CICA 8 produced more grain and was able to suppress *E. colona* better than other rice varieties. In the absence of *E. colona* competition, the less competitive varieties yielded similarly to CICA 8. Competitiveness of the rice cultivars tested was correlated with increased leaf area index, tiller number, and canopy light interception. Competitive cultivars could help in decreasing herbicide use, especially those associated with applications against late-emerging *E. colona* flushes.[181] Research is also being carried out to better characterize and exploit competitive traits, including allelopathy, for future incorporation into rice varieties.[182-184] Promising rice varieties are able to suppress growth of *E. crus-galli* by about 40 to 50% under field conditions. If such highly competitive rice varieties are introduced we expect that they would reduce the dependence on herbicides for weed control.[185,186]

A relatively simple method of resistance management is increasing seeding rates above current practice. Of course, seeding rates must be optimized according to local conditions and varieties. Increasing seeding rates may confer a competitive advantage to the crop in early stages of growth as more crop seedlings occupy the available space. Seeding rate, however, can only be increased to limits determined by economic and agronomic reality. As rice-seeding rate increases, rice tiller number per plant decreases, bringing about a compensatory effect in the number of rice panicles per unit area.[187] At high rice panicle density, grain filling also decreases and thus final yields are not improved.[188] High seeding rates also increase the risk of lodging. There is no doubt that a competitive advantage to rice is also provided by transplanting. Hand-transplanting predominates in South and Southeast Asia and machine-transplanting in East Asia. Transplanting of 30- to 35-day-old seedlings and adequate water management supplemented by some additional control practices (hand- or rotary-weeding or a pre-emergence herbicide application or both) normally suffice in flooded rice fields.[168]

An aspect to be considered in the relationship between the rice crop and its weeds is the possibility of differential fitness between herbicide-resistant and -susceptible biotypes. Very limited information is available on fitness of herbicide-resistant weeds associated with rice. Fischer et al.[53] found considerable variability in leaf area, aboveground biomass, and growth among propanil-resistant and -susceptible biotypes of *E. colona* from Colombia under greenhouse conditions, but only a lower reproductive capacity could be associated with the more resistant biotypes. There was no direct relation between competitiveness and response to propanil. Garro et al.[189] did not observe differences in growth (plant height, tiller number, and biomass) or inflorescence number between a propanil-resistant and -susceptible biotype of *E. colona* from Costa Rica. A field study has also shown that a susceptible and a resistant biotype of *E. crus-galli* equally reduced rice grain yield at weed densities ranging from 2 to 20 plants ha^{-1}. The resistant biotype, however, produced fewer panicles and less aboveground biomass than the susceptible one at each density.[190] Seed germination and initial seedling growth were also similar among propanil-resistant and -susceptible *E. crus-galli* biotypes from Sri Lanka.[191]

6.4.2.1.6 Water management

Water management influences substantially the weed flora in rice fields. Saturated soil conditions allow germination of *E. crus-galli*, *S. zeylandica*, and *C. difformis* but a more moderate soil moisture content would favor establishment of *E. colona*, *Leptochloa chinensis*, and *Cyperus iria*. A few species, such as *M. vaginalis*, *S. juncoides*, and *L. flava*, can germinate under flooded conditions. Time, duration, and depth of flooding water can be manipulated to suppress weed growth.[168] Flooding can be a very effective weed management practice as weed populations decrease as the water depth increases. Increasing depth of submergence reduces emergence and growth of *E. crus-galli*.[33] Water management in laser-leveled fields has been very successful for weed control in the United States but herbicide use remains. Deep water tables are conducive to high infestations of *S. mucronatus*.[23] Careful water management is also important in relation to herbicide activity, crop growth, and snail management.[168] In Japan, deep weed flooding that would control several weed species is not possible since levees only allow a water table of 3 to 5 cm; shallow flooding is practiced to maximize tillering for increased yields.[41]

6.4.2.1.7 Other cultural practices

Other crop husbandry practices that place rice at competitive advantage over weeds can help in decreasing the need for repeated or sequential herbicide applications. Balanced and timely fertilization regimes improve rice vigor. Nitrogen application is frequently split during the cropping season to balance the nutrition needs of the plant. Under conditions of competition for mobile nutrients such as nitrogen, supplementary applications of fertilizer cannot substitute for effective weed control and may actually aggravate weed interference.[192,193] Weeds should be eliminated before applying nitrogen. In Central and South America, when propanil fails to control *E. colona* because of resistance, growers frequently use ACCase herbicides late in the season, some of which are only marginally selective to rice. To stimulate the crop to recover from partial phytotoxicity caused by the herbicide, growers delay application of nitrogen until the weed begins to show toxicity symptoms. Nitrogen nutrition status of the rice plant influences its response to fenoxaprop.[194,195]

6.4.2.2 Chemical Control of Resistant Populations

6.4.2.2.1 Alternative herbicides

A key component of integrated weed management is the judicious use of herbicides. This is especially relevant for high-input systems, where selection of sequences and mixtures of herbicides with different modes of action and contrasting chemistry are important for resistance management.[196] The argument for this strategy is that the chances of a weed becoming simultaneously resistant to herbicides with several modes of action are low, although exceptions may occur as in cases of multiple-resistance such as reported for *A. myosuroides*[162] and *L. rigidum*.[163] For instance, the tubulin biosynthesis inhibitor pendimethalin effectively controls *E. colona* resistant to the PS II-inhibiting herbicide propanil because resistance is due to an elevated activity of AAA that metabolizes propanil but has no effect on the pendimethalin molecule. In contrast, resistance to the PS II-inhibiting herbicide chlortoluron by *A. myosuroides* is due to enhanced mixed function oxidase activity that results in rapid metabolism of both chlortoluron and pendimethalin.[197] Therefore, in this case pendimethalin is not an effective herbicide to use in mixture or a sequence with other herbicides.

Several herbicides belonging to different chemistries are available for selective weed control in rice (Table 6.4). In Japan, ALS-herbicide-resistant annual paddy weeds are controlled by sequential application of pretilachlor followed by a formulated mixture of MCPB, simetryn, and thiobencarb.[95,198,199] Pentoxasone, a new oxazolidinedione herbicide, is also effective against resistant broadleaf annuals such as *Lindernia* spp.[200] Perennial *S. juncoides* can be controlled with one-shot formulations containing several herbicides.[88,201,202]

In Australia, alternative herbicides to control ALS-resistant weeds are quite limited. In addition to the old herbicide MCPA, the pyrazole herbicide benzofenap has been recently registered as an option for the control of ALS-resistant aquatic weeds.[207] California rice growers also face serious difficulties with ALS-herbicide resistance as agrochemical companies voluntarily removed and regulatory authorities further restricted the phenoxy herbicides (MCPA and 2,4-D) in 1998 because of damage to nontarget crops, especially cotton.[111,112] Thiobencarb controls ALS-herbicide-resistant *C. difformis*[23] and triclopyr is replacing other phenoxy herbicides (Figure 6.2), especially to control *S. mucronatus*.[47] Californian authorities granted an emergency registration, under highly regulated conditions, to use carfentrazone-ethyl to specifically control ALS-resistant *S. montevidensis* and *S. mucronatus*.* Growers in Italy use oxadiazon before planting followed by a postemergence application of MCPA, usually in combination with propanil, to control ALS-resistant weeds.[89] In Portugal, growers still do not adopt specific measures to manage ALS resistance; in fact, because bensulfuron still controls *Cyperaceae* it continues to be used in combination with other products.[117] Similarly, despite resistance, bensulfuron remains the most widely used broadleaf herbicide in Australia because it is cheap and still controls several weed species.[207]

Many commercially used herbicides control propanil-resistant *E. colona* and *E. crus-galli*, including anilofos, bifenox, bispyribac, butachlor, cyhalofop, clomazone, fenoxaprop, molinate, oxadiazon, pendimethalin, pyribenzoxim, pretilachlor, quinclorac, and thiobencarb.[59,121,208,209] However, we should take into consideration that both species have already evolved resistance to most of them (Table 6.2). Some of these herbicides (anilofos, bispyribac, butachlor, clomazone, pendimethalin, pretilachlor, pyribenzoxim, pyriminobac, quinclorac, or thiobencarb) alone or in tank mixtures with propanil also control propanil- and fenoxaprop-resistant *E. colona*.[59,121,210] In Central America, propanil plus clomazone controls propanil-resistant *E. colona*.[211] Farmers would prefer to maintain propanil use if adding a second herbicide to the tank mixture could control *Echinochloa* spp., as propanil selectively controls several dicotyledonous weeds. In modified weed control regimes

---

* California Department of Pesticide Regulation. 1999. California authorization for pesticide use under USEPA Section 18 No. 99-13.

TABLE 6.4
Herbicides Registered or in Development for Selective Weed Control in Rice

Mode of Action	WSSA Group	Herbicide	Chemical Group	Company	Activity	Resistance Risk
Cell division inhibitor	15	Anilofos	Organophosphorus	Aventis	Pre-emergence control of grasses; synergist to propanil to control propanil-resistant E. colona	Low
ALS inhibitor	2	Azimsulfuron	Sulfonylurea	Du Pont	Postemergence control of Echinochloa spp. and annual and perennial broadleaf weeds and sedges	High
Lipid synthesis inhibitor but not ACCase inhibitor	16	Benfuresate	Benzofurane	Aventis	Pre-emergence control of grasses and sedges	Low
ALS inhibitor	2	Bensulfuron-methyl	Sulfonylurea	Du Pont	Pre- and postemergence control of many emerged and submerged broadleaf weeds and sedges	High
PS II inhibitor	6	Bentazon	Benzothiadiazinone	BASF	Postemergence control of Cyperaceae and aquatic and some annual broadleaf weeds	Low
Bleaching, 4-HPPD inhibitor	28	Benzofenap	Pyrazole	Mitsubishi	In one-shot formulations, annual and perennial broadleaf weeds	Low
Protox inhibitor	14	Bifenox	Diphenyl-ether	Aventis	Pre- and postemergence control of broadleaf weeds and some grasses in transplanted rice	Low
ALS inhibitor	2	Bispyribac-sodium	Pyrimidinyl oxybenzoate	Kumiai/Bayer	Control of a wide range of grasses, including Echinochloa spp., broadleaf weeds, and sedges	High
Cell division inhibitor	15	Bromobutide	Butyramide	Sumitomo	Effective against Echinochloa spp., Eleocharis spp., S. juncoides, and some broadleaf weeds	Unknown
Cell division inhibitor	15	Butachlor	Chloroacetanilide	Monsanto	Pre- and early postemergence control of annual grasses and Murdannia nudiflora	Low
Microtubule assembly inhibitor	3	Butamifos	Phosphoroamidate	Sumitomo	Pre-emergence, annual grasses and some broadleaf species	Low
Tubulin polymerization inhibitor	3	Butralin	Dinitroaniline	Aventis	Pre-emergence, annual broadleaf weeds and sedges	Low

TABLE 6.4 (CONTINUED)
Herbicides Registered or in Development for Selective Weed Control in Rice

Mode of Action	WSSA Group	Herbicide	Chemical Group	Company	Activity	Resistance Risk
Cell division inhibitor	15	Cafenstrole	Triazole	Eikoi Kasei	Pre- and postemergence control of *Echinochloa* spp., *C. difformis*, and other annual weeds in paddy rice; discontinued 1994	Low
Protox inhibitor	14	Carfentrazone-ethyl	Triazolinone	FMC	Postemergence control of broadleaf weeds	Low
Protox inhibitor	14	Chlomethoxyfen	Diphenyl-ether	Ishihara Sangyo	Controls *Echinochloa* spp., *Scirpus* spp., and other annual weeds in transplanted rice	Low
Unknown	27	Cinmethylin	Cineole	Cyanamid	Controls *Echinochloa* spp., *M. vaginalis*, and *C. difformis* by post-transplanting application	Low
ALS inhibitor	2	Cinosulfuron	Sulfonylurea	Novartis	Postemergence control of broadleaf weeds and sedges	High
ACCase inhibitor	1	Clefoxydim, BAS 625 H	Cyclohexanedione	BASF	Postemergence grass control	High
ACCase inhibitor	1	Clodinafop-propargyl	Aryloxyphenoxy propanoate	Novartis	Postemergence grass control	High
Carotenoid synthesis inhibitor	11	Clomazone	Isoxazolidinone	FMC	Pre-emergence control of grasses	Low
Synthetic auxin	4	Clomeprop	Phenoxycarboxylate	Mitsubishi	Pre- to early postemergence control of broadleaf weeds and sedges in paddy rice (in combination with pretilachlor)	Low
Unknown	27	Cumyluron, Jc-940	Phenylethyl urea	Nihon Carlit/Gamyla	Pre-emergence control of grasses	Unknown
ALS inhibitor	2	Cyclosulfamuron	Sulfamoylurea	Cyanamid	Pre- and postemergence control of broadleaf weeds and sedges	High
ACCase inhibitor	1	Cyhalofop-butyl	Aryloxyphenoxy propanoate	Dow AgroSciences	Postemergence control of grasses	High
Synthetic auxin	4	2,4-D	Phenoxycarboxylate	Several companies	Postemergence control of broadleaf weeds and some sedges	Medium

No.	Mode of action	Common name	Chemical family	Company	Use	Resistance
27	Unknown, inhibits cell division	Daimuron	Phenylethyl urea	SDS Biotech	Pre- and early postemergence control of sedges and annual grasses in paddy rice; used as safener for sulfonylureas in Japan	Unknown
8	Lipid synthesis inhibitor but not ACCase inhibitor	Dimepiperate	Thiocarbamate	Mitsubishi	Postemergence control of *Echinochloa* spp. in flooded rice	Low
5	PS II inhibitor	Dimethametryn	Triazine	Novartis	Control of annual broadleaf weeds, usually in combination with other herbicides	Low
3	Tubulin formation disruptor	Dithiopyr	Pyridine	Rohm and Haas	Pre- and postemergence control of annual grasses and some broadleaf weeds in direct-seeded and transplanted rice	Medium
8	Lipid synthesis inhibitor but not ACCase inhibitor	Esprocarb	Thiocarbamate	Zeneca	Pre- and postemergence control of annual weeds and *Echinochloa* spp. in paddy rice	Low
2	ALS inhibitor	Ethoxysulfuron	Sulfonylurea	Aventis	Pre- and postemergence control of broadleaf weeds and sedges	High
27	Unknown	Etobenzanid	Dichloroaniline	Hodogaya	Pre- and postemergence control of grasses and some broadleaf weeds in direct-seeded rice	Unknown
	Safener	Fenclorim	Pyrimidine	Novartis	Herbicide safener against damage from pretilachlor	Not applicable
1	ACCase inhibitor	Fenoxaprop-p-ethyl	Aryloxyphenoxy propanoate	Aventis	Postemergence control of grasses	High
15	Cell division inhibitor	Fentrazamide, BAY YRC 2388	Carbamylurea or tetrazolinone	Bayer	Control of grasses and annual sedges; under registration in Japan and Korea in mixture with sulfonylureas for transplanted rice, and in mixture with propanil for direct-seeded rice in Southeast Asia	Unknown
3	Tubulin polymerization inhibitor	Flucnloralin	Dinitroaniline	BASF	Annual grasses, sedges, and broadleaf weeds in transplanted rice	Medium
14	Protox inhibitor	Fluoroglycofen-ethyl	Diphenyl-ether	Novartis	Control of broadleaf weeds and grasses	Low
Synergist	Growth regulator	Flurenol-butyl	Morphactin	Cyanamid	Synergist for phenoxy herbicides	Not applicable
15	Cell division inhibitor	Fluthiamide, BAY FOE 5043	Oxyacetamide	Bayer	Preplant incorporated, pre- and early postemergence control of annual grasses and some broadleaf weeds	Unknown
2	ALS inhibitor	Halosulfuron-methyl	Sulfonylurea	Nissan/Monsanto	Pre- and postemergence broadleaf control	High

TABLE 6.4 (CONTINUED)
Herbicides Registered or in Development for Selective Weed Control in Rice

Mode of Action	WSSA Group	Herbicide	Chemical Group	Company	Activity	Resistance Risk
ALS inhibitor	2	Imazosulfuron	Sulfonylurea	Takeda	Pre- and postemergence control of broadleaf weeds and sedges	High
Unknown	27	Indanofan, MK-243	Indandione	Mitsubishi	Pre- and postemergence control of *Echinochloa* spp. and aquatic weeds in transplanted rice	Unknown
ALS inhibitor	2	Iodosulfuron	Sulfonylurea	Aventis	Pre- and postemergence control of broadleaf weeds and some grasses; under development in mixture with trifensulfuron	High
PS II inhibitor	6	Ioxynil	Nitrile	Aventis	Postemergence control of Cyperaceae and broadleaf weeds in combination with phenoxy herbicides	Low
Growth regulator	Not classified	Maleic hydrazide	Pyridazinedione	Uniroyal	Late postemergence suppression of red rice seed production	Low
Synthetic auxin	4	MCPA	Phenoxycarboxylate	Several companies	Postemergence control of broadleaf weeds and some sedges	Low
Synthetic auxin	4	MCPA-thioethyl	Phenoxycarboxylate	Hokko	Postemergence control of sedges and broadleaf weeds	Low
Cell division inhibitor	15	Mefenacet	Oxyacetamide	Bayer/Nihon Tokunho	Pre- and early postemergence control of grasses, especially *Echinochloa* spp. in transplanted rice; preferably used in mixtures	Unknown
ALS inhibitor	2	Metsulfuron-methyl	Sulfonylurea	Du Pont	Postemergence control of broadleaf weeds	High
Lipid synthesis/mitosis inhibitor	8	Molinate	Thiocarbamate	Zeneca	Pre-plant incorporated, pre- and early postemergence control of annual grasses and some broadleaf weeds	Low
Unknown	Not classified	MT-147	Not disclosed	Mitsui Chemical	Pre- and early postemergence control of *Echinochloa* spp.	Unknown
Cell division inhibitor	15	Naproanilide	Amide	Mitsui Toatsu	Control of annual and some perennial grasses, but not *E. crus-galli*, in paddy rice; controls *Sagittaria pygmaea* seedlings	Unknown

Mode of action	Group	Common name	Chemical family	Company	Description	Resistance risk
Unknown	Not classified	OK-9701	Triazole	Otsuka	Pre- and postemergence; under development	Unknown
Protox inhibitor	14	Oxadiargyl	Oxadiazole	Aventis	Pre-emergence control of grasses, broadleaf weeds, and sedges	Low
Protox inhibitor	14	Oxadiazon	Oxadiazole	Aventis	Pre- and postemergence control of grasses and broadleaf weeds	Low
Growth inhibitor, target unknown	Not classified	Oxaziclomefone, MY-100	Oxazinone	Aventis	Pre- and early postemergence control of *Echinochloa* spp., sedges, and some broadleaf weeds	Unknown
Protox inhibitor	14	Oxyfluorfen	Diphenyl-ether	Rohm and Haas	Pre-plant control of red and weedy rice and broadleaf weeds	Low
Tubulin polymerization inhibitor	3	Pendimethalin	Dinitroaniline	Cyanamid	Pre- and early postemergence control of *Echinochloa* spp. and other grasses	Medium
Unknown	Not classified	Pentoxazone	Oxazolidinedione	Kaken	Pre-emergence control of *Echinochloa* spp.; usually in mixture with other herbicides	Unknown
Cell division inhibitor	15	Piperophos	Organophosphorus	Novartis/Rohm and Haas	Pre-emergence control of grasses and some sedges, usually combined with other herbicides; synergist to propanil to control propanil-resistant *E. colona*	Low
Cell division inhibitor	15	Pretilachlor	Chloroacetamilde	Novartis	Primarily used to control grasses but it also controls some broadleaf weeds and sedges	Low
PS II inhibitor	7	Propanil	Anilide	Several companies	Postemergence control of *Echinochloa* spp. and other grasses, broadleaf weeds, and sedges emerging from seeds	High
Bleaching, 4-HPPD inhibitor	28	Pyrazolynate	Pyrazole	Sankyo	Control of grasses, sedges, and aquatic weeds in paddy rice	Unknown
ALS inhibitor	2	Pyrazosulfuron-ethyl	Sulfonylurea	Nissan	Pre- and postemergence control of annual and perennial broadleaf weeds and sedges in dry-seeded and paddy rice	High
Bleaching, 4-HPPD inhibitor	28	Pyrazoxyfen	Pyrazole	Ishihara Sangyo	Pre- and postemergence control of annual and perennial weeds in transplanted rice	Unknown
ALS inhibitor	2	Pyribenzoxim	Pyrimidinylbenzoate	LG Chemicals	Control of a wide range of grasses, including *Echinochloa* spp., broadleaf weeds, and sedges	High
Unknown	27	Pyributicarb	Thiocarbamate	Dainippon	Pre- and early postemergence control of grasses, especially *Echinochloa* spp., and some sedges and aquatic weeds in paddy rice	Unknown

TABLE 6.4 (CONTINUED)
Herbicides Registered or in Development for Selective Weed Control in Rice

Mode of Action	WSSA Group	Herbicide	Chemical Group	Company	Activity	Resistance Risk
ALS inhibitor	2	Pyriftalid, CGA 279233	Pyrimidinylthiobenzoate	Novartis	Control of grasses and some broadleaf weeds in both transplanted and direct-seeded rice	High
ALS inhibitor	2	Pyriminobac-methyl	Pyrimidinylbenzoate	Kumiai	Pre- to late postemergence control of *Echinochloa* spp. in paddy rice	High
Auxin-type activity— target unknown	4	Quinclorac	Quinolinecarboxylate	BASF	Pre- and postemergence control of *Echinochloa* spp. and some broadleaf weeds	Medium
ACCase inhibitor	1	Sethoxydim	Cyclohexanedione	BASF	Postemergence control of grasses	High
PS II inhibitor	5	Simetryn	Triazine	Nippon Kayaku, Novartis	Used in combination with thiobencarb for broadleaf weeds and algae	Low
Cell division inhibitor	15	Thenylchlor	Chloroacetanilide	Tokuyama	Pre-emergence control of annual grasses, especially *Echinochloa* spp., and broadleaf weeds in paddy rice	Unknown
Microtubule assembly inhibitor	3	Thiazopyr	Pyridine	Rohm and Haas	Pre-emergence control of *Echinochloa* spp. and other species	Medium
Lipid synthesis inhibitor but not ACCase inhibitor	8	Thiobencarb	Thiocarbamate	Kumiai, Valent	Pre-emergence to early postemergence control of *Echinochloa* spp. and other grasses, sedges, and broadleaf weeds in direct-seeded and transplanted rice	Medium
Lipid synthesis inhibitor but not ACCase inhibitor	8	Tiocarbazil	Thiocarbamate	Isagro	Pre- and postemergence control of *Echinochloa* spp. and other monocotyledonous weeds in paddy rice	Unknown
Unknown	Not classified	Triaziflam, IDH-1105	Triazine	Idemitsu Kosan	Pre- and postemergence control of broadleaf and grass weeds	Unknown
Synthetic auxin	4	Triclopyr	Pyridine carboxylic acid	Dow AgroSciences	Foliar-applied herbicide for broadleaf control	Low
ALS inhibitor	2	Triofensulfuron	Sulfonylurea	Aventis	Pre- and postemergence control of broadleaf weeds and some grasses; under development in mixture with iodosulfuron	High

Source: Compiled from information in recognized herbicide and agrochemical manuals,[203-206] information provided by agrochemical companies, and authors' experiences.

that include delayed planting, to allow for control of early flushes of *E. colona*, further benefits can be obtained by substituting propanil with pendimethalin or other herbicides with alternative modes of action.[172] The substitution or alternation of propanil-based control regimes by pendimethalin also is financially attractive.[212] In Sri Lanka, especially in the Polonnaruwa district of the north central province, farmers have switched to other herbicides to control propanil-resistant *E. crus-galli*. Quinclorac was heavily used in 1997–1998; bispyribac has gained popularity in the last two cropping seasons especially because it is also effective on many broadleaf weeds and some sedges. In areas where resistance is not currently a problem, propanil remains the preferred herbicide.[191]

A key aspect of successful resistance management is the easy and rapid determination of the response of a weed population to a herbicide to which there is suspicion of evolved resistance and to herbicides with alternative modes of action. A rapid test for sulfonylurea resistance in *Lindernia* spp. was developed in Japan using thifensulfuron methyl as the challenging herbicide.[213] The method is based on a colorimetric assessment of differential accumulation of acetoin in resistant and susceptible plants treated with an ALS herbicide.[214] It is also important to emphasize the need for testing *E. colona* for resistance to both propanil and ACCase herbicides, particularly where there has been a history of fenoxaprop use. Rapid tests are now available for this purpose.[215]

6.4.2.2.2 Use of synergists

Synergists, compounds that when mixed with herbicides result in a level of biological activity substantially higher than that of the added efficacy of each chemical separately, may be used as components of a tank mixture or as part of a formulation to overcome resistance. Synergists are useful when resistance is due to enhanced metabolism but are ineffective against target-site resistance. It is well known that some organophosphate and carbamate insecticides block the action of AAA, responsible for propanil hydrolysis in *Echinochloa* spp. and rice. Thus, the insecticide carbaryl could be used in mixture with propanil to provide effective control of propanil-resistant *Echinochloa* spp. This is not a practical strategy, however, as rice tolerance to propanil is reduced. Similarly, a formulation of propanil containing carbaryl sold in the United States causes excessive damage to the crop.[208] When organophosphate and carbamate insecticides are applied soon before or after propanil, rice can be damaged by the herbicide because the insecticide prevents its metabolism. The organophosphate herbicides anilofos and piperophos are selective in rice, however, and mixtures with propanil are not more phytotoxic to the crop than propanil alone but successfully overcome resistance in *E. colona*.[152,216,217] Piperophos is translocated in the apoplast after absorption by rice roots and it is subject to metabolic degradation.[218] Mixtures of synergist plus a low dose of propanil control propanil-resistant *E. colona*.[121]

Combination of either piperophos or anilofos with propanil, first used commercially in Costa Rica and then in other areas of Latin America to control propanil-resistant *E. colona*, is a unique example of the practical use of synergists in managing herbicide resistance. In 1995, a formulated mixture containing propanil plus piperophos was commercialized in Costa Rica.[149] The mixture was widely accepted and by 1997 it accounted for 30% of the total propanil market. The cost of the formulated mixture to the grower is about 35 to 40% more than the propanil alone but the decrease in the dose required to control *E. colona* and the improved efficacy of the formulation make it profitable. Additionally, many growers began spraying the mixture of propanil plus piperophos only once in the cropping cycle instead of the customary repeated application of propanil without synergist.[121] In Colombia, anilofos also was introduced as a synergist to propanil in 1996. In 1998 the tank mixture of propanil plus anilofos was used over 24,000 ha.[219] In 1998 co-formulated propanil plus piperophos was also introduced in Colombia. In 1999 about 35% of the rice area was treated with propanil plus a synergist (either anilofos or piperophos) and it was expected to grow to about 50% of the area in 2000. In Colombia, the per-hectare application cost (only herbicides) at recommended doses for propanil alone, co-formulated propanil plus piperophos, and propanil plus anilofos (tank mixture) are U.S. $39.00, 51.50, and 37.00, respectively.[121]

6.4.2.3   Herbicide-Resistant Rice Cultivars

Transgenic (glyphosate, glufosinate) or tissue-mutation-mediated (IMI) rice cultivars resistant to these herbicides have been developed and are being introduced to world agriculture. Commercial introduction of these cultivars will provide farmers with new chemical alternatives to control red rice and other weedy *Oryza* species and herbicide-resistant weeds. To date, most research aimed to optimize weed management in herbicide-resistant rice has been conducted with cultivars resistant to glufosinate and imidazolinones. Almost complete control of red rice and other grasses, including *E. crus-galli,* was achieved in glufosinate-resistant rice in Arkansas (U.S.A.) by sequential applications of glufosinate in early pre-emergence and at flooding. Weed response and glufosinate selectivity to rice depend on environmental conditions, such as depth of flooding water, growth stage, and other herbicides in tank mixture.[220–224] For example, increased glufosinate toxicity to transgenic rice was observed when it was applied in combination with triclopyr and resulted in a yield reduction of up to 76% compared to glufosinate alone.[222] In initial studies on weed control in IMI rice, imazethapyr controlled red rice and other grass weeds especially in sequential applications. Crop injury was less than 5%,[132] but control of *Eclipta prostrata* and *Sesbania exaltata* was inadequate.[225] In most field studies, sequential application of imazethapyr or imazaquin alone or in mixture with other herbicides is required for satisfactory weed control but crop injury is also common.[226–228] First release of commercial varieties of IMI-resistant rice is anticipated for 2001.[132] There is limited information about the development of glyphosate-resistant rice but field-testing is already in progress.*

The main concerns with this new technology are the possibility of gene flow from the commercial herbicide-resistant varieties to weedy *Oryza* species, the selection pressure imposed by sequential application of the same chemical, and the difficulty to control rice volunteers in subsequent crops. In a study of hybridization between cultivated rice and red rice Langevin et al.[229] found substantial

---

* http://www.isb.vt.edu/cfdocs/fieldtests3.cfm

hybridization. Safeguards to prevent the negative effects of gene flow from rice to wild or weedy relatives have been suggested[230] but with the imminent release of IMI rice they may arrive too late. Genes for resistance to ALS herbicides already exist in rice populations. In the development of IMI rice, 300 million rice plants of different rice accessions were screened to find two mutants resistant to imidazolinone herbicides.[231] Sequential applications, especially of imidazolinones, which also would be used in rotational crops (soybean), may subject weed populations to excessive selection pressure with a group of chemicals that has proven resistant-prone. The situation would be aggravated if other herbicides with the same ALS mode of action (sulfonylureas or pyrimidyl-oxybenzoates) were used against other grasses, broadleaf weeds, and sedges. Volunteer rice is already a problem in rotation crops and in succeeding rice crops, especially when there is a variety change and in production of certified seed. Herbicide-resistant volunteer rice will further limit the choice of chemicals for its control, especially for glyphosate, which is currently one of the most effective, inexpensive, nonresidual herbicides.

6.4.2.4   Biological Weed Control

The need for biological control solutions is increasingly promoted as an alternative tactic to control prevalent weeds in rice, especially after they have evolved resistance to the most widely used herbicides. However, there are few successful examples of classical biological control agents or bioherbicides. Rice is one of the few crops where a commercial mycoherbicide has been registered and used. In the United States a mycoherbicide containing an endemic strain of *Colletotrichum gloeosporioides* f. sp. *aeschynomene* has been used to selectively control *Aeschynomene virginica* in rice.[232] This weed, however, has not evolved resistance to herbicides.

Efforts are under way in Australia to develop biocontrol for *Alismataceae* weeds. The endemic fungus *Rhynchosporium alismatis* has been tested in several species of weeds, related families, and crops. In *A. lanceolatum* and *D. minus* it causes necrotic lesions on leaves and scapes.[171] The fungus also infects other species and the pathogen has been recovered from crops such as cucurbits and tomato, but the risk to crops grown in proximity to rice in southern Australia is regarded as negligible.[233] The fungus is intended to suppress both *A. lanceolatum* and *D. minus*.

Echinochloa spp. are also receiving consideration for possible biological control. A strain of *C. graminicola* was found highly pathogenic to *E. crus-galli* and innocuous to rice.[234] The fungus *Exserohilum monoceras* is also a promising biocontrol agent for *E. crus-galli*.[235] Biological control of rice weeds, however, is still primarily at the research level, with very few practical uses, especially in relation to management of herbicide-resistant species.

6.4.3   Feasibility and Impact of Integrated Management of Herbicide-Resistant Weeds

There is ample opportunity to adapt a range of control tactics to local conditions and to incorporate them into scientifically sound, flexible, and realistic integrated management programs for herbicide-resistance management. It is also of paramount importance to ensure that the farmers realize the importance of herbicide resistance and understand the rationale for integrating control tactics. In this regard, Ho[236] reviews an interesting example in the Muda area of Malaysia. Based on identification of weed shifts resulting from changing agricultural practices, the Muda Agricultural Development Authority (MADA) prioritized research and extension activities, considering the possible effects of future changes in cultural practices on weed population dynamics and identifying current efficient practices for weed control. A traditional approach of developing weed management techniques and delivering them as a "package" through training and field demonstrations was unsatisfactory. A program was then initiated by conducting a thorough survey to understand farmers' knowledge and attitudes related to weed management and using group interviews to determine how farmers decided upon management practices and how they valued their effectiveness. MADA's extension program emphasized group work among farmers, adoption of cultural practices including use of certified seed and manual weeding, weed identification at early growth stages, and judicious herbicide use. Pilot projects covering 10 to 15 ha and involving 5 to 10 growers were established at different locations in close consultation and with active participation of growers. Two clear objectives were stated for a campaign launched in 1989 inspired by the Farmers School principles:[237,238] to reduce grass weed infestations in direct-seeded rice fields by 70% and to reduce rice crop losses due to weeds by 25%. By 1994, substantial progress had been made. Rice yields in the main cropping season increased from 4.3 t ha^{-1} to about 5.4 t ha^{-1}; successful control of *Echinochloa* spp. was obtained and, interestingly, herbicide use declined, probably as a result of improved land preparation and leveling. There is also ample experience in implementation of integrated insect management in rice in Asia[238–240] upon which we can model an integrated weed management approach for herbicide resistance.

For sustainable herbicide-resistance management, control measures must deplete the weed seedbank. For example, in the case of *E. colona*, the seedbank is short-lived.[121] *E. colona*, however, produces vast amounts of seed and unless seed production is almost eliminated it will only be the management practices used in the previous season that will have an immediate impact on seedling emergence in the following year.[121] *A. plantago-aquatica* is self-compatible and disseminates in rice environments predominantly by seeds that have hard-coat dormancy and form a persistent seedbank.[89] Management of resistant perennial species may be complicated by the persistence of their seeds. More than 90% of seeds of *S. juncoides* subsp. *hotarui* buried for up to 7 years germinated and emerged from the soil.[241] *S. maritimus* spreads extensively by creeping rhizomes and stolons with tubers at the nodes; *S. mucronatus* is a perennial sedge with short rhizomes but under rice field conditions reproduces mostly by seeds.[89]

6.5   SUMMARY AND FUTURE OUTLOOK

World rice ecosystems are quite variable, as are the weeds associated with the rice crop. Nevertheless, there are similarities in weed floras, especially within production systems and specific geographical areas. Worldwide, where herbicide control is the favored tool for weed management, herbicide resistance has evolved. Herbicide-resistant weeds in rice are becoming more frequent, especially those resistant to ALS-inhibiting herbicides and especially within the genus *Echinochloa*. Increased reliance on herbicides and the huge shifts toward direct seeding will impose additional selection pressure on weed populations that will respond by evolving herbicide resistance. Resistance is now documented for herbicides representing the most important modes of action available in rice: ALS, PS-II, ACCase, lipid and protein synthesis, and auxinic herbicides. Multiple herbicide resistance is also emerging, imposing a greater limitation to weed management, especially in areas where herbicide use is highly regulated in relation to water pollution, spray drift, and residues.

Herbicide usage is increasing in the developing world. While the developed economies account for approximately 70% of herbicide use,[242] in recent years, Latin America, which accounts for almost 11% of the market, has had the largest increases in sales (16% in U.S. dollar terms in 1996), mostly driven by heavy herbicide consumption in Brazil and Argentina.[243,244] In 1996, pesticide sales dropped in Japan by 12% largely because of shrinking area of paddy rice, but the market continued to grow in other Asian countries, especially China and South Korea.[243–245] The rice-herbicide market is heavily concentrated in Asia and is dominated by Japan, with half of the total global rice-herbicide sales revenue. Other important markets are those of South Korea, Brazil, Colombia, and China.[94] According to industry estimates, the rice-herbicide market in Japan is expected to decrease as a result of liberalization of import barriers by the World Trade Organization, which would bring a reduction in rice prices with a concomitant decrease in farm inputs, and planted area reductions promoted by the Japanese government since 1994. Herbicide sales in the rest of Asia, including South Korea, and emerging new markets in the Indian subcontinent are expected to grow.[247]

It is expected that ALS-inhibiting herbicides will continue to play an important role in the rice-herbicide market, especially in the developing world. Market developments will undoubtedly influence local herbicide-use patterns and therefore the likelihood of more weed species evolving resistance. Only through a coherent integrated approach to weed management can resistance problems be prevented or managed. A clear understanding of the biology and ecology of weeds associated with each rice ecosystem, the aspects that determine selection pressure, and consideration of the experience accumulated in areas where resistance has evolved are essential to better anticipate and manage resistance evolution. We hope that the information presented in this chapter helps scientists, field advisors, and growers to achieve this goal.

ACKNOWLEDGMENTS

We are indebted to several colleagues who provided useful information on current status of herbicide-resistance problems in rice throughout the world either as personal communications or by providing manuscripts still in preparation: Octavio Almario (AgrEvo, S.A., Colombia); Michael Carriere (University of California at Davis, U.S.A.); Isabel M. Calha (Direcção-Geral de Protecção das Culturas, Portugal);

Poon-Min Chang (DuPont, Malaysia); Jean-Pierre Claude (DuPont de Nemours, France); Gustavo Cobo (Compañía Molinera San Cristóbal, Chile); Albert J. Fischer (University of California at Davis, U.S.A.); Israel Garita (DuWest Central America, Costa Rica); Diego Gómez de Barreda (Instituto Valenciano de Investigaciones Agrarias, Spain); Klaus Grossmann (BASF, Germany); Bob Harris (Aventis, Australia); James E. Hill (International Rice Research Institute, Philippines); Kil-Ung Kim (Kyungpook National University, Korea); Akira Koarai (Kyusyu National Agricultural Experiment Station, Japan); Christoph Labhart (Novartis, Switzerland); Chanya Maneechote (Department of Agriculture, Thailand); Buddhi Marambe (University of Peradeniya, Sri Lanka); Valmir Menezes (Instituto Riograndense do Arroz-IRGA, Brazil); Soichi Nakayama (National Agriculture Research Center, Japan), Jose Alberto Noldin (Epagri-Estação Experimental de Itajai, Brazil); Aida Ortiz (Universidad Central de Venezuela, Venezuela); Arturo Rangel Muñoz (Agropecuaria Internacional, S.A. de C. V., Mexico); Dearl Sanders (Louisiana State University, U.S.A.); Maurizio Sattin (Centro Biologia e Controllo Piante Infestanti - CNR, Italy); Oskar Schmidt (BASF, Germany); and Joe Street (Mississippi State University, U.S.A.). Mr. D. Okumura, California Department of Pesticide Regulation, kindly provided information regarding the regulatory status of rice herbicides in California. B. E. Valverde was supported by a grant from the Danish Council for Development Research (Project 91011).

REFERENCES

1. International Rice Research Institute (IRRI), *Rice Almanac*, International Rice Research Institute, Los Baños, Philippines, 1997, 181.
2. Food and Agriculture Organization (FAO), Statistical Database, Agricultural Production; www.fao.org, 2000.
3. Fischer, K. S., Research approaches for variable rainfed systems — thinking globally, acting locally, in *Plant Adaptation and Crop Improvement*, Cooper, M. and Hammer, G. L., Eds., CAB International, Wallingford, U.K., 1996, 25.
4. Pinstrup-Andersen, P., Pandya-Lorch, R., and Rosegrant, M. W., *World Food Prospects: Critical Issues for the Early Twenty-First Century*, 2020 Vision Food Policy Report, International Food Policy Research Institute, Washington, D.C., 1999, 32.
5. Garriti, D. P., Singh, V. P., and Hossain, M., Ecosystems analysis for research priorities, in *Rice Research in Asia: Progress and Priorities*, Evenson, R. E., Herdt, R. W., and Hossain, M., Eds., CAB International, Oxon, U.K., 1996, 35.
6. Khush, G. S., Terminology for rice growing environments, in *Terminology for Rice Growing Environments*, International Rice Research Institute, Los Baños, Laguna, Philippines, 1984, 5.
7. Mackill, D. J., Coffman, W. R., and Garrity, D. P., *Rainfed Lowland Rice Improvement*, International Rice Research Institute, Manila, Philippines, 1996, 242.
8. Zeigler, R. S. and Puckridge, D. W., Improving sustainable productivity in rice-based rainfed lowland systems of South and Southeast Asia, *GeoJournal*, 35, 307, 1995.
9. Fukai, S., Cooper, M., and Wade, L. J., Adaptation of rainfed lowland rice, *Field Crops Res.*, 64, 1, 1999.

10. Wade, L. J., George, T., Ladha, J. K., Singh, U., Bhuiyan, S. I., and Pandey, S., Opportunities to manipulate nutrient by water interactions in rainfed lowland rice systems, *Field Crops Res.*, 56, 93, 1998.

11. Widawsky, D. A. and O'Toole, J. C., *Prioritizing the Rice Biotechnology Research Agenda for Eastern India*, The Rockefeller Foundation, New York, 1990.

12. Wade, L. J., Fukai, S., Samson, B. K., Ali, A., and Mazid, M. A., Rainfed lowland rice: physical environment and cultivar requirements, *Field Crops Res.*, 64, 3, 1999.

13. Savary, S., Willocquet, L., Elazegui, F. A., Teng, P. S., Du, P. V., Zhu, D., Tang, Q., Huang, S., Lin, X., Singh, H. M., and Srivastava, R. K., Rice pest constraints in tropical Asia: characterization of injury profiles in relation to production situations, *Plant Dis.*, 84, 341, 2000.

14. Fujisaka, J. S., Moody, K., and Ingram, K. T., A descriptive study of farming practices for dry-seeded rainfed lowland rice in India, Indonesia and Myanmar, *Agric. Ecosyst. Environ.*, 45, 115, 1993.

15. Pandey, S. and Velasco, L., Economics of direct seeding in Asia: patterns of adoption and research priorities, *Int. Rice Res. News*, 24, 6, 1999.

16. Naklang, K., Direct seeding for rainfed lowland rice in Thailand, in *Breeding Strategies for Rainfed Lowland Rice in Drought-Prone Environments*, Fukai, S., Cooper, M., and Salisbury, J., Eds., Proc. Int. Workshop, Ubon Ratchathani, Thailand, *ACIAR Proc.*, 77, 126, 1997.

17. Vongsaroj, P., Weed management in deepwater rice, in *Weed Management in Rice*, Auld, B. A. and Kim, K.-U., Eds., Paper 139, FAO Plant Production and Protection, Food and Agriculture Organisation of the United Nations, Rome, 1996, 113.

18. Huang, B. and Wang, X., Studies on resistance of barnyardgrass in the rice planting regions in China and its management, in *Proc. Conf. Integrated Pest Management (IPM) — Theory and Practice, Develop. Sustainable Agric.*, Guanzhou, China, 1998, 210.

19. Guang, X., The status of weed control in Jiangsu province of China, in *Proc. 2nd Int. Weed Control Congr.*, Copenhagen, Denmark, 1996, 699.

20. Su, S. Q. and Ahrens, W. H., Weed management in northeast China, *Weed Technol.*, 11, 817, 1997.

21. Hossain, M., Economic prosperity in Asia: implications for rice research., in *Proc. 3rd Int. Rice Genetics Symp.*, *Rice Genetics III*, International Rice Research Institute, Manila, Philippines, 1996, 3.

22. Fischer, A. J., Integrated red rice management in Latin American rice fields, in *Proc. 2nd Int. Weed Control Congr.*, Copenhagen, Denmark, 1996, 653.

23. Hill, J. E. and Hawkins, L. S., Herbicides in United States rice production: lessons for Asia, in *Herbicides in Asian Rice: Transitions in Weed Management*, Naylor, R., Ed., Institute for International Studies, Stanford University, Palo Alto, CA, and International Rice Research Institute, Manila, Philippines, 1996, 37.

24. Hassan, S. M. and Rao, A. N., Weed management in rice in the Near East, in *Weed Management in Rice*, Auld, B. A. and Kim, K.-U., Eds., Paper 139, FAO Plant Production and Protection, Food and Agriculture Organisation of the United Nations, Rome, 1996, 143.

25. Johnson, D. E., Weed management in small holder rice production in the tropics, in *Radcliffe's IPM World Textbook*, Radcliffe, E. B. and Hutchison, W. D., Eds., University of Minnesota, St. Paul, Online, Internet, URL: http://ipmworld.umn.edu, 2000, 8.

26. Haefele, S. M., Johnson, D. E., Diallo, S., Wopereis, M. C. S., and Janin, I., Improved soil fertility and weed management is profitable for irrigated rice farmers in Sahelian West Africa, *Field Crops Res.*, 66, 101, 2000.

27. McDonald, J. D., Temperate rice technology for the 21st century: an Australian example, *Aust. J. Exp. Agric.*, 34, 877, 1994.

28. Lattimore, M. E., Pastures in temperate rice rotations of south-eastern Australia, *Aust. J. Exp. Agric.*, 34, 959, 1994.

29. Sanders, B. A., The life cycle and ecology of *Cyperus difformis* (rice weed) in temperate Australia: a review, *Aust. J. Exp. Agric.*, 34, 1031, 1994.

30. Taylor, M., Management herbicide resistance for water-seeded rice in New South Wales, in *Proc. 16th Asian-Pacific Weed Sci. Soc. Conf.*, Kuala Lumpur, Malaysia, 1997, 215.

31. Faure, J. and Mazaud, F., Rice quality in the European Union, *Agric. Dev.*, 2, 12, 1995.

32. Herdt, R., Research priorities for rice biotechnology, in *Rice Biotechnology*, Klush, G. S. and Toenniessen, G. J., Eds., International Rice Research Institute, Manila, Philippines, and CAB International, Wallingford, U.K., 1991, 19.

33. Moody, K., Weed management in rice, in *Handbook of Pest Management in Agriculture*, Pimentel, D., Ed., CRC Press, Boca Raton, FL, 1991, 301.

34. Savary, S., Willocquet, L., Elazegui, F. A., Castilla, N. P., and Teng, P. S., Rice pest constraints in tropical Asia: quantification of yield losses due to rice pests in a range of production situations, *Plant Dis.*, 84, 357, 2000.

35. Rojas, M. and Agüero, R., Malezas asociadas a canales de riego y terrenos colindantes de arroz anegado en la finca El Cerrito, Guanacaste, Costa Rica, *Agric. Mesoam.*, 7, 9, 1996.

36. Shibayama, H., Experience with rice herbicides in Japan, in *Herbicides in Asian Rice: Transitions in Weed Management*, Naylor, R., Ed., Institute for International Studies, Stanford University, Palo Alto, CA, and International Rice Research Institute, Manila, Philippines, 1996, 243.

37. Ho, N. K., Itoh, K. I., and Othman, A., Weed management and changes in rice production practices, in *Rice Pest Science and Management*, Teng, P. S., Heong, K. L., and Moody, K., Eds., International Rice Research Institute, Manila, Philippines, 1994, 217.

38. Kim, K. U., Ecological forces influencing weed competition and herbicide resistance, in *Herbicides in Asian Rice: Transitions in Weed Management*, Naylor, R., Ed., Institute for International Studies, Stanford University, Palo Alto, CA, and International Rice Research Institute, Manila, Philippines, 1996, 130.

39. Moody, K., Weed community dynamics in rice fields, in *Herbicides in Asian Rice: Transitions in Weed Management*, Naylor, R., Ed., Institute for International Studies, Stanford University, Palo Alto, CA, and International Rice Research Institute, Manila, Philippines, 1996, 27.

40. Mortimer, A. M. and Hill, J. E., Weed species shifts in response to broad spectrum herbicides in sub-tropical and tropical crops, in *Proc. Brighton Crop Prot. Conf. – Weeds*, 1999, 425.

41. Shibayama, H., Integrated management of paddy weeds in Japan: current status and prospects for improvement, in *Integrated Management of Paddy and Aquatic Weeds in Asia*, FFTC Book Series No. 45, Food and Fertilizer Technology Center, Taipei City, 1994, 78.

42. Corporación Colombiana de Investigación Agropecuaria (CORPOICA), *Principales avances en investigación y desarrollo tecnológico por sistemas de producción agrícola*, CORPOICA, Bogotá, 1998, 451.

43. Calha, I. M., Machado, C., and Rocha, F., Resistance of *Alisma plantago-aquatica* (waterplantain) to bensulfuron-methyl, in *Programme and Abstracts Resistance 97*, Herts, U.K., 1997, 46.

44. Calha, I. M., Machado, C., and Rocha, F., Resistance of *Alisma plantago aquatica* to sulfonylurea herbicides in Portuguese rice fields, *Hydrobiologia*, 415, 289, 1999.

45. Sattin, M., Airoldi, M., Barotti, R., Marchi, A., Salomone, M. C., Trainini, G., Tabacchi, M., and Zanin, G., La resistenzaagli erbicidi ALS, fenomeno da gestire, *Terra Vita*, 8(Suppl.), 30, 1999.

46. Claude, J. P., personal communication, 2000.

47. Hill, J. E., personal communication, 2000.

48. Chang, P. M., personal communication, 2000.

49. Graham, R. J., Pratley, J. E., and Slater, P. D., Herbicide resistance in *Cyperus difformis*, a weed of New South Wales rice crops, in *Proc. Conf. Temperate Rice — Achievements and Potential*, Leeton, New South Wales, Australia, 1994, 433.

50. Pappas-Fader, T., Cook, J. F., Butler, T., Lana, P. J., and Hare, J., Resistance of California arrowhead and smallflower umbrella plant to sulfonylurea herbicides, in *Proc. Western Weed Sci. Soc.*, 1993, 76.

51. Taylor, M., Rice weed control developments, *Farmers' Newsl. Large-Area*, 145, 10, 1995.

52. Garita, I., Valverde, B. E., Vargas, E., Chacón, L. A., de la Cruz, R., Riches, C. R., and Caseley, J. C., Occurrence of propanil resistance in *Echinochloa colona* in Central America, in *Proc. Brighton Crop Prot. Conf. — Weeds*, Brighton, U.K., 1995, 193.

53. Fischer, A. J., Granados, E., and Trujillo, D., Propanil resistance in populations of junglerice (*Echinochloa colona*) in Colombia rice fields, *Weed Sci.*, 41, 201, 1993.

54. Villa-Casarez, J. T., Repuesta de *Echinochloa colona* (L.) Link a propanil en el cultivo de arroz (*Oryza sativa* L.) en áreas selectas de México, M.Sc. thesis, Universidad Autónoma Chapingo, Chapingo, Mexico, 1998, 140.

55. Carey, V. F., III, Hoagland, R. E., and Talbert, R. E., Verification and distribution of propanil-resistant barnyardgrass (*Echinochloa crus-galli*) in Arkansas, *Weed Technol.* 9, 366, 1995.

56. Ortiz, A., Pacheco, M., Pérez, V., Ramos, R., and Sejías, E., Identificación de biotipos de *Echinochloa colona* (L.) Link. Potencialmente resistentes al propanil en Venezuela, *Rev. COMALFI* (Colombia), 26, 28, 2000.

57. Riches, C. R., Caseley, J. C., Valverde, B. E., and Down, V. M., Resistance of *Echinochloa colona* to ACCase inhibiting herbicides, in *Proc. Int. Symp. Weed and Crop Resistance to Herbicides*, De Prado, R., Jorrín, J., García-Torres, L., and Marshall, G., Eds., University of Cordoba, Spain, 1996, 14.

58. Almario, O., personal communication, 2000.

59. Caseley, J. C., Palgrave, C., Haas, E., Riches, C. R., and Valverde, B. E., Herbicides with alternative modes of action for the control of propanil- and fenoxaprop-*p*-resistant *Echinochloa colona*, in *Proc. Brighton Crop Prot. Conf. – Weeds*, 1997, 215.

60. Schmidt, O., personal communication, 2000.

61. Giannopolitis, C. N. and Vassiliou, G., Propanil tolerance in *Echinochloa crus-galli* (L.) Beauv, *Trop. Pest Manage.*, 35, 6, 1989.

62. Smith, R. J., Talbert, R. E., and Baltazar, A. M., Control, biology and ecology of propanil-tolerant barnyardgrass, *Arkansas Rice Res. Studies*, University of Arkansas, Fayetteville, 1992, 46.

63. Marambe, B., Amarasinghe, L., and Senaratne, G. R., Propanil resistant barnyardgrass (*Echinochloa crus-galli* L. Beauv.) in Sri Lanka, in *Proc. 16th Asian-Pacific Weed Sci. Soc. Conf.*, Kuala Lumpur, Malaysia, 1997, 222.

64. Maneechote, C. and Krasaesindhu, P., Propanil resistance in barnyardgrass (*Echinochloa crus-galli* (L.) Beauv.), in *Proc. 17th Asian-Pacific Weed Sci. Soc. Conf.*, Abstracts, Bangkok, Thailand, 1999, 97.

65. Huang, B. and Lin, S., Study on the resistance of barnyardgrass to butachlor in paddy fields in China, *J. South China Agric. Univ.*, 14, 103, 1993.

66. Maneechote, C., personal communication, 2000.

67. Sanders, D. E., personal communication, 2000.

68. Street, J., personal communication, 2000.

69. Schmidt, O., Aurich, O., Lopez, N., de Prado, R., and Walter, H., Botanical identification of Spanish *Echinochloa* biotypes with differential responses to quinclorac, in *Proc. 6th EWRS Mediterranean Symp.*, *Montpellier, France*, European Weed Research Society, 1998, 232.

70. Noldin, J. A., personal communication, 2000.

71. Menezes, V., personal communication, 2000.

72. Gómez de Barreda, D., Carretero, J. L., del Busto, A., Asins, M. J., Carbonell, E., and Lorenzo, E., Response of *Echinochloa* spp. (barnyardgrass) populations to quinclorac, in *Proc. Int. Symp. Weed Crop Resistance Herbicides*, University of Cordoba, Spain, 1996, 157.

73. Fischer, A. J., Ateh, C. M., Bayer, D. E., and Hill, J. E., Herbicide-resistant *Echinochloa oryzoides* and *E. phyllopogon* in California *Oryza sativa* fields, *Weed Sci.*, 48, 225, 2000.

74. Hata, K., Okuda, K., Aoki, M., and Kuramochi, H., Occurrence of *Elatine triandra* Sckk., resistant to sulfonylurea herbicides, *J. Weed Sci. Technol.*, 43(Suppl.), 28, 1998.

75. Watanabe, H., Ismail, M. Z., and Ho, N. A., Response of 2,4-D resistant biotype of *Fimbristylis miliacea* (L.) Vahl. to 2,4-D dimethylamine and its distribution in the Muda Plain, peninsular Malaysia, *Weed Res.* (Japan), 42, 240, 1997.

76. Heap, I. M., International Survey of Herbicide Resistant Weeds, Online, Internet, available www.weedscience.com, 2000.

77. Nakayama, S., Azmi, M., and Ruslan, A. G., A biotype of *Limnocharis flava* multiple-resistant to 2,4-D and bensulfuron-methyl in Malaysia, in *Proc. 17th Asian-Pacific Weed Sci. Soc. Conf.*, Bangkok, Thailand, 1999, 827.

78. Itoh, K. and Wang, G. X., An outbreak of sulfonylurea herbicide resistance in Scrophulariaceae paddy weeds in Japan, in *Proc. 16th Asian-Pacific Weed Sci. Soc. Conf.*, Kuala Lumpur, Malaysia, 1997, 219.

79. Nakayama, S., personal communication, 2000.

80. Kohara, H., Yamashita, H., and Yamazaki, N., Resistance to sulfonylurea herbicides in *Monochoria korsakowii* Regel te Maack in Hokkaido, *Weed Res.* (Japan), 41(Suppl.), 236, 1996.

81. Wang, G. X., Kohara, H., and Itoh, K., Sulfonylurea resistance in a biotype of *Monochoria korsakowii*, an annual paddy weed in Japan, in *Proc. Brighton Crop Prot. Conf. — Weeds*, 1997, 311.

82. Park, T. S., Kim, C. S., Park, J. P., Oh, Y. K., and Kim, K. U., Resistant biotype of *Monochoria korsakowii* against sulfonylurea herbicides in the reclaimed paddy fields in Korea, in *Proc. 17th Asian-Pacific Weed Sci. Soc. Conf.*, Bangkok, Thailand, 1999, 251.

83. Kim, K. U., Park, T. S., and Park, J. E., Present situation of herbicide resistant paddy weeds in Korea, in Proc. *Int. Workshop, Biology and Management of Noxious Weeds for Sustainable and Labor Saving Rice Production*, Ibaraki, Japan, National Agriculture Research Center, Japan, 2000, 160.

84. Koarai, A., Diagnosis of susceptibility of sulfonylurea herbicides on *Monochoria vaginalis*, an annual paddy weed in Japan, *J. Weed Sci. Technol.*, 45(Suppl.), 40, 2000.

85. Lee, D. J., Kwon, O. D., Lee, H. J., and Guh, J. A., Control of sulfonylurea-resistant biotype of *Monochoria vaginalis* in paddy fields in Chinnam province, Korea, in *Proc. Int. Workshop, Biology and Management of Noxious Weeds for Sustainable and Labor Saving Rice Production*, Ibaraki, Japan, National Agriculture Research Center, Japan, 2000, 169.

86. Itoh, K., Uchino, A., and Watanabe, H., A resistant biotype to sulfonylureas in *Rotala indica* (Wild.) Koehn, in Omagari, Akita Prefecture, *J. Weed Sci. Technol.*, 43(Suppl.), 40, 1998.

87. Itoh, K., Takagai, Y., Blancaver, M. E., Odan, H., and Chang, P. M., A sulfonylurea resistant biotype of *Sagittaria guyanensis* H.B.K. Val., in paddy fields of Malaysia, *J. Weed Sci. Technol.*, 45(Suppl.), 102, 2000.

88. Kohara, H., Konno, K., and Takekawa, M., Occurrence of sulfonylurea-resistant biotypes of *Scripus juncoides* Roxb. var. *ohwianus* T. Koyama in paddy fields of Hokkaido prefecture, Japan, *J. Weed Sci. Technol.*, 44(Suppl.), 228, 1999.

89. Sattin, M., Berto, D., Zanin, G., and Tabacchi, M., Resistance to ALS inhibitors in weeds of rice in north-western Italy, in *Proc. Brighton Crop Prot. Conf. — Weeds*, 3, 783, 1999.

90. Itoh, K., Ed., *Life Cycles of Rice Field Weeds and Their Management in Malaysia*, Trop. Agric. Res. Center MAFF, Japan, 1991, 92.

91. Sy, S. J. and Mercado, B. L., Note: Comparative response to 2,4-D of *Sphenoclea zeylandica* collected from two locations, *Philippines J. Weed Sci.*, 10, 90, 1983.

92. Migo, T. R., Mercado, B. L., and de Datta, S. K., Response of *Sphenoclea zeylandica* to 2,4-D and other recommended herbicides for weed control in lowland rice, *Philippines J. Weed Sci.*, 13, 28, 1986.

93. Saari, L. L., Cotterman, J. C., and Thill, D. C., Resistance to acetolactate synthase inhibiting herbicides, in *Herbicide Resistance in Plants*, Powles, S. B. and Holtum, J. A. M., Eds., CRC Press, Boca Raton, FL, 1994, 83.

94. Allison, D., ALS/AHAS inhibitor herbicides, *Agrow Rep.*, DS 176, 1999, 167.

95. Itoh, K., Wang, G. X., and Ohba, S., Sulfonylurea resistance in *Lindernia micrantha*, an annual paddy weed in Japan, *Weed Res.*, 39, 413, 1999.

96. Itoh, K., Uchino, A., Wang, G. X., and Yamakawa, S., Distribution of *Lindernia* spp. resistant biotypes to sulfonylurea herbicides in Yuza Town, Yamagata Prefecture, *J. Weed Sci. Technol.*, 42(Suppl.), 22, 1997.

97. Itoh, K., Wang, G. X., and Uchino, A., Non-effective problems of *Lindernia* weeds to one-shot application herbicides including sulfonylureas in Tohoku area, *J. Weed Sci. Technol.*, 42(Suppl.), 12, 1997.

98. Morita, A. and Motojima, K., Paddy field weeds, *Lindernia* spp. in Tohoku area, *J. Weed Sci. Technol.*, 42(Suppl.), 164, 1997.

99. Seitoh, F., Tazawa, F., and Kawamura, M., Recent situation of remaining of *Lindernia dubia* subsp. *major* (L.) Penn. and past use records of herbicide in a town, Tsgaru region, Aomori Prefecture, *J. Weed Sci. Technol.*, 42(Suppl.), 166, 1997.

100. Kohara, H., Uchino, A., and Watanabe, H., Seed germination of sulfonylurea-resistant *Scirpus juncoides* Roxb. var. *ohwianus* T. Koyama at a low temperature, *J. Weed Sci. Technol.*, 44(Suppl.), 74, 1999.

101. Yoshida, S., Onodera, K., Soeda, T., Takeda, Y., Sasaki, S., and Watanabe, H., Occurrence of *Scirpus juncoides* subsp. *juncoides*, resistant to sulfonylurea herbicides in Miyagi Prefecture, *J. Weed Sci. Technol.*, 44(Suppl.), 70, 1999.

102. Ormeño, J., Prospección de las principales malezas asociadas al cultivo del arroz (*Oryza sativa* L.), *Agric. Téc.* (Chile), 43, 285, 1983.
103. Cobo, G., personal communication, 2000.
104. Alverde, B. E., Chaves, L., González, J., and Garita, I., Field-evolved imazapyr resistance in *Ixophorus unisetus* and *Eleusine indica* in Costa Rica, in *Proc. Brighton Crop Prot. Conf. — Weeds*, Brighton, U.K., 1993, 1189.
105. Aoki, M., Kuramochi, H., Hata, K., and Otsuka, K., Distribution of weed biotypes resistant to sulfonylurea herbicides in Kazo City, Saitama Pref., *J. Weed Sci. Technol.*, 43(Suppl.), 34, 1998.
106. Sugimoto, M., Ohashi, Y., and Kawase, K., An occurrence of paddy weeds resistant to sulfonylurea herbicides in Kyoto and an effect of several herbicides on them, *J. Kinki-Sakuiku-Kenkyu*, 44, 43, 1999.
107. Graham, R. J., Pratley, J. E., Slater, P. D., and Baines, P. R., Herbicide resistant aquatic weeds, a problem in New South Wales rice crops, in *Proc. 11th Australian Weeds Conf.*, Melbourne, Australia, 1996, 598.
108. Barbe, G. D. and Hills, J. C., *Effects of the Herbicide Propanil on Stone Fruit Trees in California*, Special Publication No. 69-1, Bureau of Plant Pathology, California Department of Agriculture, Sacramento, 1969.
109. Hill, J. E., Smith, R. J., Jr., and Bayer, D. E., Rice weed control: current technology and emerging issues in temperate rice, *Aust. J. Exp. Agric.*, 34, 1021, 1994.
110. Hill, J. E., Weed control in rice: Where have we been? Where are we going? *Proc. Calif. Weed Sci. Soc.*, 50, 114, 1998.
111. Okumura, D., personal communication, 2000.
112. Carriere, M., personal communication, 2000.
113. [CalEPA] State of California Environmental Protection Agency, Department of Pesticide Regulation, Pesticide use report, Annual 1995, California Environmental Protection Agency, Sacramento, 1996, 334.
114. [CalEPA] State of California Environmental Protection Agency, Department of Pesticide Regulation, Summary of pesticide use report data 1996, California Environmental Protection Agency, Sacramento, 1999, 360.
115. [CalEPA] State of California Environmental Protection Agency, Department of Pesticide Regulation, Summary of pesticide use report data 1997, California Environmental Protection Agency, Sacramento, 1999, 362.
116. [CalEPA] State of California Environmental Protection Agency, Department of Pesticide Regulation, Summary of pesticide use report data 1998, Preliminary data. California Environmental Protection Agency, Sacramento, 1999, 413.
117. Calha, I. M., personal communication, 2000.
118. Wang, G. X., Watanabe, H., Uchino, A., and Itoh, K., A biotype of kikumo (*Limnophila sessiliflora*) resistant to sulfonylurea herbicides and its control in Akita, Japan, *J. Weed Sci. Technol.*, 43(Suppl.), 38, 1998.
119. Uchino, A., Itoh, K., Wang, G. X., and Tachibana, M., Sulfonylurea resistant biotypes of *Lindernia* species in the Tohoku region and their response to several herbicides, *J. Weed Sci. Technol.*, 45, 13, 2000.
120. Garro, J. E., de la Cruz, R., and Shannon, P., Propanil resistance in *Echinochloa colona* populations with different herbicide use histories, in *Proc. Brighton Crop Prot. Conf. — Weeds*, Brighton, U.K., 1991, 1079.
121. Valverde, B. E., Riches, C. R., and Caseley, J. C., *Prevention and Management of Herbicide Resistant Weeds in Rice: Experiences from Central America with Echinochloa colona*, Cámara de Insumos Agropecuarios, Costa Rica, 2000, 123.

122. Valverde, B. E., Bolaños, A., Villa, J. T., Chaves, L., and Ramírez, F., Resistencia a herbicidas en malezas del arroz: situación actual y estrategias de manejo, *AgroSíntesis* (Mexico), April 30, 1999, 31.

123. Paez, O. E. and Almeida, N. C., Control integrado de malezas en arroz bajo riego en el estado Portuguesa, *Agron. Trop.*, 2, 245, 1994.

124. Baltazar, A. M. and Smith, R. J., Propanil-resistant barnyardgrass (*Echinochloa crus-galli*) control in rice (*Oryza sativa*), *Weed Technol.*, 8, 575, 1994.

125. Gressel, J. and Baltazar, A., Herbicide resistance in rice: status, causes and prevention, in *Weed Management in Rice*, Auld, B. A. and Kim, K.-U., Eds., Paper 139, FAO Plant Production and Protection, Food and Agriculture Organisation of the United Nations, Rome, 1996, 195.

126. Huang, B. and Gressel, J., Barnyardgrass (*Echinochloa crus-galli*) resistance to both butachlor and thiobencarb in China, *Resistant Pest Manage.*, 9, 5, 1997.

127. Knights, J. S., The activity of graminicide alone and in mixtures for the control of herbicide resistant junglerice (*Echinochloa colona* (L.) Link), M.Sc. project, University of Bath, U.K., 1995, 135.

128. Lopez-Martinez, N., De Prado, R., Finch, R. P., and Marshall, G., A molecular assessment of genetic diversity in *Echinochloa* spp., in *Proc. Brighton Crop Prot. Conf. Weeds*, 1995, 445.

129. Lopez-Martinez, N., Hidalgo, N. J., Pujadas, A., Marshall, G., and De Prado, R., Resistance of *Echinochloa* spp. to atrazine and quinclorac, in *Proc. Int. Symp. Weed Crop Resistance Herbicides*, University of Cordoba, Spain, 1995, 23.

130. Lopez-Martinez, N., Marshall, G., and De Prado, R., Resistance of barnyardgrass (*Echinochloa crus-galli*) to atrazine and quinclorac, *Pestic. Sci.*, 51, 171, 1997.

131. DeWitt, T. C., Vickery, C., and Heier, J., Control of herbicide resistant watergrass in northern California rice with Regiment herbicide, in *Proc. California Weed Sci. Soc.*, 51, 182, 1999.

132. Hackworth, H. M., Sarokin, L. P., and White, R. H., 1997-field evaluation of imidazolinone tolerant rice, in *Proc. South. Weed Sci. Soc.*, 1998, 221.

133. Thompson, C., Thill, D., and Shafii, V. B., Growth and competitiveness of sulfonylurea-resistant and -susceptible kochia (*Kochia scoparia*), *Weed Sci.*, 42, 172, 1994.

134. Mallory-Smith, C., Thill, D., and Dial, M., Identification of sulfonylurea herbicide-resistant prickly lettuce (*Lactuca serriola*), *Weed Technol.*, 4, 163, 1990.

135. Lee, C. D., Martin, A. R., Roeth, F. W., Johnson, B. E., and Lee, D. J., Comparison of ALS inhibitor resistance and allelic interactions in shattercane accessions, *Weed Sci.*, 47, 275, 1999.

136. Shibaike, H., Uchino, A., and Itoh, K., Genetic variation and relationships of herbicide-resistant and -susceptible biotypes of *Lindernia micrantha*, in *Proc. Brighton Crop Prot. Conf. – Weeds*, 1999, 197.

137. Uchino, A. and Watanabe, H., Mutation in the acetolactate synthase genes of the biotypes of *Lindernia* spp. resistant to sulfonylurea herbicide, *J. Weed Sci. Technol.*, 44(Suppl.), 80, 1999.

138. Shibuya, K., Yoshioka, T., Saitoh, S., Yoshida, S., and Hashiba, T., Analysis of acetolactate synthase genes of sulfonylurea herbicides-resistant and -susceptible biotypes in *Scirpus juncoides*, *J. Weed Sci. Technol.*, 44(Suppl.), 72, 1999.

139. Matsumura, T., Wang, G. X., Goukon, K., and Itoh, K., Fauna and activity of insects visiting *Lindernia* flowers in paddy fields in northern Japan, in *Proc. 17th Asian-Pacific Weed Sci. Soc. Conf.*, Bangkok, Thailand, 1999, 850.

140. Wang, G. X., Watanabe, H., Uchino, A., and Itoh, K., Gene flow in an experimental population of sulfonylurea resistant *Monochoria korsakowii*, *J. Weed Sci. Technol.*, 43(Suppl.), 42, 1998.

141. Frear, D. S. and Still, G. G., The metabolism of 3,4-dichloropropionanilide in plants. Partial purification and properties of an aryl acylamidase from rice, *Phytochemistry*, 7, 913, 1968.

142. Still, C. C. and Kuzirian, O., Enzyme detoxication of 3',4'-dichloropropionanilide in rice and barnyard grass, a factor in herbicide selectivity, *Nature*, 216, 799, 1967.

143. Yih, R. Y., McRae, D. H., and Wilson, H. F., Mechanism of selective action of 3',4'-dichloropropionanilide, *Plant Physiol.*, 43, 1291, 1968.

144. Winkler, R. and Sanderman, H., Plant metabolism of chlorinated anilines: isolation and identification of *N*-glucosyl and *N*-malonyl conjugates, *Pestic. Biochem. Physiol.*, 33, 239, 1989.

145. Sanderman, H., Schmidt, R., Eckley, H., and Bauknecht, T., Plant biochemistry of xenobiotics: isolation and properties of soybean *O*- and *N*-glucosyl and *O*- and *N*-malonyltransferases for chlorinated phenols and anilines, *Arch. Biochem. Biophys.*, 287, 341, 1991.

146. Schmidt, B., Thiede, B., and Rivero, C., Metabolism of the pesticide metabolites 4-nitrophenol and 3,4-dichloroaniline in carrot (*Daucus carota*) cell suspension cultures, *Pestic. Sci.*, 40, 231, 1994.

147. Matsunaka, S., Propanil hydrolysis: inhibition in rice plants by insecticides, *Science*, 60, 1360, 1968.

148. Leah, J. M., Caseley, J. C., Riches, C. R., and Valverde, B. E., Association between elevated activity of aryl acylamidase and propanil resistance in jungle-rice, *Echinochloa colona*, *Pestic. Sci.*, 42, 281, 1994.

149. Valverde, B. E., Management of herbicide resistant weeds in Latin America: the case of propanil-resistant *Echinochloa colona* in rice, in *Proc. 2nd Int. Weed Control Congr.*, Copenhagen, Denmark, 1996, 415.

150. Hoagland, R. E. and Graf, G., Enzymatic hydrolysis of herbicides in plants, *Weed Sci.*, 20, 303, 1972.

151. Smith, R. J., 3,4-Dichloropropionanilide for control of barnyardgrass in rice, *Weeds*, 3, 318, 1961.

152. Caseley, J. C., Leah, J. M., Riches C. R., and Valverde, B. E., Combating propanil resistance in *Echinochloa colona* with synergists that inhibit acylamidase and oxygenases, in *Proc. 2nd Int. Weed Control Congr.*, Copenhagen, Denmark, 1996, 455.

153. Gronwald, J. W., Resistance to Photosystem II inhibiting herbicides, in *Herbicide Resistance in Plants*, Powles, S. B. and Holtum, J. A. M., Eds., CRC Press, Boca Raton, FL, 1994, 27.

154. Leah, J. M., Caseley, J. C., Riches, C. R., and Valverde, B. E., Age-related mechanisms of propanil tolerance in jungle-rice, *Echinochloa colona*, *Pestic. Sci.*, 43, 347, 1995.

155. Carey, V. F., III, Duke, S. O., Hoagland, R. E., and Talbert, R. E., Resistance mechanism of propanil-resistant barnyardgrass. I. Absorption, translocation and site of action studies, *Pestic. Biochem. Physiol.*, 52, 182, 1995.

156. Carey, V. F., III, Hoagland, R. E., and Talbert, R. E., Resistance mechanism of propanil-resistant barnyardgrass. II. *In-vivo* metabolism of the propanil molecule, *Pestic. Sci.*, 49, 333, 1997.

157. Wu, J., Omokawa, H., and Hatzios, K. K., Glutathione *S*-transferase activity in unsafened and fenclorim-safened rice (*Oryza sativa*), *Pestic. Biochem. Physiol.*, 54, 220, 1996.

158. Usui, K., Deng, F., Shim, I. S., and Kobayashi, K., Differential contents of pretilachlor, fenclorim and their metabolites between rice and early watergrass (*Echinochloa oryzicola*) seedlings leading to selectivity and safening action, *J. Weed Sci. Technol.*, 44, 37, 1999.

159. Hsieh, Y.-N., Liu, L.-F., and Wang, Y.-S., Uptake, translocation and metabolism of the herbicide molinate in tobacco and rice, *Pestic. Sci.*, 53, 149, 1998.

160. Coupland, D., Resistance to the auxin analog herbicides, in *Herbicide Resistance in Plants*, Powles, S. B. and Holtum, J. A. M., Eds., CRC Press, Boca Raton, FL, 1994, 171.

161. Deshpande, S. and Hall, J. C., Auxin herbicide resistance may be modulated at the auxin-binding site in wild mustard (*Sinapsis arvenses* L.): a light scattering study, *Pestic. Biochem. Physiol.*, 66, 41, 2000.

162. Moss, S., Herbicide resistance in the weed *Alopecurus myosuroides* (Blackgrass): the current situation, in *Achievements and Developments in Combating Pesticide Resistance*, Denholm, I., Devonshire, A. L., and Hollomon, D., Eds., Elsevier, London, 1992, 28.

163. Burnet, M. W. M., Hart, Q., Holtum, J. A. M., and Powles, S. B., Resistance to nine herbicide classes in a population of rigid ryegrass (*Lolium rigidum*), *Weed Sci.*, 42, 369, 1994.

164. Grossmann, K., Quinclorac belongs to a new class of highly selective auxin herbicides, *Weed Sci.*, 46, 707, 1998.

165. Grossmann, K. and Kwiatkowski, J., The mechanism of quinclorac selectivity in grasses, *Pestic. Biochem. Physiol.*, 66, 83, 2000.

166. Fischer, A. J., personal communication, 2000.

167. Preston, C. and Powles, S. B., Amitrole inhibits diclofop metabolism and synergises diclofop-methyl in a diclofop-methyl-resistant biotype of *Lolium rigidum*, *Pestic. Biochem. Physiol.*, 62, 179, 1998.

168. De Datta, S. K. and Baltazar, A. M., Integrated weed management in rice in Asia, in *Herbicides in Asian Rice: Transitions in Weed Management*, Naylor, R., Ed., Institute for International Studies, Stanford University, Palo Alto, CA, and International Rice Research Institute, Manila, Philippines, 1996, 145.

169. Naylor, R., Herbicide use in Asian rice production: perspectives from economics, ecology, and the agricultural sciences, in *Herbicides in Asian Rice: Transitions in Weed Management*, Naylor, R., Ed., Institute for International Studies, Stanford University, Palo Alto, CA, and International Rice Research Institute, Manila, Philippines, 1996, 3.

170. Wrubel, R. P. and Gressel, J., Are herbicide mixtures useful for delaying the rapid evolution of resistance? A case study, *Weed Technol.*, 8, 635, 1994.

171. Cother, E. J. and Gilbert, R. L., Efficacy of a potential mycoherbicide for control of *Alisma lanceolatum* and *Damasonium minus*, *Aust. J. Exp. Agric.*, 34, 1043, 1994.

172. Valverde, B. E., Chaves, L., Garita, I., Ramírez, F., Vargas, E., Carmiol, J., Riches, C. R., and Caseley, J. C., Modified herbicide regimes for propanil-resistant *Echinochloa colona* control in rainfed rice, *Weed Sci.*, 2000, in press.

173. Khush, G. S., Origin, dispersal, cultivation and variation of rice, *Plant Mol. Biol.*, 35, 25, 1997.

174. Eastin, E. F., Weed management for Southern U.S. rice, in *Handbook of Pest Management in Agriculture*, Pimentel, D., Ed., CRC Press, Boca Raton, FL, 1991, 329.

175. Foloni, L. L., Adjuvant effects on sulfosate and glyphosate for control of red-rice in rice, in *Proc. Brighton Crop Prot. Conf. — Weeds*, Brighton, U.K., 1995, 743.

176. Mello, I., Plantio direto de arroz irrigado no sul do Brasil, *Lavoura Arrozeira*, 48, 3, 1995.

177. Franco, P., Caseley, J. C., Kim, D. S., Riches, C. R., and Miyasato, Y., Evaluation of response of fluazifop-*p* resistant carib grass (*Eriochloa punctata*) to other ACCase inhibitors and herbicides with other modes of action, *WSSA Abstr.*, 39, 106, 1999.

178. Garriti, D. P., Movillon, M., and Moody, K., Differential weed suppression ability in upland rice cultivars, *Agron. J.*, 84, 586, 1992.

179. Puckridge, D. W., Processes determining crop yield, in *Breeding Strategies for Rainfed Lowland Rice in Drought-Prone Environments*, Fukai, S., Cooper, M., and Salisbury, J., Eds., Proc. Int. Workshop, Ubon Ratchathani, Thailand, *ACIAR Proc.*, 77, 23, 1997.

180. Fischer, A. J., Ramírez, H. V., and Lozano, J., Suppression of junglerice (*Echinochloa colona*) (L.) Link by irrigated rice cultivars in Latin America, *Agron. J.*, 89, 516, 1997.

181. Fischer, A. J. and Ramírez, A., Mixed-weed infestations: prediction of crop losses for economic weed management in rice, *Int. J. Pest Manage.*, 39, 354, 1993.

182. Gealy, D. R., Dilday, R. H., and Rutger, J. N., Interaction of flush irrigation timing and suppression of barnyardgrass with potentially allelopathic rice lines, *Res. Ser. Ark. Agric. Exp. Stn.*, 460, 49, 1998.

183. Dilday, R. H., Lin, J., and Yan, W., Identification of allelopathy in the USDA-ARS rice germplasm collection, *Aust. J. Exp. Agric.*, 34, 907, 1994.

184. Dingkuhn, M., Johnson, D. E., Sow, A., and Audebert, A. Y., Relationships between upland rice canopy characteristics and weed competitiveness, *Field Crops Res.*, 61, 79, 1999.

185. Navarez, D. and Olofsdotter, M., Relay seeding technique for screening allelopathic rice (*Oryza sativa*), in *Proc. 2nd Int. Weed Control Congr.*, Copenhagen, Denmark, 1996, 1285.

186. Olofsdotter, M. and Navarez, D., Allelopathic rice for *Echinochloa crus-galli* control, in *Proc. 2nd Int. Weed Control Congr.*, Copenhagen, Denmark, 1996, 1175.

187. Azmi, M., Weed populations in direct-seeded rice as affected by seeding rates, in *Proc. 16th Asian-Pacific Weed Sci. Soc. Conf.*, Kuala Lumpur, Malaysia, 1997, 251.

188. Gravois, K. A. and Helms, R. S., Path analysis of rice yield and yield components as affected by seeding rate, *Agron. J.*, 84, 1, 1992.

189. Garro, J. E., de la Cruz, R., and Merayo, A., Estudio del crecimiento de materiales de *Echinochloa colona* (L.) Link. susceptibles y tolerantes al propanil, *Manejo Integrado Plagas*, 26, 39, 1993.

190. Carey, V. F., III and Talbert, R. E., Investigation of propanil-resistant barnyardgrass (*Echinochloa crus-galli* L. Beauv.) in Arkansas, *WSSA Abstr.*, 34, 38, 1994.

191. Marambe, B., personal communication, 2000.

192. Ampong-Nyarko, K. and de Datta, S. K., Effects of light and nitrogen and their interaction on the dynamics of rice-weed competition, *Weed Res.*, 33, 1, 1993.

193. Cavero, J., Hill, J. E., Lestrange, M., and Plant, R. E., Efecto de la dosis de nitrógeno en la competencia de *Echinochloa oryzoides* con arroz de siembra directa, in *Proc. 1997 Congr. Spanish Weed Sci. Soc.*, Valencia, Spain, 1997, 55.

194. Moya, J. C., Influencia de la época de aplicación y dos fuentes de nitrógeno sobre la tolerancia del arroz (*Oryza sativa*) y la caminadora (*Rottboellia cochinchinensis*) a los herbicidas fenoxaprop-etil y haloxyfop-metil, Tesis Ingeniero Agrónomo, Facultad de Agronomía, Universidad de Costa Rica, 1990, 53.

195. Acuña, A., Efecto del nitrógeno y azufre en la tolerancia del arroz (*Oryza sativa*) al fenoxaprop-etilo, M.S. thesis, Universidad de Costa Rica, Costa Rica, 1995, 67.

196. Jutsum, A. R. and Graham, J. C., Managing weed resistance: the role of the agrochemical industry, in *Proc. Brighton Crop Prot. Conf. — Weeds*, 1995, 557.
197. James, E. H., Kemp, M. S., and Moss, S. R., Phytotoxicity of trifluoromethyl- and methyl-substituted dinitroaniline herbicides on resistant and susceptible populations of black-grass (*Alopecurus myosuroides*), *Pestic. Sci.*, 43, 273, 1995.
198. Itoh, K., Wang, G. X., Uchino, A., and Tachibana, M., Control of *Lindernia* weeds resistant to sulfonylureas, *J. Weed Sci. Technol.*, 42(Suppl.), 24, 1997.
199. Itoh, K., Ed., *Proc. Symp. Invasion of Herbicide Resistant Weeds in Agroecosystems in Japan*, NIAES, Tsukuba, 1998, 82.
200. Watanabe, H., Wang, G. X., Uchino, A., and Itoh, K., Control effect of several herbicides on biotypes of *Lindernia* spp., *Limnophila sessiliflora* and *Monochoria korsakowii*, lowland broadleaved weeds, resistant to sulfonylurea herbicides, *J. Weed Sci. Technol.*, 43(Suppl.), 46, 1998.
201. Fukumoto, T., Kagawa, M., and Shirakura, S., Development of the one shot herbicides which have efficacy on SU resistant *Scirpus juncoides*, *J. Weed Sci. Technol.*, 44(Suppl.), 90, 1999.
202. Sugiura, K., Watanabe, S., and Soeda, T., Herbicidal efficacy of ethoxysulfuron/pyrazolate/pretilachlor mixture against sulfonylurea resistant *Scirpus juncoides*, *J. Weed Sci. Technol.*, 4(Suppl.), 86, 1999.
203. Hopkins, W. L., *Ag Chem New Compound Review*, Ag Chem Information Services, Indianapolis, IN, 1998, 120.
204. Tomlin, C., Ed., *The Pesticide Manual*, British Crop Protection Council and The Royal Society of Chemistry, U.K., 1998, 1606.
205. WSSA, Weed Science Society of America, *Herbicide Handbook*, Ahrens, W. H., Ed., Weed Science Society of America, Champaign, IL, 1994, 352.
206. WSSA, Weed Science Society of America, *Herbicide Handbook, Supplement to Seventh Edition*, Hatzios, K. K., Ed., Weed Science Society of America, Champaign, IL, 1998, 104.
207. Harris, B., personal communication, 2000.
208. Baldwin, F. L., Talbert, R. E., Carey, V. F., III, Kitt, M. J., Helms, R. S., Black, H. L., and Smith, R. J., Jr., A review of propanil resistant *Echinochloa crus-galli* in Arkansas and field advice for its management in dry seeded rice, in *Proc. Brighton Crop Prot. Conf. — Weeds*, 1995, 577.
209. Norsworthy, J. K., Rutledge, J. S., Talbert, R. E., and Hoagland, R. E., Agrichemical interactions with propanil on propanil-resistant barnyardgrass (*Echinochloa crus-galli*), *Weed Technol.*, 13, 296, 1999.
210. Carmiol-Zúñiga, J. A., Opciones químicas para el control de *Echinochloa colona* resistente a propanil en arroz de secano, Tesis de Licenciatura, Universidad de Costa Rica, Sede Regional del Atlántico, Turrialba, Costa Rica, 1999, 41.
211. Garita, I., personal communication, 2000.
212. Riches, C. R., Knights, J. S., Chaves, L., Caseley, J. C., and Valverde, B. E., The role of pendimethalin in the integrated management of propanil-resistant *Echinochloa colona* in Central America, *Pestic. Sci.*, 51, 341, 1997.
213. Uchino, A., Watanabe, H., Wang, G. X., and Itoh, K., Light requirement in rapid diagnosis of sulfonylurea-resistant weeds of *Lindernia* spp. (Scrophulariaceae), *Weed Technol.*, 13, 680, 1999.
214. Gerwick, B. C., Mireles, L. C., and Eilers, R. J., Rapid diagnosis of ALS/AHAS-resistant weeds, *Weed Technol.*, 7, 519, 1993.

215. Do-Soon, K., Caseley, J. C., Brain, P., Riches, C. R., and Valverde, B. E., Rapid detection of propanil and fenoxaprop resistance in *Echinochloa colona*, *Weed Sci.*, 48, 695, 2000.

216. Valverde, B. E., Chaves, P., Garita, I., Vargas, E., Riches, C. R., and Caseley, J. C., From theory to practice: development of piperophos as a synergist to propanil to combat herbicide propanil resistance in junglerice, *Echinochloa colona*, *WSSA Abstr.*, 37, 33, 1997.

217. Valverde, B. E., Garita, I., Vargas, E., Chaves, L., Ramírez, F., Fischer, A. J., and Pabón, H., Anilofos as a synergist to propanil for controlling propanil-resistant junglerice, *Echinochloa colona*, *WSSA Abstr.*, 39, 318, 1999.

218. Mayer, P., Kriemler, H.-P., Hamböck, H., and Laanio, T. L., Metabolism of O,O-dipropyl S-[2-(2′-methyl-1′-piperidinyl)-2-oxo-ethyl] phosphorodithioate (C 19 490) in paddy rice, *Agric. Biol. Chem.*, 45, 355, 1981.

219. Almario, O., personal communication, 1999.

220. Sankula, S., Braverman, M. P., Jodari, F., Lindscombe, S. D., and Oard, J. H., Evaluation of glufosinate on rice (*Oryza sativa*) transformed with the BAR gene and red rice (*Oryza sativa*), *Weed Technol.*, 11, 70, 1997.

221. Sankula, S., Braverman, M. P., and Linscomb, S. D., Response of BAR-transformed rice (*Oryza sativa*) and red rice (*Oryza sativa*) to glufosinate application timing, *Weed Technol.*, 11, 303, 1997.

222. Sankula, S., Braverman, M. P., and Linscomb, S. D., Glufosinate-resistant, BAR-transformed rice (*Oryza sativa*) and red rice (*Oryza sativa*) response to glufosinate alone and in mixtures, *Weed Technol.*, 11, 662, 1997.

223. Wheeler, C. C., Baldwin, F. L., Talbert, R. E., and Webster, E. P., Weed control in glufosinate-resistant rice, in *Proc. Southern Weed Sci. Soc.*, 51, 34, 1998.

224. Griffin, J. L., Linscombe, S. D., and Zhang, W., Tolerance of glufosinate-resistant rice lines to glufosinate, *WSSA Abst.*, 39, 11, 1999.

225. Dillon, T. L., Baldwin, F. L., and Webster, E. P., Weed control in IMI-tolerant rice, in *Proc. Southern Weed Sci. Soc.*, 51, 268, 1998.

226. Sanders, D. E., Strahan, R. E., Linscombe, S. D., and Croughan, T. P., Control of red rice (*Oryza sativa*) in imidazolinone tolerant rice, in *Proc. Southern Weed Sci. Soc.*, 51, 36, 1998.

227. Webster, E. P., Masson, J. A., and Zhang, W., Imazethapyr rates and timings in water-seeded rice, *WSSA Abstr.*, 40, 271, 2000.

228. Williams, B. J., Barnyardgrass (*Echinochloa crus-galli*) control in Clearfield rice, *WSSA Abstr.*, 40, 273, 2000.

229. Langevin, S. A., Clay, K., and Grace, J. B., The incidence and effects of hybridization between cultivated rice and its related weed red rice (*Oryza sativa* L.), *Evolution*, 40, 1000, 1990.

230. Gressel, J., Herbicide resistant tropical maize and rice: needs and biosafety considerations, in *Proc. Brighton Crop Prot. Conf. — Weeds*, 1999, 637.

231. Croughan, T., Utomo, H. S., Sanders, D. E., and Braverman, M. P., Herbicide resistant rice offers potential solution to red rice problem, *Louisiana Agric.*, 39, 10, 1996.

232. Auld, B. A. and Morin, L., Constraints in the development of bioherbicides, *Weed Technol.*, 9, 638, 1995.

233. Cother, E. J., Host range studies of the herbistat fungus *Rhyncosporium alsimatis*, *Australasian Plant Pathol.*, 28, 149, 1999.

234. Yang, Y. K., Kim, S. O., Chung, H. S., and Lee, Y. H., Use of *Colletotrichum graminicola* KA001 to control barnyardgrass, *Plant Dis.*, 84, 55, 2000.

235. Gohbara, M., Use of plant pathogens for the control of weeds in Japan, in *Proc. Int. Weed Control Congr.*, Copenhagen, Denmark, 1996, 1205.
236. Ho, N. K., Introducing integrated weed management in Malaysia, in *Herbicides in Asian Rice: Transitions in Weed Management*, Naylor, R., Ed., Institute for International Studies, Stanford University, Palo Alto, CA, and International Rice Research Institute, Manila, Philippines, 1996, 167.
237. Matteson, P. C., Implementing IPM: policy and institutional revolution, *J. Agric. Entomol.*, 13, 173, 1996.
238. Ooi, P. A. C., Experiences in educating rice farmers to understand biological control, *Entomophaga*, 41, 375, 1996.
239. Heong, K. L., Escalada, M. M., Huan, N. H., and Mai, V., Use of communication media in changing rice farmers' pest management in the Mekong delta, Vietnam, *Crop Prot.*, 17, 413, 1998.
240. Mangan, J. and Mangan, M. S., A comparison of two IPM training strategies in China: the importance of concepts of the rice ecosystem for sustainable insect pest management, *Agric. Human Values*, 15, 209, 1998.
241. Sumiyoshi, T., Seed germination and emergence of three *Scirpus* weeds, *Tohoku J. Crop Sci.*, 40, 61, 1997.
242. Bryant, R., Agrochemicals in perspective: analysis of the worldwide demand of agrochemical active ingredients, *Fine Chem. Conf.*, 1999.
243. Agrow, *World Crop Protection News*, December 13, 1996.
244. Agrow, *World Crop Protection News*, February 14 and 28, 1997.
245. Agrow, *World Crop Protection News*, July 11, 1997.
246. Young, I., Rhone-Poulenc in R&D accord to develop genetically modified rice, *Chem. Week*, April 7, 1999, 17.
247. Labhart, C., personal communication, 2000.

7 Economic and Sociological Factors Affecting Growers' Decision Making on Herbicide Resistance

David J. Pannell and David Zilberman

CONTENTS

0-8493-2219-7/01/$0.00+$.50
© 2001 by CRC Press LLC

7.1   INTRODUCTION

Economic and sociological aspects of herbicide resistance have received little attention in the published literature; the only published studies to date are those of Schmidt and Pannell,[1,2] Gorddard et al.,[3,4] and Orson.[5] The published literature on economic and social aspects of insecticide and fungicide resistance is a little larger[e.g., 6–13] but, apart from general concepts, contains little that can readily be applied to the case of herbicide resistance.

The management problem for herbicide resistance differs in several respects from these other types of resistance. One important practical difference is that, in most cases, a greater range of control methods is available for weeds than for insects and microorganisms, so that substitution out of herbicide control options is much more likely to be economically viable. Indeed, many farmers already practice some nonherbicide methods, in tandem with selective herbicides. Another difference is the importance of spread. Resistant insects and diseases are commonly much more mobile than resistant weeds, and so have much greater problems of communal resistance management, whereas herbicide resistance is mainly a problem for private, individual farmers.

Our chapter proceeds in three parts. First, we discuss issues relating to the phase when farmers are considering adoption of changed farming practices, either to delay the onset of resistance or to deal with its arrival. We review the key factors affecting the speed and nature of farmers' responses to possible changed management practices and discuss the implications of this information for adoption of management practices for herbicide resistance.

Second, we employ a detailed bioeconomic model to examine herbicide resistance in the context of a complex mixed farming system. We examine the impact of resistance on the profitability of farming in a case study, and discuss the changes in farm management that resistance requires or encourages.

Third, we examine the policy implications of herbicide resistance. Is there a need for government intervention to manage the resistance problem for the common good? In what ways might this intervention occur?

7.2 HERBICIDE RESISTANCE AND THE ADOPTION OF CHANGED FARMING PRACTICES

Before addressing some of the specific issues relating to resistance we will present some of the major concepts related to technological change and agriculture. In the discussion of these concepts, we will illustrate key points with historical examples relating to adoption of mechanical and herbicide weed control methods. We will then discuss specific issues relating to the adoption of herbicide-resistance management practices.

7.2.1 Introduction of New Technologies

It is useful to distinguish between two processes: innovation, which is the introduction or development of new technologies, and adoption, which is actual use of these new technologies. The induced innovation hypothesis, originated by Hayami and Ruttan[14] and formulated by Binswanger,[15] argues that innovations are strongly influenced by profitability considerations, particularly "demand-driven" innovations. Demand-driven innovations, in the context of farming, are innovations generated by, or at the behest of, farmers in direct response to farming problems perceived or recognized by farmers. Such innovations may save scarce resources, increase yields, or enable production of new products that are more profitable. "Supply-driven" innovations, on the other hand, are generated from scientific research. They may not necessarily address a problem recognized in advance by farmers, but may nevertheless generate benefits to farmers and/or others in the community, for example, by revealing a hitherto unrecognized constraint on yields.

In traditional society, manual weeding was (and often still is) the dominant form of weed control. The development of mechanized weeding through disks and cultivators was driven both by demand factors, namely, the increased scarcity of labor, and by supply factors, such as the invention of the tractor. Mechanical weeding, however, still requires substantial machinery and labor time and entails energy costs. During the 20th century, costs of inputs such as labor, machinery, and energy increased relative to the prices of the major agricultural commodities, so mechanical weeding became more expensive per unit value of production. This created a potential for benefits from alternative weed control technologies, which was met through the development of chemical herbicides. Environmental factors also drove the demand for herbicide solutions, as concerns about soil erosion made cultivation less desirable. The availability of chemical herbicides made it feasible to introduce no-tillage and minimal-tillage systems, allowing earlier seeding of crops. Recently concerns also have risen that heavy tillage releases carbon dioxide into the atmosphere.[16] The attractiveness of low-tillage activities may be enhanced by policies that provide economic incentives for soil carbon sequestration, as part of an effort to curtail climate change.[16]

The introduction of chemical herbicides was also driven by supply factors. First, developments in weed science and chemistry made chemical herbicides cheaper and safer. Furthermore, the use of chemicals to control other pests provided opportunities for complementarity. Marginal application costs of herbicides are low if farmers already use spraying machinery to protect against other pests, such as insects.

While the use of chemicals for control of other pests has not grown significantly since 1973, the use of herbicides has increased dramatically in many nations[17] (Figure 7.1). This growth in herbicide use occurred first in most industrialized countries and is now occurring in developing countries.

The creation of an innovation is only part of the process of technological change. For change to occur, the product must be adopted by the producer.

7.2.2   ADOPTION AND DIFFUSION

We can distinguish between adoption, which is the extent to which an individual producer is using the new technology, and diffusion, which is the percentage of farmers that have adopted a new technology or the percentage of land on which the innovation is used. Adoption and diffusion have been studied by both sociologists and economists. The empirical evidence shows that the pattern of diffusion usually follows a sigmoidal or S-shaped function over time[18] (Figure 7.2). That is, diffusion may be represented by a function of the form $P(t) = K/[1 + \exp(-a - bt)]$, where $P(t)$ is the diffusion level at time t; K is the long-run upper limit on the level of diffusion; the slope coefficient, b, reflects the speed of uptake of the new technology; and the intercept, a, reflects aggregate adoption at the start of the period. K, in most cases, is smaller than 1 since diffusion of an innovation is generally not complete.

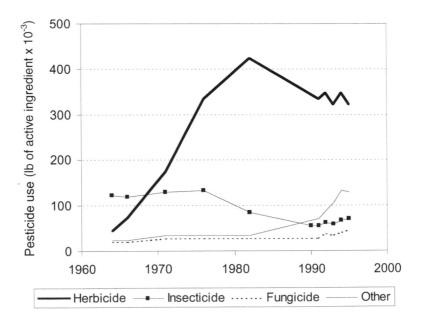

FIGURE 7.1   Pesticide use on major crops in the United States. (From USDA, Economic Research Service estimates.)

In a landmark study Griliches[19] demonstrated that the three parameters are strongly affected by economic profitability. Producers or regions for which adoption of a new technology is relatively profitable are likely to adopt the technology earlier and to a greater extent. For example, in the case of herbicides, regions that have relatively high costs of energy or labor are likely to adopt herbicides more rapidly, which saves on these inputs.

Nonfinancial influences also play important roles in the diffusion process, including factors such as the quality of communication channels, social acceptability, social infrastructure, and farming subcultures or farming styles.[20] These are often conceptualized by economists in terms of their influence on incentives for or against adoption, and in this way most sociological factors can readily be considered within the sort of economic framework normally applied to financial influences on adoption.[e.g., 21] Sociologists use different language and different conceptual frameworks to discuss these issues,[e.g., 20] but the underlying logic of the two disciplinary approaches is, in most cases, identical.

7.2.3   EXPLANATIONS OF THE SIGMOIDAL DIFFUSION CURVE

One explanation for the S-shaped diffusion curve is the process of imitation. There are a small number of early adopters but then the number of imitators grows exponentially and the diffusion process enters its take-off period. Since the population is finite beyond a certain point, the diffusion rate slows until the level stabilizes. Eventually, there may be disadoption of the technology following the introduction

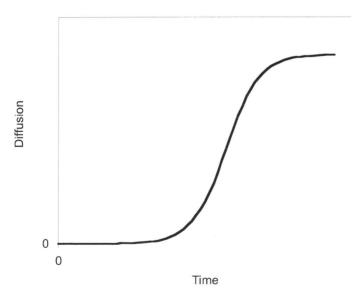

FIGURE 7.2   A typical diffusion curve, showing increasing usage of the innovation among the population of potential adopters.

of more advanced technology. This pattern of dynamic behavior has occurred with various weed control technologies. There was a transition from manual to mechanical weeding. Then there was diffusion of chemical herbicides, and within this category there has continued to be diffusion of new herbicide products at the expense of earlier herbicides. Following the failure of herbicides due to herbicide resistance, sigmoidal diffusion curves for replacement technologies and practices will again be apparent.

An alternative explanation to the S-shaped diffusion curve was presented by David[22] and Stoneman,[23] who argued that the reason for the initially gradual diffusion is heterogeneity. The causes of heterogeneity are differences in farm size, human skills and knowledge, machinery capacity, crop yields, soil types, disease risks, and other variables. Producers are motivated to choose technologies that maximize profits and, because the profitability of any new technology will vary widely from farmer to farmer, this provides different incentive levels for adoption. For example, a technology that requires indivisible investment in equipment (purchase of a tractor or computer) is more likely to be adopted earlier by larger producers because the fixed cost can be spread over a larger volume of operation. If a new technology earns α extra dollars per unit of output and requires a fixed investment of k dollars per period, only producers with a volume greater than k/α will adopt the new technology. Thus, there would be a threshold farm size separating adopters and nonadopters. Over time, the threshold level is likely to decline because of experience and learning by both the farmers and, if relevant, the manufacturers of the technology. Learning on the part of farmers results in an increased α, and learning on the part of manufacturers may reduce the cost of producing the technology, and thus decrease k. Either way, the threshold farm size for adoption declines over time, resulting in an increased level of diffusion. Tractors and other farm machinery are likely to be examples for which this threshold model is relevant. Larger farms adopted the technology first, increasing production volumes and allowing tractors to become more affordable so that smaller farms later joined the ranks of adopters.

However, the threshold model does not fully explain the dynamics of new technology diffusion, even for technologies such as tractors that conform to the model (an indivisible, high-cost item). For example, earlier adopters of such a technology may have a significant economic advantage over nonadopters. Some would have sufficient excess productive capacity to enable them to purchase extra land, leading to an increase in average farm size. On the other hand, Olmstead and Rhode[24] suggest that farmers form partnerships so that they can afford mechanical innovations. This enables several small farmers to adopt expensive machinery and thus accelerates the diffusion process. In some areas with small farms, private contractors established firms that provided mechanical services to farmers who could not afford to purchase the machinery.

Both of the above explanations undoubtedly apply simultaneously to some extent for virtually all innovations. Overlaid on these explanations (and consistent with them) is the approach that conceptualizes adoption as being primarily a process of dynamic learning and refinement of decision making over time.[e.g., 21,25] This approach recognizes that growers vary in their risk preferences and their perceptions of riskiness of an innovation, and that their perceptions are affected by information from various sources at different times. Sources of information can include experience gained from on-farm trials of the innovation, but also off-farm sources such as extension agents, neighboring farmers, or the media. Heterogeneity of perceptions, perhaps due to different levels of experience or different sources of information, is another factor that would contribute to the sigmoid shape of the diffusion curve.

7.2.4   INFLUENCES ON THE SPEED AND LEVEL OF DIFFUSION

The incentives for or against adoption may be influenced by many factors, including those outlined below.

7.2.4.1   Land Quality

Caswell and Zilberman[26] argued that differences in land quality and soil conditions (e.g., slope of land and water-holding capacity of soil) are also important in explaining adoption of modern technologies. There is strong evidence to support this empirically.[27] Similarly, adoption of reduced-tillage practices occurred at locations with lower soil quality (measured by depth of topsoil).[28] Adoption of herbicides was complementary to the reduced-tillage practices and thus was particularly favored in situations where soil erosion was a problem and topsoil was shallow.

7.2.4.2   Prices

Increases in the price of fuel and labor tend to increase the adoption of herbicides.[17,29] Increases in commodity prices or support programs that are based on yield tend to increase the intensification of farming and thus increase adoption and use of herbicides.[30,31] Higher sale prices also tend to increase cultivation of land in locations of lower soil quality[32] and this may further increase use of herbicides.

7.2.4.3   Government Policies

Government policies that aim to improve environmental policies also serve as inducements for adoption of alternative practices. Since there has been an increased emphasis on soil erosion prevention, the U.S. government has required certain environmental practices as conditions for participation in commodity programs that raise product prices. This has led to increases in the area of land with no or minimum tillage, and thus increased the reliance on herbicides.[28,33] Note, however, that these programs also include provisions to encourage reductions in the use of some herbicides, such as atrazine. Thus, their overall effect is to increase the use of herbicides but to modify them toward herbicides that are judged to be more environmentally benign. The global tendency toward freer agricultural markets and reduced government supports for agriculture may actually lead to a reduction in the area of reduced-tillage practices. On the other hand, the introduction of policies that subsidize soil carbon sequestration may increase the use of soil conserving practices and, consequently, the use of herbicides.

7.2.4.4   Location

Initial studies of diffusion linked diffusion rates to locations.[18] For most technologies the evidence points to higher adoption rates by producers who are closer to the nearest urban center than producers who are farther away.[18] That may be due to a lower transportation cost and/or closer contact with dealers and distributors of the technology or extension agents. In some cases, however, remote regions are more likely to adopt technologies such as herbicides if they save inputs (labor and energy) that are more costly in these remote locations.

7.2.4.5   Uncertainty and Risk

Uncertainty and risk are major obstacles for the adoption of new technologies. Farmers to a large extent are risk averse,[34] meaning that they are willing to give up some average profit in order to reduce fluctuations in, or uncertainty about, profit. In many cases adoption of new technologies is gradual so farmers can experiment with them and have more reliable assessments of their impacts. One of the main activities of marketing specialists and extension agents is to reduce farmers' uncertainties about new technologies. This is achieved, for example, by the establishment of demonstration plots, demonstration activities, money-back guarantees, or warranties. At the time of the introduction of herbicides, there must have been substantial uncertainty among farmers regarding their performance both in terms of their effectiveness and their possible side effects, and this would have affected the pace of their diffusion. Effective marketing plans that educate farmers about the properties of a new technology and reduce their uncertainty about it will likely result in higher adoption rates. The high initial adoption rates of transgenic glyphosate-resistant crops in the United States suggest that Monsanto was able to reduce much of farmers' uncertainty about the transgenic crop by its marketing activities. There would also have been relatively low uncertainty about the performance of the herbicide due to extensive prior experience with it by most farmers.

7.2.4.6   Observability and Trialability

These issues are closely related to the issue of uncertainty. If a technology is highly observable, the diffusion process via the imitation mechanism is likely to proceed more rapidly. One can well imagine that the high observability of tractors enhanced their rate of spread through farming communities (although it clearly does not override other considerations in a farmer's final decision to adopt or reject an innovation). Perhaps even more important is the potential to trial the innovation on a small scale, so that information can be obtained, uncertainties reduced, and skills developed without the risk of large financial costs if the technology turns out to be uneconomic or fails due to inexperience. Innovations such as a new herbicide are highly trialable; its impacts on weeds are readily observable, it can be readily compared to existing herbicides, and it can be used effectively on a very small scale.

7.2.4.7   Adjustment Costs

The rapid and high rate of adoption of genetically modified herbicide-resistant crop varieties in North America and, now, South America can partly be explained by the fact that these technologies require very little adjustment cost. They do not require modification of the production system or a significant amount of learning by the farmer. Farmers simply replace one seed variety with another and continue to use herbicides with which they are familiar (glyphosate in most cases). One of the key obstacles to the adoption of new technology is that it requires investment of time and effort from the farmers. Farmers do not consider only the direct financial cost of adoption but also any nonmonetary costs. One reason that the adoption of IPM has been relatively slow is the relatively high adjustment cost it entails.[35] High adjustment cost also explains the slow rate of adoption of modern irrigation technologies.[36] Because of the need to invest in adjustment, older farmers are less likely to invest in learning for adoption of new technologies than are younger farmers.[27] One rationalization of the observation that individuals with higher human capital tend to adopt new technologies earlier is that they probably have relatively low learning costs. The adoption of mechanized farming practices and mechanized weeding was slow because of farmers' needs to learn to operate new machinery, with the result that complete adoption of tractors took about 40 years in the United States.[37] On the other hand, adoption of herbicides has been less problematic, especially for farmers who already sprayed against insects and therefore required little up-front investment in machinery. This helps explain the rapid increase in herbicide use in the 1980s. Adoption of complex resistance-management strategies may be slowed by the adjustment cost that they entail, including purchase costs of any new machinery that is necessary.

7.2.4.8   Public Acceptance

Adoption of new technologies may also be affected by public acceptance considerations. Adoption rate may be lowered if farmers perceive that by adopting a new technology they may endanger their capacity to sell their product at a high price, or that they may encounter legal problems or protests. Some of the controversy regarding genetically modified crop varieties is likely to reduce the area planted to such varieties. Farmers will engage in extra activities that will enable them to adopt new technologies in a manner that will be more socially acceptable, as long as the overall benefit is greater than the cost.

7.2.4.9   Education

More educated farmers tend to adopt advanced technology earlier. Putler and Zilberman[38] have shown that education was a key explanatory variable for the adoption of computers by farmers in Tulare County, California. Education has also been shown to be a key factor explaining adoption of various forms of IPM[35] and soil testing.[28] However, farmers with less education may adopt more sophisticated technologies when they rely on private consultants or cooperative extension.

Huffman[39] has shown that the productivity of extension is higher for farmers who are less educated. Wolf[40] documented the evolution and growth of private consulting in pest management.

7.2.4.10   Extension and Consulting Advice

Zilberman et al.[41] have argued that, increasingly, decisions regarding pest management are made based on significant input from professionals, particularly in industrialized agricultural nations. Smaller farms rely on dealers or independent consultants and some of the bigger farms have in-house experts. Furthermore, government extension is shifting much of its efforts toward education and certification of consultants. In the United States, these experts are mostly concerned about insect control but, conversely, in Australia, they are mostly focused on weed control. The role of consultants in weed control undoubtedly will increase. This trend suggests that educational activities by companies and universities regarding herbicide resistance and weed control in general should be increasingly directed at consultants. Indeed, in Australia, where herbicide resistance is widespread, conscious decisions have been made by government agencies and universities to target consultants as information providers.

7.2.4.11   Marketing

A key to the diffusion of a new technology is providing information about its existence and properties. A significant amount of the budget of herbicide manufacturers is dedicated to marketing their products through activities such as advertisements, demonstrations and trade shows, money-back guarantees and warranties, and product stewardship through salespeople.[42] The marketing literature views adoption as a hierarchical and gradual process with several stages. The first stage is product awareness, where farmers may become aware of a new product through advertisements and discussions with other farmers. The second stage is information search, where farmers may go to a retail outlet to learn about new products. That stage is followed by evaluation, purchasing decision, and post-purchase evaluation. Advertising is crucial for product awareness to trigger interest in a new product. Kotler[42] divided potential adopters into two segments: (1) early adopters who are influenced by advertisements and (2) imitators who are influenced by word of mouth of previous adopters.

Successful marketing of a new herbicide entails identifying the market segments where this pesticide is most appropriate, developing a marketing strategy using the right pitch to gain potential buyers' attention, and training dealers to recognize and deal with buyers' concerns and to help them adjust to the new product. Activities of salespeople must be supported by effective demonstrations, educational material, and warranty programs.[42]

This by no means exhausts the list of factors that have been found or proposed to influence adoption, but it includes the main types of factors of influence mentioned in the literature (although not necessarily with the same language).

7.2.5   RESISTANCE MANAGEMENT

While there is a significant economic literature on the adoption of integrated pest management (see survey by Carlson and Wetzstein[43]), no existing studies focus on adoption of resistance-management practices in agriculture and only a few focus on the economics of such practices. Most of the economic studies are normative and/or conceptual. They derive economically optimal pest control strategies taking resistance into account and compare them to alternative strategies that may be more pervasive. One notable example is a study by Hueth and Regev.[6] They argue that optimal resistance management must be analyzed within a dynamic context. Pesticide resistance development is a dynamic phenomenon and once resistance reaches a high level, chemical treatment will become ineffective. Therefore, they modeled vulnerability to a pesticide as a nonrenewable resource and argued that overuse of a pesticide would lead to depletion of this vulnerability. They argued that concern about resistance should reduce pesticide usage.

An observation made in the pesticide resistance literature is that farmers may apply pesticide suboptimally, disregarding resistance considerations, resulting in overapplication of pesticides. There are several possible reasons for high chemical usage. First, decision makers may be unsophisticated, using myopic or short-term decision-making rules rather that dynamic, long-term ones. Second, producers may have high discount rates because of their financial situations and thus tend to give a relatively low weighting to long-term benefits relative to short-term costs of their present activities. In particular, this may be the case with poorer farmers whose economic survival is potentially under threat. It may also be the situation in an economy where access to credit is very limited and interest rates faced by farmers are very high. Third, long-term planning may actually result in little difference in management compared to short-term planning. A case study presented later shows that this is a realistic possibility for herbicide resistance in at least one farming system. Fourth, the adoption of strategies to slow resistance will be low when farmers expect substitute pesticides to become available over time, replacing those lost to resistance.

Extension and other outreach activities may educate farmers to modify their behavior toward more sustainable practices when ignorance is the cause for disregarding resistance considerations. In agriculture of the developed world, the previously noted increase in reliance on consultants for pesticide advice is likely to modify behavior so that resistance will be taken into account. In some farming regions, however, it seems possible that farmers may use consultants and be exposed to sophisticated advice on insect control problems but may be less well informed about weed control and herbicide resistance. On the other hand, the converse is more likely to be true in other environments (e.g., Australia). Therefore, it is important to develop educational opportunities that will enable farmers to consider the resistance problem for all types of pesticides.

Education will not modify pest application practices, however, if farmers' behaviors are dictated by high discount rates. In these cases, pest control strategies will best be modified if informed farmers have easier access to credit under better terms so that their discount rates will decline and they will give more weight to future consequences of their activities. Thus, in many cases, improvement of credit channels is essential for more emphasis on long-term perspectives in pest management. Neither education nor cheap credit may be sufficient if the prime cause for nonadoption of herbicide-resistance prevention strategies is that the long-run economically optimal strategy is to exploit the herbicide resource to exhaustion.

It has been noted in the literature on insecticide resistance that a lack of attention to resistance may also occur when it develops as a result of the joint activities of many producers. In cases where there is sufficiently rapid movement of pests between farms, individual farmers are aware that their individual actions will not prevent resistance (which depends on the overall behavior of the community). Therefore, individual investments to prevent resistance will incur costs without generating benefits that can be captured by the investor. In this case there is likely to be a "tragedy of the commons." Namely, each individual considers only his/her private, short-term interests and over-uses chemicals such that an excessive level of resistance occurs. Possible policy mechanisms for preventing such over-application are discussed in Section 7.4.

It seems that regional dependency is less important for herbicide resistance than some cases of insecticide resistance because of the relatively low mobility of most weeds relative to some insects. Furthermore, much of the use of herbicides occurs on large operations where herbicide-resistance problems are contained within the boundaries of the operation. In these cases, there is less interdependence among operators and more prospect of private herbicide decision being consistent with the interests of the community.

Most studies of resistance have considered only one chemical and developed optimal resistance strategies, assuming that this chemical will be the only one available in the long run. However, if producers expect to see new technologies for pest control appearing over time, then every such category can be viewed as a backstop technology that provides relief from the problem associated with resistance to existing chemicals. The potential development of new pest treatment technologies in the future leads producers to reduce their emphasis on resistance considerations in the present. This is analogous to a decision to ignore the exhaustibility of fossil fuels because of a perception that there is a high likelihood of developing alternative energy sources. A major issue here is the speed of development of alternative treatment types relative to the speed of resistance development. Furthermore, uncertainty regarding the development of alternative pest control technologies provides another reason for adoption of a resistance management strategy. It may be that at a certain point we run out of new forms of pest management. This risk is enhanced by the occurrence of resistance to multiple herbicides in particular weeds, particularly where there are multiple mechanisms for endowing multiple-resistance.[e.g., 44] In the case of herbicides, the rate of discovery of chemicals with new modes of herbicidal action is low. Obviously, assessment of the relative rates of resistance enrichment vs. development of new treatments is an empirical issue.

That suggests that whenever a new form of treatment is being introduced (e.g., new technologies developed using biotechnology) it is important to consider the potential for resistance and, if possible, develop management strategies that will economically slow the onset of resistance. This may not always be technically possible, as illustrated in the next section. There may also be difficulties obtaining cooperation from chemical producers. For example, delay of resistance may require the integration of chemicals (and other approaches) that are promoted by different chemical companies. Sellers of a new chemical, especially in the early stages of production and patent protection, have every incentive to underemphasize the drawbacks of their product and to aggressively capture a significant market share that will trigger the diffusion process. On the other hand, the sellers of incumbent products may not wish to be accommodating to the introduction of new chemicals that may displace their own products. This suggests an important role for extension specialists and independent consultants in developing resistance-management strategies.

7.2.6 ADOPTION OF RESISTANCE-PREVENTION STRATEGIES

Considering the nature of herbicide resistance and management, it is possible to make some observations about adoption of resistance-prevention strategies based on lessons from the general adoption literature.

In particular, we note that herbicide resistance, as a farm management issue, is

- complex
- potentially expensive or difficult to prevent
- a source of considerable uncertainty and
- difficult to observe until herbicide resistance is advanced

Complexity increases uncertainty, adjustment costs, and, possibly, direct costs. Complexity in herbicide resistance management arises from several sources. First, farmers faced with resistance can benefit from understanding the ecological and biochemical theories that explain the occurrence of resistance, but these theories may be difficult to fully understand for nonscientists. Second, the weed treatment options that can be used to substitute for herbicides and delay resistance are themselves complex to use relative to herbicides, and normally must be used in combinations in order to achieve adequate weed control. Third, the number of potential combinations of treatments to be considered is vast, especially when the issue is considered as a dynamic problem spanning a number of years. In this case, the options include not just changes in weed kill methods but also, potentially, changes in land use; for example, pasture may be included in place of a crop to broaden the available range of weed control options.

The cost-effectiveness of herbicides is revealed by their ubiquitous use through cropping systems of agriculture in the developed world. If a farmer is considering changing to a lower intensity of herbicide use in order to delay the development of herbicide resistance, he or she will be very conscious that in most cases such a move will increase short-term costs. The change will therefore be unattractive unless the farmer is convinced that there are sufficient offsetting benefits. There may actually not be sufficient offsetting benefits in some situations.

Even if there *are* sufficient offsetting benefits to justify early adoption, it may be very difficult for the farmer to determine this with sufficient certainty in time for preventative strategies to be put into place. For some examples of herbicide-resistance development, full resistance is apparent within a small number of years (e.g., 5 years or less in many recorded cases for annual ryegrass, *Lolium rigidum*, in Australia). Even if the innovative technologies are trialable, few years of trialing are available before the potential for preventative action has passed.

This problem is worsened by the difficulty of trialing some types of treatment recommended to delay resistance. Difficulties include the following:

1. Where treatments are implemented in combinations (as they generally must be) it can be very difficult to determine from field observations alone the individual impacts of each element of the combination in order to assess their individual worthiness for inclusion in an integrated management system.
2. Impacts of some treatments are relatively difficult to observe even if implemented in isolation. For example, increasing the crop seeding rate affects the seed production of both crops and weeds (in opposite directions), and the latter is rather difficult to observe quantitatively in the field without tedious collection and counting of weed seeds in both standard and high seeding rate plots.
3. The effectiveness of some alternative weed treatments is very sensitive to weather conditions or the quality of implementation, and so trials give highly variable results from time to time. Even if a treatment is beneficial in the long run, it may not appear so in a short-term trial, or it may take a long time before its value can be determined with adequate confidence.

For all of these reasons, rapid adoption of integrated weed management systems, involving combinations of unfamiliar, complex, and expensive treatments that are difficult to trial, is unlikely to occur until it is essential.

On the other hand, the nature of herbicide resistance is that, once it has developed, farmers have no choice but to alter their weed management systems. Thus, following the onset of resistance, most of the problems involved in encouraging farmers to change to some alternative system evaporate. The problem for farmers then becomes which of the many possible alternative systems should best be adopted? The decision support system outlined in the next section is the outcome of one attempt to provide guidance on this question.

7.3　HERBICIDE RESISTANCE AS A FARMING SYSTEMS/RESOURCE MANAGEMENT PROBLEM

There has been increasing interest in nonherbicide weed control options in response to the threat of herbicide resistance, [e.g., 45,46] but little work on their optimal economic integration into farming systems. Stewart[47] and Schmidt and Pannell[1] applied early versions of the Ryegrass Integrated Management (RIM) model[48] to explore the economic implications of herbicide resistance in Western Australia. RIM has since been substantially enhanced and expanded, and the 1999 version is the basis for the analyses presented here. We will first briefly describe the model (readers are referred to Pannell et al.[48] for more details) and then a selection of model results that illustrate key points about the economics of resistance management. Results are all based on a set of parameters specified for the wheat belt of Western Australia.

7.3.1 Model Description

RIM is a multiperiod simulation model representing the biology, technology, and farm-level financial aspects of ryegrass weed management in dryland farming systems of southern Australia. General features include

- a 20-year time frame
- representation of up to seven different crop- or pasture-based enterprises in a user-specified sequence
- 35 different weed control options that can be selected in any technically feasible combination and in any technically feasible sequence
- detailed representation of relevant biological relationships
- representation of relevant financial details, including input costs, output prices, machinery costs, interest rates, yields, and environmental costs for certain treatments
- potential for the user to adjust all biological and economic parameters
- ability to specify the herbicide-resistance status of the ryegrass weeds with respect to each of seven herbicide groups

The treatment options are shown in Table 7.1. There are 12 selective herbicide options, 5 nonselective herbicide options, and 18 nonchemical options. They are listed in chronological order, based on the time when farmers would implement them. Some of these treatment descriptions may not be clear. Detailed explanations for each are provided by Pannell et al.[48] Chemical names for those chemicals for which trade names are used are provided in an appendix.

The biological processes, variables, and relationships represented include

- the density of living, ungerminated seeds in or on the soil through each year
- a germination pattern of weed seeds over time within each year
- mortality or removal of seeds, seedlings, or adult plants or prevention of seed production through various processes, including both natural and human-applied mechanisms
- inter- and intra-species competition impacts on crop yields and weed seed production
- damage to crop production from weed control treatments, including phytotoxic damage from herbicides, and mechanical losses due to treatments such as green manuring

TABLE 7.1
Weed Treatment Options Included in the RIM Model

	Treatment	Type[a]
1	Knockdown option 1 – glyphosate (Group M)	N
2	Knockdown option 2 – paraquat + diquat (Group L)	N
3	2 knocks: glyphosate + paraquat + diquat (Groups M & L)	N
4	Trifluralin (Group D)	S
5	Simazine pre-emergence (Group C)	S
6	Atrazine pre-emergence (Group C)	S
7	Chlorsulfuron pre-emergence (Group B)	S
8	Use high crop seeding rate	B
9	Seed at first chance (default)	B
10	Tickle, wait 10 days, seed	B
11	Tickle, wait 20 days, seed	B
12	Simazine postemergence (Group C)	S
13	Atrazine postemergence (Group C)	S
14	Chlorsulfuron postemergence (Group B)	S
15	Diclofop methyl (Group A)	S
16	Fluazifop-butyl (Group A)	S
17	Clethodim (Group A)	S
18	Other Dim for lupins or canola (Group A)	S
19	Other selective herbicide	S
20	Grazing (selected automatically if pasture)	B
21	High intensity grazing winter/spring	B
22	Glyphosate top pasture (Group M)	N
23	Paraquat top lupins/pasture (Group L)	N
24	Green manure	B
25	Cut for hay, then glyphosate (Group M)	B
26	Cut for silage, then glyphosate (Group M)	B
27	Swathe	B
28	Mow pasture, then glyphosate (Group M)	B
29	User-defined option A (Spring)	B
30	Seed catch - burn dumps	B
31	Seed catch - total burn	B
32	Windrow - burn windrow	B
33	Windrow - total burn	B
34	Burn crop stubble or pasture residues	B
35	User-defined option B (at or after harvest)	B

[a] N = Nonselective herbicide, S = Selective herbicide, B = "Biological" treatment (nonchemical).

7.3.2 Results

7.3.2.1 Impact of Herbicide Resistance on Profit and Optimal Treatment Strategies

Consider a comparison between two otherwise similar fields of arable land that differ in their herbicide-resistance status. One has no resistance and, for illustrative purposes, is assumed to be immune from resistance development in future regardless of the pattern of herbicide use. The other has a severe herbicide-resistance problem. Table 7.2 shows a comparison of economic returns and optimal treatment strategies for the two cases. In deriving these results we note the following:

- It has been assumed that the ryegrass weeds in the second case are multiple-resistant to all "fop," "dim," triazine, and sulfonyl-urea herbicides.
- The sequence of crop types is the same for both cases — a lupin–wheat–wheat rotation that is widely practiced in the modeled region.
- The "optimal" treatment strategies have been identified by a process of trial and error. They may not be globally optimal, but they are at least close to this.

Several aspects of these results are notable.

- The onset of severe resistance results in substantial economic losses.
- After the onset of resistance, it is possible to maintain the continuous cropping rotation with a substantially altered weed management system.
- The altered system involves a greater diversity of treatment types. Each type is individually less effective than herbicides, so a greater number of treatments must be employed. The "with resistance" strategy is more strikingly consistent with what would be considered "Integrated Weed Management" (IWM).
- The "optimal" strategy with resistance present involves only a slightly greater average density of weeds. This is consistent with survey results that have found weed densities in farmers' fields with herbicide resistance are, on average, no greater than in nonresistant fields. Thus the economic difference in the scenarios is not primarily due to differences in weed density, but to differences in total treatment costs.

7.3.2.2 Benefits and Costs of Conserving the Herbicide Resource

Evidence for herbicide resistance in ryegrass in Western Australia is that use of herbicides on ryegrass is similar in nature to extraction of a nonrenewable resource. There is a limited stock of "herbicide susceptibility" and each usage of herbicide moves the farmer inexorably toward herbicide resistance, from which there is no return. Because resistant ryegrass plants are apparently no less fit than susceptible

TABLE 7.2
**RIM Model Results with and without Multiple Herbicide Resistance
for Lupin–Wheat–Wheat Crop Rotation**

	Without Resistance	With Resistance
Profit[a]	$114	$75
Treatment strategy	Simazine on first 3 lupin crops	Trifluralin on 6 of the first 7 crops
	Diclofop-methyl on most wheat crops	High seeding rates for all crops
	Clethodim on all lupin crops	Seeding delayed by 20 days for all
	High seeding rates for first 5 crops	crops, with glyphosate applied
	Seed catch all wheat crops	Paraquat top all lupin crops
	Total burn residues in first wheat	Windrow/burn windrow all lupins
	of each rotation, and burn dumps	Seed catch all wheat crops
	from seed catching for second wheat	Total burn residues in first wheat
		of each rotation, and burn dumps
		from seed catching for second wheat
Weed density[b]	3	13

[a] Australian dollars (A$), equivalent annual value over 20 years.
[b] Plants m^{-2}, average over 20 years.

ryegrass, there is no reason to expect that cessation of herbicide use will reverse the process of resistance development to any extent.

Given the high cost of resistance development (Table 7.1) the question arises whether it is economically preferable to preempt the onset of resistance by adopting a greater range of treatments at an early stage, thereby reducing the speed at which the stock of herbicide susceptibility is used up. Several aspects of this question must be considered.

Just because the cost of resistance is high, it does not necessarily follow that early adoption of IWM is economically preferred.

Factors in favor of early adoption of IWM include: (a) preserving herbicides would help to avert a major increase in weed numbers late in the planning period, and (b) if the farmer obtains experience and expertise with IWM prior to the onset of resistance, the risk of losing control of the weed population at the time of resistance onset is reduced.

Factors against early adoption include: (a) if the weed population density is low at the time resistance occurs, it is easier to maintain the weed population at low levels with the use of nonchemical treatments (e.g., increasing the crop seeding rate is more effective if the weed population is low), and (b) farmers can earn more interest from income that is earned sooner rather than later (i.e., the value of later benefits must be discounted relative to early benefits).

Existing evidence indicates that early adoption of IWM does not increase the total number of herbicide applications that are possible before resistance is fully developed. It just allows the farmer to spread out the fixed number of applications over a longer period.

Considering all these factors together, it appears unlikely that there would be a strong economic incentive either for or against early adoption of IWM. The results in Table 7.3 confirm that this is the case. Assumptions behind these results include the following:

- The allowed numbers of applications of herbicides from different groups prior to onset of resistance were "fops" and "dims," 4; sulfonyl ureas, 2; triazines, 4.
- The lupin–wheat–wheat rotation is maintained in both scenarios.
- In order to avoid a strategy involving neglect of weed management at the end of the planning period, a constraint is imposed that final weed seed density must not exceed the starting weed seed density.

TABLE 7.3
RIM Results with and without Early Adoption of Integrated Weed Management

	Early Adoption	No Early Adoption
Profit[a]	$85	$89
Treatment strategy	Available herbicide applications distributed approximately evenly over the 20-year period	Available herbicide applications used up in the first several years of the 20-year period
	Other treatments selected to provide the "optimal" strategy over the whole period	Other treatments selected to provide the "optimal" strategy over the whole period
Weed density[b]	3	13

[a] Australian dollars (A$), equivalent annual value over 20 years.
[b] Plants m^{-2}, average over 20 years.

One consideration mentioned earlier was not captured in these model runs: there may be reductions in risk if the farmer develops experience with IWM prior to onset of resistance onset. Were this factor to be properly valued, it is likely that there would be little difference in the economic values of early and late adoption of IWM.

This result should be considered illustrative only. It applies to a particular example and may not be transferable to others, depending on biology and economics. Nevertheless, it does illustrate the potential for exploitation, rather than conservation, of the herbicide resource to be the optimal long-term strategy for farmers.

7.3.2.3 Implications of Herbicide Resistance for Optimal Land Use

Results presented thus far have all been for a particular cropping rotation. Severe resistance raises the prospect that changes in land use may be beneficial. In particular, in Western Australia, it has been argued that inclusion of an occasional pasture phase can be of benefit to weed management by allowing a greater range of weed treatments, employing methods that do not rely on selective herbicides, such as grazing by livestock. These alternatives offer the opportunity to reduce the weed seedbank to very low levels. Of course, the economics of such a strategy depend not just on weed management considerations, but also on relative costs and returns of the alternative enterprises. Table 7.4 shows results for four additional strategies involving inclusion of pasture phases. They are directly comparable to the strategies shown in Table 7.3 except for the inclusion of the pasture phases. In each case, the pasture phases last for 3 years, which is considered the time necessary to reduce ryegrass seedbank densities substantially. The scenarios with two pasture phases include two groups of 3 years within the 20-year planning horizon.

TABLE 7.4
Profit[a] Estimated by the RIM Model Following Inclusion of Pasture Phases

Number of Pasture Phases	Early Adoption of IWM	No Early Adoption of IWM
0	$85	$89
1	$85	$86
2	$79	$82

[a] Australian dollars (A$), equivalent annual value over 20 years.

Given the costs and prices included in the model, profit from a strategy involving a single pasture phase is either similar to or slightly less than a strategy involving no pasture. Two pasture phases are less profitable for both scenarios regardless of the timing of IWM adoption.

Of course, these results are dependent on product sale prices. The main product of pastures in the farming system modeled is wool, harvested from merino sheep. The 1999 wool price used in the analysis was a relatively low level historically. If it were to increase, the economic attractiveness of pasture phases would rise accordingly.

7.3.2.4   Implications for Farmer Decision Making

The following observations highlight the implications of these modeling results for likely farmer decision making.

The management of herbicide-resistant weeds is complex, involving a greater range of more costly and unfamiliar weed treatments than farmers are initially used to. One would expect farmers to take some time to establish their preferred weed management strategy after the onset of resistance, following a period of trial and, inevitably, error.

There appears to be no compelling case for reducing the reliance on herbicides in order to delay the time when they will be lost to resistance. Even in cases where early adoption is in fact beneficial, it will be very difficult for farmers to determine that this is true. In the face of major uncertainties about such a change, and no compelling arguments in favor of it, most farmers are likely to maintain a more or less traditional, herbicide-based weed management system until forced to change by the full development of resistance. There is evidence that this is indeed what farmers tend to do.[e.g., 49]

7.4 POLICY IMPLICATIONS
OF HERBICIDE RESISTANCE

The threat of herbicide resistance, given its potentially severe economic consequences, raises the question of what government policies should be implemented in response. Economists recognize two morally respectable categories of possible reasons for governments to intervene in markets: market failure (efficiency related) and unacceptable distributional consequences (fairness related).

Market failure refers to several clearly defined circumstances where the operation of a free market may not lead to the most efficient outcome. Pannell[50] reviews the full range of market failures and outlines their relevance to weeds in general. Several types of market failure may be relevant to the problem of herbicide resistance.

7.4.1 EXTERNALITIES FROM SPREAD

Spread of resistance from farm to farm or from farm to the nonfarm environment may occur by several mechanisms, causing "externalities," costs external to the farmer who has the power to take action to reduce the risk of spread. Spread of resistant weed seeds can occur either naturally (e.g., by water, wind, or animals) or be carried by humans (e.g., in hay or as a contaminant in purchased seed). It is also possible for resistance to spread via pollen blown by wind[51] or carried by insect pollinators.[52] Potential government policies in response to this include regulation of seed sellers, or of movements of agricultural produce containing resistant seeds. In practice, in Australia, given that virtually all farmers are using herbicides, they are almost all developing resistance problems of their own, regardless of any imports. Thus it has been difficult to detect any cases of farm-to-farm spread, because they are masked (and rendered irrelevant) by on-farm resistance development. An exception to this is agricultural areas where animal grazing is the dominant enterprise, with the consequence that herbicides are used relatively infrequently. There are anecdotal stories of farmers using herbicide for the first time in a field and finding resistance, either because they sowed *Lolium rigidum* seed that was resistant (*Lolium* is a prized feed to animal producers[e.g., 53]) or imported resistant seed in hay. Similarly, resistant *L. rigidum* seed from Australia has been sold in Spain and introduced resistance problems (P. Boutsalis, personal communication, 1999).

For some cases, it is possible that externality problems from resistance spread may be important. Some of the issues associated with designing a mechanism to control problems with "tragedy of commons" due to spread are treated in the environmental economics literature.[54] In general, when resistance is a shared problem within a community, the solution requires some form of collective action. Some economists have designed policy mechanisms intended to provide incentives for such collective action to occur.

7.4.2   EXTERNALITIES FROM PRACTICES ADOPTED IN RESPONSE TO HERBICIDE RESISTANCE

Two weed control practices that are brought back into consideration by the onset of resistance are practices from which agriculture has been moving away: mechanical cultivation and burning.[1] Externalities are part of the reason for the loss of favor of these practices. This is particularly the case for cultivation, as concerns about erosion and soil health have driven a worldwide trend among crop producers in developed countries toward various versions of low- or no-tillage production systems.[55] Herbicides have been critical to the success of this trend, replacing previously high intensities of cultivation as the main method of weed control. Resistance clearly poses a threat to this trend. A potential government response is to ensure that farmers who do increase their level of cultivation are made to bear the burden of the external costs that result from greater movement of soils off the farm. Theoretically, an efficient way to do this is to impose a tax on soil loss. In practice, there are considerable difficulties in monitoring and evaluating the extent of such losses with sufficient accuracy to form the basis for imposing a tax. Economists recognize these difficulties as "transaction costs" and they are likely to be very high for most nonpoint sources of pollution, such as eroded soil particles. It may be necessary, then, to consider second-best approaches which are theoretically less efficient than a tax, but may be more efficient when transaction costs are considered. Possibilities include regulations to prohibit or limit use of certain farming practices, or subsidies to encourage particular practices.

7.4.3   UNCERTAINTY/INFORMATION FAILURES

Herbicide resistance is technically complex and difficult to observe until it has reached a high level, so misperceptions, ignorance, and high levels of uncertainty are the norm for farmers facing herbicide resistance for the first time. Even once resistance has occurred, farmers may find it relatively difficult to reach a full understanding of its biological nature and of the effectiveness of the many possible combinations of available treatment options. Rogers[18] identified trialability as a key characteristic affecting the speed and level of adoption of new technologies. Pannell[56] argued that the potential to learn from trials is reduced for "sustainability" problems such as herbicide resistance, relative to more traditional agricultural practices with quick response times. This suggests the need for public support for research and extension, which have been subjects of substantial investment in Australia since herbicide resistance became prominent. Government involvement in such research and extension may also be justified by its "public good" nature. Pannell[50] briefly describes the two types of public goods. Their most important feature in the current context is that part of the benefits resulting from the research and extension is of a nature that it cannot be traded in markets (consider the benefits from reduced soil particle loads in waterways).

7.4.4 DISTRIBUTIONAL CONSEQUENCES

Pannell[50] noted that

> There is nothing in economic theory which helps us to objectively evaluate the relative merits of different distributions of wealth,* although in practice redistribution is often one of the goals of government policy and action. In these cases the professional contribution of economists is limited to
>
> - quantifying the distributional effects of government action,
> - quantifying the impacts of government action on efficiency, and
> - assessing the performance of the action with regard to several rules of thumb which attempt to capture (or perhaps shape) community attitudes regarding what is fair and equitable.

It is therefore difficult to generalize about the implications of distributional impacts for policy formation, other than to observe that it further complicates the consideration of potential policy options. For example, the potential to apply a tax on soil loss to address external costs caused by herbicide resistance may be considered "fair" by some, since it allocates the cost to the person who is the source of that cost (i.e., it is consistent with the allocative rule of thumb called "polluter pays"). On the other hand, such a tax may be considered unfair, since it imposes an additional financial burden on a group of farmers who have already suffered financial losses from the development of herbicide resistance. The solution to this dilemma is necessarily a subjective judgment, made within a political context.

7.5 CONCLUSION

The existing literature on adoption of innovations by farmers provides a number of findings and insights that are highly relevant to farmers' responses to the herbicide resistance problem. We have noted that herbicide resistance has a number of characteristics that make it rather unlikely for farmers to rapidly adopt strategies that would delay the onset of full resistance. Rather, they seem more likely to wait for resistance to occur, and then adjust their management practices accordingly.

Using a detailed bioeconomic model, we have provided insights into the nature of these adjustments for a particular case study relating to the farming system, which provides the most extreme example of herbicide-resistance development in the world. The model confirms the expectation that, with resistance present, profits are lower and the optimal weed management system is more diversified. It includes a greater range of nonchemical treatments than is the case for farming systems without herbicide resistance. Details of the exact changes were provided, although in practice these will vary from case to case.

---

* In fact, one of the most celebrated theorems in economics, Arrow's Impossibility Theorem, purports to prove that it would be impossible for society to agree on a method for judging the merits of different distributions.

Finally we discussed a range of policy implications of herbicide resistance. It appears that the externality problem from spread is, in many cases, not as severe as for resistance to insecticides or fungicides, but it may be an issue in some cases. The problem of uncertainty and information failures was discussed as an issue to which governments might justifiably respond, and the distributional implications of resistance were also considered as a policy issue.

Clearly herbicide resistance poses very serious challenges not only to farmers and scientists, but also to extension agents, consultants, and policy makers. Sensible responses to many of these challenges will require a good appreciation of the social and economic issues that we have presented here.

REFERENCES

1. Schmidt, C. and Pannell, D. J., Economic issues in management of herbicide resistant weeds, *Rev. Market. Agric. Econ.,* 64, 301, 1996.
2. Schmidt, C. and Pannell, D. J., The role and value of herbicide-resistant lupins in Western Australian agriculture, *Crop Prot.,* 5, 539, 1996.
3. Gorddard, R. J., Pannell, D. J., and Hertzler, G. L., An optimal control model of integrated weed management under herbicide resistance, *Aust. J. Agric. Econ.,* 39, 71, 1995.
4. Gorddard, R. J., Pannell, D. J., and Hertzler, G. L., Economic evaluation of strategies for management of herbicide resistance, *Agric. Syst.,* 51, 281, 1996.
5. Orson, J., The cost to the farmer of herbicide resistance, *Weed Technol.,* 13, 607, 1999.
6. Hueth, D. and Regev, U., Optimal agricultural pest management with increasing pest resistance, *Am. J. Agric. Econ.,* 56(3), 543, 1974.
7. Taylor, C. R. and Headley, J. C., Insecticide resistance and the evaluation of control strategies for an insect population, *Can. Entomol.,* 107(3), 237, 1975.
8. Moffitt, L. J. and Farnsworth, R. L., Bioeconomic analysis of pesticide demand, *Agric. Econ. Res.,* 33(4), 12, 1981.
9. Regev, U., Shalit, H., and Gutierrez, A. P., On the optimal allocation of pesticides with increasing resistance: the case of alfalfa weevil, *J. Environ. Econ. Manage.,* 10, 86, 1983.
10. Knight, A. L. and Norton, G. W., Economics of agricultural pesticide resistance in arthropods, *Annu. Rev. Entomol.,* 34, 293, 1989.
11. Clark, S. J. and Carlson, G. A., Testing for common versus private property: the case of pesticide resistance, *J. Environ. Econ. Manage.,* 19(1), 45, 1990.
12. Peck, S. L. and Ellner, S. P., The effect of economic thresholds and life-history parameters on the evolution of pesticide resistance in a regional setting, *Am. Nat.,* 149(1), 43, 1997.
13. Smale, M., Singh, R. P., and Dubin, H. J., Estimating the economic impact of breeding nonspecific resistance to leaf rust in modern bread wheats, *Plant Dis.,* 82(9), 1055, 1998.
14. Hayami, Y. and Ruttan, V. W., Factor prices and technical change in agricultural development: the United States and Japan 1880–1960, *J. Polit. Econ.,* 78, 1115, 1970.
15. Binswanger, H. P., A microeconomic approach to induced innovation, *Econ. J.,* 84, 940, 1974.

16. Marland, G., McCarl, B., and Schneider, U., Soil carbon: policy and economics, in *Carbon Sequestration in Soils: Science, Monitoring, and Beyond,* Rosenberg, N. J., Izaurralde, R. C., and Malone, E. L., Eds., Battelle Press, Columbus, OH, 1999, 153.

17. Miranowski, J. A. and Carlson, G. A., Agricultural resource economics: an overview, in *Agricultural and Environmental Resource Economics*, Carlson, G. A., Zilberman, D., and Miranowski, J. A., Eds., Oxford University Press, New York, 1993, 3.

18. Rogers, E. M., *Diffusion of Innovations*, Free Press, New York, 1995.

19. Griliches, Z., Hybrid corn: an exploration in the economics of technological change, *Econometrica,* 25, 501, 1957.

20. Vanclay, F., The social basis of environmental management in agriculture: a background for understanding landcare, in *Critical Landcare*, Key Papers Series 5, Lockie, S. and Vanclay, F., Eds., Centre for Rural Social Research, Charles Sturt University, Wagga Wagga, 1997, 9.

21. Lindner, R. K., Adoption and diffusion of technology: an overview, in *Technological Change in Postharvest Handling and Transportation of Grains in the Humid Tropics*, Champ, B. R., Highley, E., and Remenyi, J. V., Eds., ACIAR Proceedings No. 19, 1987, 144.

22. David, P. A., A contribution to the theory of diffusion, *Stanford Center for Research in Economic Growth,* Memorandum No. 7, 1969.

23. Stoneman, P., Intra firm diffusion, Bayesian learning and profitability, *Econ. J.,* 91, 375, 1981.

24. Olmstead, A. L. and Rhode, R., Induced innovation in American agriculture: a reconsideration, *J. Polit. Econ.,* 101, 100, 1993.

25. Abadi Ghadim, A. K. and Pannell, D. J., A conceptual framework of adoption of an agricultural innovation, *Agric. Econ.,* 21, 145, 1999.

26. Caswell, M. F. and Zilberman, D., The effects of well depth and land quality on the choice of irrigation technology, *Am. Agric. Econ.,* 68, 798, 1986.

27. Green, G. P., Technology Adoption and Water Management in Irrigated Agriculture, Ph.D. thesis, Agricultural and Resource Economics, University of California, Berkeley, 1995.

28. Wu, J. and Babcock, B. A., The choice of tillage, rotation, and soil testing practices: economic and environmental implications, *Am. J. Agric. Econ.,* 80, 494, 1998.

29. Carlson, G. A. and Wetzstein, M. E., Firm decisions and behavior in pest management on a regional level, in *Agricultural and Environmental Resource Economics,* Carlson, G. A., Zilberman, D., and Miranowski, J. A., Eds., Oxford University Press, Oxford, 1993, 273.

30. Helms, G. L., Bailey, D., and Glover, T. F., Government programs and adoption of conservation tillage practices on nonirrigated wheat farms, *Am. J. Agric. Econ.,* 69, 786, 1987.

31. Miranowski, J. A., Hrubovak, J., and Sutton, J., The effects of commodity programs on resource use, in *Commodity and Resource Policies in Agricultural Systems,* Bockstael, N. and Just, R., Eds., Springer-Verlag, New York, 1991.

32. de Gorter, H. and Fisher, E. O., The dynamic effects of agricultural subsidies in the United States, *J. Agric. Res. Econ.,* 18, 147, 1993.

33. Uri, N. D., Conservation tillage and the use of energy and other inputs in U.S. agriculture, *Ener. Econ.,* 20, 389, 1998.

34. Bar-Shira, Z., Just, R., and Zilberman, D., Estimation of farmers' risk attitude: an econometric approach, *Agric. Econ.,* 17, 211, 1997.

35. Wiebers, U. C., Economic and Environmental Effects of Pest Management Information and Pesticides: The Case of Processing Tomatoes in California, Ph.D. dissertation, Technische Universität Berlin, Fachbereich Internationale Agrarentwicklung, Institut für Agrarbetriebs- und Standortsökonomie, 1992.
36. Green, G., Sunding, D., Zilberman, D., Parker, D., Trotter, C., and Collup, S., How does water price influence irrigation technology adoption? *Calif. Agric.*, 50, 36, 1996.
37. Cochrane, W. W., *The Development of American Agriculture: A Historical Analysis*, University of Minnesota Press, Minneapolis, 1979.
38. Putler, D. S. and Zilberman, D., Computer use in agriculture: evidence from Tulare County, California, *Am. J. Agric. Econ.*, 70, 790, 1988.
39. Huffman, W. E., Decision making: the role of education, *Am. J. Agric. Econ.*, 56, 85, 1974.
40. Wolf, S. A., Privatization of Crop Production Information Service Markets: Spatial Variation and Policy Implications, Ph.D. thesis, University of Wisconsin-Madison, 1996.
41. Zilberman, D., Sunding, S., Dobler, M., Campbell, M., and Manale, A., Who makes pesticide use decisions: implications for policymakers, in *Pesticide Use and Product Quality*, Armbruster, W., Ed., Farm Foundation, Glenbrook, IL, 1994, 23.
42. Kotler, P., *Marketing Management: Analysis, Planning, Implementation, and Control*, 9th ed., Prentice-Hall, Upper Saddle River, NJ, 1997, 789.
43. Carlson, G. A. and Wetzstein, M. E., Pesticides and pest management, in *Agricultural and Environmental Resource Economics*, Carlson, G. A., Zilberman, D., and Miranowski, J. A., Eds., Oxford University Press, New York, 1993, 268.
44. Tardif, F. J., Preston, C., and Powles, S. B., Mechanisms of herbicide multiple resistance in *Lolium rigidum*, in *Weed and Crop Resistance to Herbicides*, Kluwer, Dordrecht, the Netherlands, 1997.
45. Boerboom, C. M., Nonchemical options for delaying weed resistance to herbicides in Midwest cropping systems, *Weed Technol.*, 13, 636, 1999.
46. Nalewaja, J. D., Cultural practices for weed resistance management, *Weed Technol.*, 13, 643, 1999.
47. Stewart, V. A. M., Economics of Integrated Strategies for Managing Herbicide Resistant Ryegrass under Continuous Cropping, Undergraduate dissertation, Faculty of Agriculture, University of Western Australia, Nedlands, 1993.
48. Pannell, D. J., Stewart, V. A. M., Bennett, A. L., Monjardino, M., Schmidt, C. P., and Powles, S. B., RIM 99 User's Manual, A Decision Tool for Integrated Management of Herbicide-Resistant Annual Ryegrass, University of Western Australia, Nedlands, 1999.
49. Peterson, D. E., The impact of herbicide-resistant weeds on Kansas agriculture, *Weed Technol.*, 13, 632, 1999.
50. Pannell, D. J., Economic justifications for government involvement in weed management: a catalogue of market failures, *Plant Prot. Q.*, 9(4), 131, 1994.
51. Timmons, A. M., O'Brien, E. T., Charters, Y. M., Dubbels, S. J., and Wilkinson, M. J., Assessing the risks of wind pollination from fields of genetically modified *Brassica napus*, *Euphytica*, 85, 417, 1995.
52. Scheffler, J., Parkinson, R., and Dale, P. J., Frequence and distance of pollen dispersal from transgenic oilseed rape (*Brassica napus*), *Transgenet. Res.*, 2, 356, 1993.
53. Abadi Ghadim, A. K. and Pannell, D. J., The economic trade-off between pasture production and crop weed control, *Agric. Syst.*, 36(1), 1, 1991.

54. Segerson, K., Flexible incentives: a unifying framework for policy analysis, in *Flexible Incentives for the Adoption of Environmental Technologies in Agriculture*, Casey, F., Schmitz, A., Swinton, S., and Zilberman, D., Eds., Kluwer, Boston, 1999, 79.

55. Cornish, P. S. and Pratley, J. E., Eds., *Tillage: New Directions in Australian Agriculture,* Inkata Press, Melbourne, 1987.

56. Pannell, D. J., Uncertainty and Adoption of Sustainable Farming Systems, Paper presented at the 43rd Annual Conference of the Australian Agricultural and Resource Economics Society, Christchurch, New Zealand, January 20–22, 1999.

8 Regulatory Aspects of Resistance Management for Herbicides and Other Crop Protection Products

Dale L. Shaner and Paul Leonard

CONTENTS

0-8493-2219-7/01/$0.00+$.50
© 2001 by CRC Press LLC

8.1   INTRODUCTION

Countries regulate the use of herbicides to ensure that they are applied properly and safely with minimal risk to the environment. These regulations include how the herbicides will be applied, where they will be applied, and at what rates they will be used. These regulations may have a profound impact on the effectiveness of different resistant management strategies, particularly if the regulations prohibit certain practices such as mixtures, frequency of application, etc. Resistance management has always been a consideration in the registration of herbicides, but it has not been a major aspect of the registration or labeling process. However, in the 1990s regulators increased their attention to resistance management of herbicides as the cases of resistance increased. New labeling and registration requirements that deal specifically with the risk of selecting for resistance are being discussed and new laws may be implemented in the near future. It is important to understand how these new laws will impact resistance management. This chapter addresses the history of the regulation of herbicides in terms of resistance management and how newly proposed changes may affect resistance practices in Europe and the United States.

8.2   RESISTANCE REGULATION IN EUROPE

8.2.1   BACKGROUND OF THE LEGISLATION AND INVOLVEMENT OF THE EUROPEAN AND MEDITERRANEAN PLANT PROTECTION ORGANIZATION

8.2.1.1   Council Directive 91/414

The European Union Council Directive 91/414/EEC, "concerning the placing of plant protection products on the market," aimed to establish a harmonized framework for the registration of crop protection products in the European Union. The directive requires that applicants who wish to register plant protection products evaluate the risk of resistance developing and propose management strategies to address such risks (Commission Directive 93/71/EEC of 27 July 1993 amending Council Directive 91/414/EEC). At the time Council Directive 91/414 was drafted, a guideline for addressing resistance risk and resistance management did not exist. However, Commission Directive 93/71/EEC indicated that European Plant Protection Organization (EPPO) guidelines should be used for fulfilling efficacy data requirements. The Dutch Plant Protection Service (PPS), concerned that herbicide resistance could have a significant impact on horticulture in the Netherlands, decided to address this anomaly.[1]

8.2.1.1.1   Dutch Plant Protection Service resistance initiative

At the time the Dutch PPS started work on guideline documentation, a number of national authorities in Europe were already concerned about the lack of regulatory control of pest "susceptibility." This notion was based on the realization that the susceptibility of naturally occurring pest populations to crop protection products is a valuable and often nonrenewable natural resource. The PPS regarded the need to manage susceptibility as particularly acute for the Dutch horticultural industry where high value crops, such as ornamentals and flower bulbs, constitute important components of the agricultural economy.

In order to market produce internationally and to gain access to premium prices, growers are forced to achieve cosmetically clean crops with no visible signs of pest infestation. In addition, high levels of phytosanitary control must be maintained to meet the demands of export markets. Crop protection products are relied on to a great extent, especially in the production of high value ornamentals. However, due to the limited size of the market and increasingly stringent regulatory requirements, the number of new active ingredients entering the market is declining. At the same time, as the availability of new active ingredients is on the decline, the diversity of existing crop protection products is being eroded by the de-registration of products as a result of national re-registration programs.

The Dutch PPS concluded that, when combined, these factors produce a scenario of greater reliance on fewer crop protection compounds. They saw this as a significant threat to the long-term viability of the Dutch horticultural industry.

In response, the PPS drafted a resistance risk evaluation scheme in line with Council Directive 911/414 EEC, to evaluate resistance risk and to prescribe management strategies. A copy of this draft was submitted to the Netherlands National Crop Protection Association (NEPHYTO) for consultation. NEPHYTO in turn referred the matter to the Global Crop Protection Federation (GCPF) resistance action committees (Insecticide Resistance Action Committee [IRAC], Fungicide Resistance Action Committee [FRAC], and the Herbicide Resistance Action Committee [HRAC]).

Industry's initial response was cautious, as such a prescriptive approach would not be applicable in all situations and could be used out of context, resulting in denied or delayed registrations. However, it was recognized that the PPS shared the objective of maintaining the activity of valuable crop protection products. The Resistance Action Committees (RACs) therefore welcomed the chance to work with the PPS to develop and improve the draft decision-making scheme. A series of productive meetings resulted in a final draft that industry recognized as much improved[2] although it still caused some concern.

Overall, the scheme offered the great advantage of making registration decisions more objective, but industry remained concerned that there was potential for misinterpretation by regulators in countries with a limited knowledge of resistance management. If used out of context, registrations could be unnecessarily denied or delayed.

8.2.1.2 Development of the European and Mediterranean Plant Protection Organization's Resistance Risk Analysis Guideline

Although originally developed for the Netherlands, the PPS submitted the draft decision-making scheme to EPPO for possible adoption in the light of alternative guidance of how to meet the data requirements laid down in Council Directive 91/414. After careful consideration, EPPO chose not to adopt the PPS proposal, but it indicated that the proposal had real merit and would be adopted if a more appropriate guideline were not made available.

In 1997 the RACs set out to develop a resistance risk evaluation guideline that was based on the normal practice of responsible crop protection companies. Additionally, the RACs determined to develop a guideline that was practical, achievable, and relevant to the agricultural situation. A key feature would be to place responsibility for evaluating risk and proposing resistance management strategies on the registration applicant. All three committees worked on several drafts before a final draft was submitted to EPPO in December 1997. This submission included a common introduction and separate sections for fungicide, herbicide, and insecticide resistance risk evaluation. The three sections shared a common format but reflected the practicalities and differences of the three disciplines. EPPO evaluated the proposal alongside relevant documents from the U.K.'s Pesticide Safety Directorate and the Dutch PPS decision-making scheme.

In June 1998, EPPO drafted a guideline that was based on the RAC proposal but amalgamated the fungicide, insecticide, and herbicide sections to produce a uniform approach. This document provided the starting point for EPPO's resistance risk assessment guideline. EPPO worked with the crop protection industry, represented by the RACs and experts from France, the U.K., the Netherlands, Switzerland, and Germany to further develop the guideline. While not advocating legislation for resistance management, the RACs welcomed the opportunity to contribute to the development of the guideline, as it would ultimately be the industry that would have to adopt its recommendations.

8.2.2 Current Status of the EPPO Guideline

EPPO circulated a final draft of the resistance risk analysis guideline to member countries for approval in May 1999. The draft was formally adopted with minor changes at the EPPO Working Group meeting in May 1999. The guideline was published in the EPPO Bulletin during the first quarter of 2000.[3]

8.2.3 KEY FEATURES OF THE EUROPEAN AND MEDITERRANEAN PLANT PROTECTION ORGANIZATION'S DRAFT RESISTANCE RISK ANALYSIS GUIDELINE

8.2.3.1 Objective

The draft EPPO guideline communicates to both registration authorities and applicants their obligations for assessing resistance risk and developing suitable management strategies.

The guideline presents concepts of resistance risk analysis, evaluation, management, and regulatory decision making; information on the nature and development of practical resistance; and ways in which resistance risk may be reduced to acceptable levels.

8.2.3.2 Scope of the Guideline — When Is It Relevant?

An important aspect of the guideline is that it focuses exclusively on "practical resistance." Resistance is defined as "the naturally occurring, inheritable adjustment in the ability of individuals in a population to survive a plant protection product treatment that would normally give effective control." This is a term further used to describe "loss of field control due to a shift in sensitivity." The EPPO draft resistance risks assessment guideline limits its scope to that of changes in sensitivity which may occur under agricultural conditions and which would have a measurable impact on field performance. The guideline is therefore not intended to encompass areas of resistance research which are not of direct relevance to practical field performance of crop protection products.

8.2.4 RESISTANCE RISK ANALYSIS, ASSESSMENT, MANAGEMENT, AND DECISION MAKING

The guideline provides information on the basic concepts of resistance. Terms used to describe various aspects of resistance research, measurement, and management are described. In order to provide structure, resistance risk evaluation is broken up into several distinct processes; risk analysis, risk assessment, risk management, and decision making are the most important.

8.2.4.1 Risk Analysis

Resistance risk analysis is described as an "iterative two-stage process." The process is composed of two separate phases. The first phase is to evaluate the risk, or probability of resistance developing. The second phase is to describe the impact such a resistance development may have on use of other compounds, chemistries, and agronomic practice. The guideline provides information on how to negotiate both of these phases of the risk assessment process.

8.2.4.2   Factors Considered

The applicant is required to take into consideration factors that are inherent to the crop protection product and its effect on the pest. Such inherent factors include product persistence kinetics, mode of action, pest life cycle, and cross-resistance. The guideline also requires consideration of the impact that agronomic practice may have on resistance development. These factors include the diversity of control measures available and the number of applications that would normally be required to control target pests.

8.2.4.3   Modifiers and Management Strategies

If the risk of resistance development is considered unacceptable, when a plant protection product is used according to the label but without consideration of the need to reduce selection pressure, measures must be taken to reduce this risk to acceptable levels. These measures are termed "modifiers." The process of using modifiers to reduce risk is termed "resistance management." Resistance management may take many forms. Most often these strategies involve restricting application frequency, mixing with or rotating applications with other chemistries. These management strategies share the same objective: to reduce selection pressure of the specific crop protection product to an acceptable and sustainable level.

8.2.5   Data Requirements

The EPPO draft guideline makes specific reference to the information that applicants are required to provide in order for regulatory authorities to evaluate whether the intended use pattern represents an acceptable resistance risk. In this way, the guideline places responsibility for evaluating and addressing resistance risk on the applicant and responsibility for determining acceptability of the resultant risk on regulatory authorities.

The guideline sets out what information is required regarding the modes of action, mechanisms of known resistance, evidence of related resistance from other parts of the world, cross-resistance, and methods for monitoring resistance.

8.2.6   Monitoring of Key Species

During the early stages of development the guideline required applicants to monitor susceptibility before and after registration to all pests identified on the label. While there was general agreement that this would be of theoretical value, it was quickly realized that such an all-encompassing objective would not be achievable. The draft guideline therefore focuses attention on key species where there is reason to believe the risk of resistance development is relatively great and where validated monitoring methods are available. In cases where suitable methods are not available, information from efficacy dossiers can be used as a practical indication of susceptibility at the time of submission. In order to make this relevant the applicant should present efficacy information in this context.

Following registration, the draft guideline indicates that the continued efficacy of the product should be monitored to establish whether additional management strategies may be required. However, it is recognized that this is not always possible or practical. Registration holders are therefore required to report any unexpected changes in susceptibility to registration authorities.

8.2.7 REGULATORY DECISION MAKING

Applicants for registration are required to evaluate the risk of resistance and if necessary to propose management strategies. Registration authorities are charged with the responsibility of determining whether the proposed use pattern represents acceptable levels of resistance risk. Regulatory authorities therefore need to establish if a suitable assessment has been performed and if an appropriate management process is proposed. If, accounting for modifiers proposed by the registration applicant, the risk of resistance is deemed too great, registration of that use should not be granted. In this event, provision is made for the registration applicant to propose additional modifiers in order to reduce the risk of resistance development to an acceptable level.

8.2.8 "REALITY CHECK"

In cases where registrations are not granted because of an unacceptable resistance risk, the guideline makes provision for one final "reality check." In order to avoid the unreasonable restriction of new or existing crop protection products, registration authorities are required to determine whether the proposed or modified use is consistent with that of other commercially available products with the same resistance mechanism. In cases where the proposed registration decision would restrict use which is not in context with other products having the same resistance mechanism, registration authorities are required to consider how best to resolve such an inconsistency. Consultation with the applicant, and if necessary with independent experts, is encouraged.

8.2.9 IMPLICATIONS OF THE EPPO GUIDELINE
FOR CROP PROTECTION PRODUCT DEVELOPMENT

Once published, EPPO's resistance risk analysis guideline will mark the start of a new era in resistance management in countries where EPPO guidelines are used.

From the crop protection industry's perspective, the implementation of EPPO's resistance risk assessment guideline means that companies will need to analyze resistance risk and modify use patterns or develop management strategies at an early stage in product development. Practically speaking, this means that resistance risk assessment must be performed as soon as a decision is made to develop crop protection products.

Resistance risk evaluation will become an integral part of the registration decision-making process. Registration authorities in EPPO countries will be required to consider whether proposed uses are of acceptable resistance risk. If they are not acceptable, then it must be expected that registrations will not be granted, even if in all other aspects the applications would normally be considered acceptable.

Registration dossiers must be developed in line with good agricultural practices (GAPs) that have acceptable levels of resistance risk. This approach will have little impact on responsible manufacturers taking a long-term view to protect an increasingly significant investment in research and development. The guideline will have more significant consequences for organizations with a shorter-term focus and which fail to address adequately the threat of resistance development.

Implementation of EPPO's resistance risk assessment guideline will have a profound impact on organizations that fail to take account of its recommendations suitably early in the development process.

If a crop protection company fails to develop registration dossiers that reflect GAPs with acceptable levels of resistance risk, registrations will ultimately be refused. To achieve registrations in these situations, applicants may have to modify GAPs in order to reduce potential selection pressure. This action may in itself create significant regulatory problems for applicants, as proposed use patterns may no longer be consistent with GAPs on which the rest of the dossier is based. For example, if a proposal to use a fungicide for a maximum of eight times a season is rejected on grounds of resistance risk, applications may need to be restricted to a maximum of four in order to present an acceptable risk. If this hypothetical situation were to occur, residue studies that had been conducted to support eight applications would no longer be relevant to the supported GAP. As a result, these would need to be repeated and MRL would have to be modified accordingly. There is therefore a powerful incentive for registration applicants to analyze resistance risk early in the development process and to develop regulatory dossiers that are consistent with use patterns having acceptable resistance risk.

8.2.10 IMPACT OF EUROPEAN REGULATIONS ON RESISTANCE

The value that susceptibility in pest populations represents has been recognized in Europe with the adoption of Council Directive 91/414EEC. This, coupled with the recent development of EPPO's guideline, represents a significant step forward in addressing the need to maintain the performance of valuable crop protection products.

The development of the EPPO resistance risk analysis guideline was made possible by collaboration of many experts from member states and the crop protection industry. Development of the guideline was a challenging process to which many people contributed directly or indirectly over a period of several years. EPPO's recently published guideline is now recognized in other parts of the world as a benchmark in the continuing struggle against pest resistance.

Resistance remains a specialist area of research. Relatively few practitioners are familiar with the process of resistance risk evaluation and development of resistance management strategies. With the development and ultimate implementation of EPPO's resistance risk assessment guideline, there is likely to be a great need for education in this area.

In conclusion, those who apply for registration of new and existing crop protection products in Europe and other EPPO countries need to evaluate resistance risk and, where necessary, modify use at an early stage in product development.

8.3 RESISTANCE REGULATION IN THE UNITED STATES

8.3.1 GENERAL REGISTRATION REQUIREMENTS AND RESISTANCE MANAGEMENT

In the United States, regulation of herbicides falls under three major statutory authorities: the Federal Insecticide, Fungicide and Rodenticide Act (FIFRA); the Federal Food Drug and Cosmetic Act (FFDCA); and the Food Quality Protection Act (FQPA). These acts are administered by the U.S. Environmental Protection Agency (USEPA) which has the authority to regulate the development, sale, distribution, use, storage, and disposal of herbicides. One requirement for registration under FIFRA is that a herbicide will not cause "unreasonable adverse effects" to human health or the environment. To determine whether this requirement is satisfied, the USEPA considers "the economic, social, and environmental costs and benefits" of the use of the herbicide.[4]

The USEPA has considered the development of herbicide resistance in its regulatory decisions, but it does not have an official policy or standard data requirement in place.[5] The stance of the USEPA is that the development and spread of resistance is generally associated with the frequency and rate of application of a herbicide. The selection of resistance to a herbicide can have serious consequences if farmers can no longer use a particular herbicide to control one or more pests and they are forced to use a herbicide that poses higher risks to humans and the environment. Thus, preventing or managing resistance is in both the user's and public's interest.[5]

The USEPA considers herbicide resistance under a number of sections of FIFRA, including Section 3 (registration decisions); Section 6 (special review decisions based on unreasonable human health and/or environmental risks); and Section 18 (emergency exemption decisions).[4] Pesticide resistance had been a major factor in granting emergency exemptions under Section 18. In 1995 more than 30% of the requests for Section 18 exemptions involved resistant pest populations upon which registered crop protection chemicals were no longer effective. In two cases Section 18 exemptions were granted due to herbicide resistance. One was for lactofen to control paraquat- and diquat-resistant nightshades (*Solanum* spp.) in tomatoes and peppers in Florida and the other was for quinclorac to control propanil-resistant barnyard grass (*Echinochloa crus-galli*) in rice in Arkansas.[6]

The USEPA has also considered pesticide resistance when determining whether an unreasonable adverse effect would occur if registered uses of a pesticide are cancelled as part of a Special Review process (Section 6). Although this has not involved a herbicide to date, pesticide resistance was a primary consideration in assessing the benefits of the continued use of the ethylene bisdithiocarbamate fungicides (e.g., mancozeb, maneb, metriam, and nabam). There are no reports of resistance to this class of fungicides and they are widely used in fruit and vegetable crops. Had these fungicides been discontinued, a major component of fungicide resistance management would have been lost. Hence, the uses of these fungicides on a number of commodities were maintained because of the benefits in fungicide resistance management.[6]

8.3.2   RESISTANCE MANAGEMENT LABEL STATEMENTS

Herbicide resistance has not generally been a consideration of the USEPA in determining whether a new active ingredient should be registered. However, beginning in the late 1980s, there were some specific cases where language was included on the label advising users on ways to avoid or delay the onset of resistance. For example, on the label for Amber®* (triasulfuron) for use in wheat, a "General Information" section on "Weed Resistance to Sulfonylurea Herbicides" indicates that certain biotypes of several weed species are not controlled by sulfonyureas. This section recommends tank mixes with other herbicides with different mechanisms of action in fields where resistance is a problem. Another statement reads: "Do not apply Amber or other herbicides with the same mode of action within a 12-month period after an Amber application...If additional weed control is needed, use a herbicide with a different mode of action from Amber."[7]

8.3.3   GENETICALLY MODIFIED ORGANISMS

In the early 1990s the USEPA began focusing increased attention on pesticide resistance management when the first crops containing *Bacillus thuringiensis* (Bt) toxin were being registered. In 1992 the assistant administrator of the USEPA requested that the Office of Pesticide Programs form a working group to address the potential of developing resistance to Bt toxins. This led to the formation of the Pesticide Resistance Management Workgroup (PRMW), which consists of plant pathologists, microbiologists, entomologists, weed scientists, biologists, and biochemists. This group was charged with considering the role that USEPA plays in resistance management of conventional, biological, and genetically engineered plant crop protection chemicals.[8]

The PRMW group identified seven elements that must be addressed for an adequate resistance management plan: (1) knowledge of pest biology and ecology; (2) appropriate gene deployment strategy; (3) appropriate refugia (primarily for insecticides); (4) monitoring and reporting of incidents of pesticide resistance development; (5) employment of integrated pest management; (6) communication and educational strategies on use of the product; and (7) development of alternative modes of action.[8] These factors were endorsed in 1995 by a subpanel of the FIFRA Scientific Advisory Panel.

These seven factors have been used to help define and establish the resistance management plans that have been recommended for use with Bt crops.[9] The registration of all Bt corn and Bt cotton mandated specific resistance management data requirements and mitigation measures. These registrations were conditional to allow for completion of studies to determine the effectiveness of the resistance management

---

* Registered Trademark of Novartis, Greensboro, North Carolina.

strategies. When these conditional registrations expire (April 1, 2001 for Bt corn and January 1, 2001 for Bt cotton) the USEPA will have re-evaluated the effectiveness of each registrant's resistance management plans and decided whether to convert the registration to a nonexpiring registration. The USEPA's experience with the registration of Bt crops will likely guide how it will handle similar registrations in the future.

8.3.4 LABELING HERBICIDES BY MODE OF ACTION

The first cases of widespread herbicide-resistant populations began to appear in the late 1970s and the number has increased relatively constantly up to the present day.[10] Resistance to triazine herbicides accounted for most of the early cases. However, beginning in the mid-1980s, resistance to ACCase inhibitors and ALS inhibitors began to appear. In certain cropping situations, such as cereals in the United States, Canada, and Australia, resistance to these two classes of herbicides became a serious problem. To combat this situation, efforts to educate the farmer on resistance management intensified. One of the most important methods for preventing, delaying, or managing resistance is to reduce the reliance on a single herbicide mode of action. To do this, farmers must know which herbicides share the same mode of action, but the relatively complex nature of plant biochemistry makes this difficult to determine. Attempts were made to simplify the understanding of herbicide mode of action by developing a classification system that grouped herbicides by mode of action. Eventually HRAC and the Weed Science Society of America (WSSA) developed a classification system.[11,12] A mandatory classification system was implemented in Australia in 1992.

The next step in this educational process was to introduce this system to the farmers. This was done in Australia and Canada through extensive and intensive educational programs.[13] In Canada, the herbicide group classifications have been used in crop protection guides in Saskatchewan and Manitoba since 1993 and in Alberta since 1997.[14] Thus far, only in Australia is it mandatory for herbicide labels to state the mode of action of the herbicide based on the classification system. The mode of action identification symbol is positioned on the main panel immediately beneath the active ingredient statement. If a product contains two or more active ingredients with different modes of action, then classifications for all modes of action are shown on the label.[15]

In 1996 the Pest Management Regulatory Agency (PMRA) of the Government of Canada developed voluntary guidelines concerning pesticide-resistance management labeling for herbicides, fungicides, and insecticides. These guidelines include the mode of action classification of the pesticide on the label plus standardized resistance management statements in the use directions.[16] These guidelines were to be implemented in Canada in 1997; however, the initiative was expanded into a joint project of Canada, the United States, and Mexico under the auspices of the North America Free Trade Agreement (NAFTA). This initiative was made part of the work of the Risk Reduction Subcommittee of the NAFTA Technical Working Group on Crop Protection Chemicals. The goal is to implement a uniform approach across North America to help reduce the development of pesticide resistance and to achieve consistency in resistance-management labeling across NAFTA.[6] On October 6, 1999 the PMRA issued

Regulatory Directive DIR99-06 on "Voluntary Pesticide Resistance-Management Labeling Based on Target Site/Mode of Action." This document replaced Regulatory Proposal Pro96-03 "Pesticide Resistance Management Labeling" issued in December 1996.[16] The USEPA plans to issue a similar directive. A draft was published for public comment in May 2000.[17] DIR99-06 covers both new products and old products that are being re-evaluated. The guidelines are voluntary and cover all crop protection chemicals. The goal is to include the resistance-management grouping symbols and statements on the labels of all new and existing products by January 1, 2004.

In this directive, resistance is defined as a heritable and significant decrease in the sensitivity of a pest population that reduces the field performance of the pesticide. The premise of the directive is that an important proactive resistance management strategy is to avoid the repeated use of a particular pesticide or crop protection chemicals sharing the same mode of action in the same field by rotating crop protection chemicals with different modes of action. The directive also acknowledges that this approach does not address resistance that develops due to nontarget-site mechanisms such as enhanced metabolism, decreased penetration, or behavior changes. However, the goal is to provide users with easy access to information regarding target-site resistance. The site of action grouping and identification symbol for herbicides is based on the WSSA/HRAC classification system.[16]

The mode of action identification should appear on all end-use product labels in the upper right corner of the front panel surrounded by a black rectangle (Figure 8.1). If a product contains more than one active ingredient with two or more modes of actions, these will appear together on the label. The label will also include resistance management statements in the use directions under the heading "Resistance Management Recommendations" in the "General" portion of the "Use Directions" in the U.S. and under "Use Directions" in Canada. DIR99-06 recommends product-specific labeling and that these recommendations be included in any product-specific literature.

Suggested text for the resistance-management statement is

> For resistance management, (name of product) is a Group (site of action group number) herbicide. Any weed population may contain or develop plants naturally resistant to (name of product) and other Group (site of action group number) herbicides. The resistant biotypes may dominate the weed population if these herbicides are used repeatedly in the same field. Other resistance mechanisms that are not linked to site of action, but specific for individual chemicals, such as enhanced metabolism, may also exist. Appropriate resistance-management strategies should be followed.

In addition, there is suggested language on what practices could be followed. These include the following:

- Where possible, rotate the use of (name of product) or other Group (site of action group number) herbicides with different herbicide groups that control the same weeds in a field.

GROUP   █████████ 1 █████████   HERBICIDE

FIGURE 8.1 Example of proposed mode of action identification for herbicide labels in the United States and Canada.

- Use tank mixtures with herbicides from a different group when such use is permitted.
- Herbicide use should be based on an IPM program that includes scouting and historical information related to herbicide use and crop rotation, and that considers tillage (or other mechanical), cultural, biological, and other chemical control practices.
- Monitor treated weed populations for resistance development.
- Prevent movement of resistant weed seeds to other fields by cleaning harvesting and tillage equipment and planting clean seed.
- Contact your local extension specialist or certified crop advisors for any additional pesticide-resistance management and/or integrated weed-management recommendations for specific crops and weed biotypes.
- For further information or to report suspected resistance, contact (company representatives) at (toll free number) or at (Internet site).

8.3.4.1 Goals of Labeling Herbicides by MOA for Resistance Management

In the United States, the goal for adding the "Group" classification and mode of action to the herbicide product label is to provide a simple and practical approach to deliver the information to users. This information will make it easier to develop educational material that is consistent and effective. It should increase user's awareness of herbicide mode of action and provide more accurate recommendations for resistance management.[18] Another goal is to make it easier for users to keep records on which herbicide mode of actions are being used on a particular field from year to year.

Because herbicide mode of action labeling has been mandatory in Australia since 1992, the HRAC commissioned a survey in Australia to determine how effectively growers use mode of action labeling in their resistance-management programs.[19] In this survey, farmers, agricultural consultants and academic experts were asked a series of 17 questions. Focus groups were established in six locations throughout Australia where resistance was a serious problem. Of the farmers who responded to the survey, 85% said they were aware of a herbicide's MOA and considered MOA an important aspect when making a buying decision. The majority of the respondents found the single MOA letter was very or quite easy to understand while only 10% found it confusing. However, when there were multiple MOA letters (for mixtures of herbicides), 25% of the respondents said they were confused, suggesting that the farmers were not sure how to use the information. One of the purposes for MOA labeling is to make it easier for farmers to record which MOA were used from year to year. However, only 39% of the respondents indicated that they always recorded MOA used each year, 58% of the farmers never recorded the MOA or only did so sometimes.[19] Many of the agricultural consultants questioned felt that farmers were aware of the MOA label, but feelings were mixed about how the information was being used. Many of the written and oral farmer comments from the focus groups indicated that they did not pay attention to the label until they had a resistance problem. They then found the labeling very helpful in planning their weed management program.[19]

The survey concluded that MOA labeling of herbicides is beneficial, but there is some confusion in how to fully utilize this information. It is apparent that MOA labeling alone is not enough. However, if an effective educational program is coupled with MOA labeling, farmers more successfully use this information. In Australia, farmers tended to begin to use mode of action labeling after they had developed a resistance problem, but not before.

In a similar survey, conducted in Saskatchewan in 1998,[14] the mode of action classification was only in the crop guides and was not on the herbicide label. A majority (64%) of the farmers surveyed indicated that they were familiar with the herbicide group concept and a majority of the respondents could match herbicides with group number. Most of those surveyed had read or heard that the use of the same herbicide in a field over time can lead to the development of resistance. However, similar to the Australian results, only 37% of the farmers kept records of the crops they grew and the herbicides they used. Only 29% of growers believed that including the mode of action of herbicide on labels would help growers who do not currently practice herbicide group rotation to begin doing so. Although 83% of the respondents thought that including the herbicide group number on the label would be useful, 89% said that including this information on the label would not change their current rotation practices. These results further support the conclusions from the Australian study that MOA labeling alone is inadequate for implementing effective resistance-management strategies.

8.3.4.2   Other Potential Problems with MOA Labeling

The potential legal implications of MOA labeling have yet to be fully considered and addressed in the United States. This should be accomplished before EPA adopts MOA labeling under FIFIRA. Under U.S. law, for example, a herbicide is misbranded if the labeling does not contain directions for use that are complete and adequately protect human health and the environment. USEPA regulations require that directions for use must "be easily read and understood by the average person likely to use" the herbicide and USEPA requirements for MOA labeling must "prevent unreasonable adverse effects on the environment." It is a violation of U.S. law to use a herbicide in a manner inconsistent with its label. Finally, the nature and extent of the label language approved by USEPA could have an impact on the extent of product liability claims related to alleged resistance. Each of these issues requires serious consideration and review as the USEPA moves forward on MOA labeling.

8.4   ROLE OF REGULATIONS
IN HERBICIDE-RESISTANCE MANAGEMENT

As resistance to herbicides and other crop protection chemicals develops, regulatory agencies throughout the world are devoting more time and attention to this aspect in registration decisions. Europe is moving toward a requirement of resistance risk assessment and the development of a resistance-management plan before registration. North America is also moving forward in efforts to promote voluntary labeling of all crop protection chemicals for mode of action as well as resistance-management statements in the directions for use. The USEPA recently revised its adverse effects report rule (FIFRA Section 6(a)2) to require that substantiated resistance be reported to the Agency.[17] As these initiatives and developments move forward, it is important that a thorough assessment of the impact of these regulations on effective resistance management be conducted.

Mandating rotations of herbicide modes of action may, in some cases, be counterproductive if herbicide mixtures are used, particularly if the mixtures contain herbicides that share the same mode of action, but do not control the same spectrum of weeds. There are increasing cases of weed populations that are resistant to a broad range of herbicides due to enhanced metabolism. In many cases the herbicides do not share the same mode of action but rather share the same pathway of detoxification. Thus, solely rotating herbicide modes of action will be an ineffective strategy. The regulatory agencies, producers, and users share a common goal in maintaining the effective use of herbicides for weed management with minimal impact on human health and the environment. Losing a herbicide due to the development of resistance is not in anyone's interest. All of these groups need to work together to develop effective herbicide-resistance strategies.

TABLE 8.1
Acronyms and Definitions

Acronym	Definition
EEC	European Economic Community
EPPO	European and Mediterranean Plant Protection Organization
FFDCA	Federal Food, Drug and Cosmetic Act
FIFRA	Federal Insecticide, Fungicide, and Rodenticide Act
FQPA	Food Quality Protection Act
FRAC	Fungicide Resistance Action Committee
GAP	Good Agricultural Practice
GCPF	Global Crop Protection Federation
HRAC	Herbicide Resistance Action Committee
IRAC	Insecticide Resistance Action Committee
MOA	Mode of Action
MRL	Maximum Residue Limit
NAFTA	North America Free Trade Agreement
NEPHYTO	Netherlands National Crop Protection Association
PMRA	Pest Management Regulatory Agency
PPS	Plant Protection Service
PRMW	Pesticide Resistance Management Workgroup
RAC	Resistance Action Committee
USEPA	United States Environmental Protection Agency
WSSA	Weed Science Society of America

REFERENCES

1. Rotteveel, A., personal communication, 1999.
2. Rotteveel, T., de Goeij, J. W. F. M., and van Gemerden, A. F., Towards the construction of a resistance risk evaluation scheme, *Pestic. Sci.*, 51, 407, 1997.
3. McNamara, D., personal communication, 1999.
4. Matten, S. R., Pesticide resistance management activities by the U.S. Environmental Protection Agency, *Resist. Pest Manage.*, Summer 3, 1997.
5. Matten, S. R., Lewis, P. I., Tomimatsu, G., Sutherland, D. W. S., Anderson, N., and Colvin-Snyder, T. L., The U.S. Environmental Protection Agency's role in pesticide resistance management, in *Molecular Genetics and Evolution of Pesticide Resistance*, Brown, T. M., Ed., American Chemical Society, Washington, D.C., 1996, chap. 24.
6. Matten, S. R., EPA regulation of resistance management for Bt plant-pesticides, *Resist. Pest Manage.*, 10, 3, 1998.
7. Anon., *Crop Protection Reference*, 14th ed., 1998, 1422.
8. Lewis, P. I. and Matten, S. R., Consideration and management of pesticide resistance by the U.S. Environmental Protection Agency, *Resist. Pest Manage.*, 2, 10, 1995.
9. Matten, S. R., EPA regulation of plant-pesticides and Bt plant-pesticide resistance management, in *Agricultural Biotechnology and Environmental Quality: Gene Escape and Pest Resistance*, Hardy, R. W. and Segelken, J. B., Eds., NABC Report 10, National Agricultural Biotechnology Council, Ithaca, NY, 1999, 121.
10. Heap, I., International Survey of Herbicide Resistant Weeds, Online, Internet, January 6, 2000. Available www.weedscience.com. 1999.
11. Schmidt, R. R., HRAC classification of herbicides according to mode of action, *Brighton Crop Prot. Conf. — Weeds*, 3, 1133, 1997.
12. Retzinger, E. J. and Mallory-Smith, C., Classification of herbicides by site of action for weed resistance management strategies, *Weed Technol.*, 11, 384, 1997.
13. Goodwin, M., An extension program for ACCase inhibitor resistance in Manitoba: a case study, *Phytoprotection*, 75(Suppl.), 97, 1994.
14. Beckie, H. J., Chang, F., and Stevenson, F. C., The effect of labeling herbicides with their site of action: a Canadian perspective, *Weed Technol.*, 13, 655, 1999.
15. Anon., Weed Resistance Warnings on Herbicide Labels, National Registration Authority, Barton, ACT July 1, 1995, 6 pp.
16. Anon., Regulatory Directive DIR 99-06, Submission Management and Information Division, Pest Management Regulatory Agency, Ottawa, Canada, 1999, 24 pp.
17. Anon., Pesticides; draft guidance for pesticide registrants on voluntary pesticide resistance management labeling based on mode/target site of action on the pest, *Fed. Regist.*, 65, 30115, 2000.
18. Mallory-Smith, C., Impact of labeling herbicides by site of action: a university view, *Weed Technol.*, 13, 662, 1999.
19. Shaner, D. L., Howard, S., and Chalmers, I., Effectiveness of mode of action labeling for resistance management: a survey of Australian farmers, *Proc. Brighton Crop Prot. Conf. — Weeds*, 2, 797, 1999.

Index

A

295

Printed and bound by CPI Group (UK) Ltd, Croydon, CR0 4YY

23/10/2024

01778238-0007